# OXFORD MASTER SERIES IN STATISTICAL, COMPUTATIONAL, AND THEORETICAL PHYSICS

# OXFORD MASTER SERIES IN PHYSICS

The Oxford Master Series is designed for final year undergraduate and beginning graduate students in physics and related disciplines. It has been driven by a perceived gap in the literature today. While basic undergraduate physics texts often show little or no connection with the huge explosion of research over the last two decades, more advanced and specialized texts tend to be rather daunting for students. In this series, all topics and their consequences are treated at a simple level, while pointers to recent developments are provided at various stages. The emphasis in on clear physical principles like symmetry, quantum mechanics, and electromagnetism which underlie the whole of physics. At the same time, the subjects are related to real measurements and to the experimental techniques and devices currently used by physicists in academe and industry. Books in this series are written as course books, and include ample tutorial material, examples, illustrations, revision points, and problem sets. They can likewise be used as preparation for students starting a doctorate in physics and related fields, or for recent graduates starting research in one of these fields in industry.

## CONDENSED MATTER PHYSICS

1. M.T. Dove: *Structure and dynamics: an atomic view of materials*
2. J. Singleton: *Band theory and electronic properties of solids*
3. A.M. Fox: *Optical properties of solids*
4. S.J. Blundell: *Magnetism in condensed matter*
5. J.F. Annett: *Superconductivity, superfluids, and condensates*
6. R.A.L. Jones: *Soft condensed matter*

## ATOMIC, OPTICAL, AND LASER PHYSICS

7. C.J. Foot: *Atomic physics*
8. G.A. Brooker: *Modern classical optics*
9. S.M. Hooker, C.E. Webb: *Laser physics*
15. A.M. Fox: *Quantum optics: an introduction*

## PARTICLE PHYSICS, ASTROPHYSICS, AND COSMOLOGY

10. D.H. Perkins: *Particle astrophysics*
11. Ta-Pei Cheng: *Relativity, gravitation and cosmology*

## STATISTICAL, COMPUTATIONAL, AND THEORETICAL PHYSICS

12. M. Maggiore: *A modern introduction to quantum field theory*
13. W. Krauth: *Statistical mechanics: algorithms and computations*
14. J.P. Sethna: *Statistical mechanics: entropy, order parameters, and complexity*

# Statistical Mechanics
## Algorithms and Computations

Werner Krauth

*Laboratoire de Physique Statistique, Ecole Normale Supérieure, Paris*

# OXFORD

**UNIVERSITY PRESS**

Great Clarendon Street, Oxford, OX2 6DP,
United Kingdom

Oxford University Press is a department of the University of Oxford.
It furthers the University's objective of excellence in research, scholarship,
and education by publishing worldwide. Oxford is a registered trade mark of
Oxford University Press in the UK and in certain other countries

Published in the United States of America by Oxford University Press
198 Madison Avenue, New York, NY 10016, United States of America

British Library Cataloguing in Publication Data
Data available

Library of Congress Cataloging in Publication Data
Data available

ISBN 978-0-19-851536-4

*Für Silvia, Alban und Felix*

# Preface

This book is meant for students and researchers ready to plunge into statistical physics, or into computing, or both. It has grown out of my research experience, and out of courses that I have had the good fortune to give, over the years, to beginning graduate students at the Ecole Normale Supérieure and the Universities of Paris VI and VII, and also to summer school students in Drakensberg, South Africa, undergraduates in Salem, Germany, theorists and experimentalists in Lausanne, Switzerland, young physicists in Shanghai, China, among others. Hundreds of students from many different walks of life, with quite different backgrounds, listened to lectures and tried to understand, made comments, corrected me, and in short helped shape what has now been written up, for their benefit, and for the benefit of new readers that I hope to attract to this exciting, interdisciplinary field. Many of the students sat down afterwards, by themselves or in groups, to implement short programs, or to solve other problems. With programming assignments, lack of experience with computers was rarely a problem: there were always more knowledgeable students around who would help others with the first steps in computer programming. Mastering technical coding problems should also only be a secondary problem for readers of this book: all programs here have been stripped to the bare minimum. None exceed a few dozen lines of code.

We shall focus on the concepts of classical and quantum statistical physics and of computing: the meaning of sampling, random variables, ergodicity, equidistribution, pressure, temperature, quantum statistical mechanics, the path integral, enumerations, cluster algorithms, and the connections between algorithmic complexity and analytic solutions, to name but a few. These concepts built the backbone of my courses, and now form the tissue of the book. I hope that the simple language and the concrete settings chosen throughout the chapters take away none of the beauty, and only add to the clarity, of the difficult and profound subject of statistical physics.

I also hope that readers will feel challenged to implement many of the programs. Writing and debugging computer code, even for the naive programs, remains a difficult task, especially in the beginning, but it is certainly a successful strategy for learning, and for approaching the deep understanding that we must reach before we can translate the lessons of the past into our own research ideas.

This book is accompanied by a companion website at www.oup.co.uk/companion/krauth containing more than one hundred pseudocode programs and close to 300 figures, line drawings, and tables contained in the book. Readers are free to use this material for lectures and presentations, but must ask for permission if they want to include it in their own publications. For all questions, please contact me at www.lps.ens.fr/~krauth. (This website will also keep a list of misprints.) Readers of the book may want to get in contact with each other, and some may feel challenged to translate the pseudocode programs into one of the popular computer languages; I will be happy to assist initiatives in this direction, and to announce them on the above website.

**Note added to the second printing:**

A wiki site

www.smac.lps.ens.fr

now contains much additional material, including a list of all known misprints, and more than 100 programs in popular computer languages such as Python, C, and Fortran. Readers are invited to contribute to this site.

Instructors may contact me (at werner.krauth@ens.fr) for sets of exercises with worked-out solutions (in addition to the exercises contained in the book). These sets can be used (and adapted) for homework assignments.

# Contents

# Monte Carlo methods

Starting with this chapter, we embark on a journey into the fascinating realms of statistical mechanics and computational physics. We set out to study a host of classical and quantum problems, all of value as models and with numerous applications and generalizations. Many computational methods will be visited, by choice or by necessity. Not all of these methods are, however, properly speaking, computer algorithms. Nevertheless, they often help us tackle, and understand, properties of physical systems. Sometimes we can even say that computational methods give numerically exact solutions, because few questions remain unanswered.

Among all the computational techniques in this book, one stands out: the Monte Carlo method. It stems from the same roots as statistical physics itself, it is increasingly becoming part of the discipline it is meant to study, and it is widely applied in the natural sciences, mathematics, engineering, and even the social sciences. The Monte Carlo method is the first essential stop on our journey.

In the most general terms, the Monte Carlo method is a statistical—almost experimental—approach to computing integrals using random[1] positions, called samples,[1] whose distribution is carefully chosen. In this chapter, we concentrate on how to obtain these samples, how to process them in order to approximately evaluate the integral in question, and how to get good results with as few samples as possible. Starting with very simple example, we shall introduce to the basic sampling techniques for continuous and discrete variables, and discuss the specific problems of high-dimensional integrals. We shall also discuss the basic principles of statistical data analysis: how to extract results from well-behaved simulations. We shall also spend much time discussing the simulations where something goes wrong.

The Monte Carlo method is extremely general, and the basic recipes allow us—in principle—to solve any problem in statistical physics. In practice, however, much effort has to be spent in designing algorithms specifically geared to the problem at hand. The design principles are introduced in the present chapter; they will come up time and again in the real-world settings of later parts of this book.

---

[1] "Random" comes from the old French word *randon* (to run around); "sample" is derived from the Latin *exemplum* (example).

Children randomly throwing pebbles into a square, as in Fig. 1.1, illustrate a very simple direct-sampling Monte Carlo algorithm that can be adapted to a wide range of problems in science and engineering, most of them quite difficult, some of them discussed in this book. The basic principles of Monte Carlo computing are nowhere clearer than where it all started: on the beach, computing π.

**Fig. 1.1** Children computing the number π on the Monte Carlo beach.

# 1.1 Popular games in Monaco

The concept of sampling (obtaining the random positions) is truly complex, and we had better get a grasp of the idea in a simplified setting before applying it in its full power and versatility to the complicated cases of later chapters. We must clearly distinguish between two fundamentally different sampling approaches: direct sampling and Markov-chain sampling.

## 1.1.1 Direct sampling

Direct sampling is exemplified by an amusing game that we can imagine children playing on the beaches of Monaco. In the sand, they first draw a large circle and a square exactly containing it (see Fig. 1.1). They then randomly throw pebbles.[2] Each pebble falling inside the square constitutes a trial, and pebbles inside the circle are also counted as "hits".

By keeping track of the numbers of trials and hits, the children perform a direct-sampling Monte Carlo calculation: the ratio of hits to trials is close to the ratio of the areas of the circle and the square, namely $\pi/4$. The other day, in a game of 4000 trials, they threw 3156 pebbles inside the circle (see Table 1.1). This means that they got 3156 hits, and obtained the approximation $\pi \simeq 3.156$ by just shifting the decimal point.

Let us write up the children's game in a few lines of computer code (see Alg. 1.1 (direct-pi)). As it is difficult to agree on language and dialect, we use the universal *pseudocode* throughout this book. Readers can then translate the general algorithms into their favorite programming language, and are strongly encouraged to do so. Suffice it to say here that calls to the function ran $(-1, 1)$ produce uniformly distributed real random numbers between $-1$ and $1$. Subsequent calls yield independent numbers.

**procedure** direct-pi
$N_{\text{hits}} \leftarrow 0$ (initialize)
**for** $i = 1, \ldots, N$ **do**
$\begin{cases} x \leftarrow \text{ran}(-1, 1) \\ y \leftarrow \text{ran}(-1, 1) \\ \textbf{if } (x^2 + y^2 < 1) \ N_{\text{hits}} \leftarrow N_{\text{hits}} + 1 \end{cases}$
**output** $N_{\text{hits}}$
———

**Algorithm 1.1** direct-pi. Using the children's game with $N$ pebbles to compute $\pi$.

The results of several runs of Alg. 1.1 (direct-pi) are shown in Table 1.1. During each trial, $N = 4000$ pebbles were thrown, but the ran-

**Table 1.1** Results of five runs of Alg. 1.1 (direct-pi) with $N = 4000$

| Run | $N_{\text{hits}}$ | Estimate of $\pi$ |
|---|---|---|
| 1 | 3156 | 3.156 |
| 2 | 3150 | 3.150 |
| 3 | 3127 | 3.127 |
| 4 | 3171 | 3.171 |
| 5 | 3148 | 3.148 |

[2]The Latin word for "pebble" is *calculus*.

dom numbers differed, i.e. the pebbles landed at different locations in each run.

We shall return later to this table when computing the statistical errors to be expected from Monte Carlo calculations. In the meantime, we intend to show that the Monte Carlo method is a powerful approach for the calculation of integrals (in mathematics, physics, and other fields). But let us not get carried away: none of the results in Table 1.1 has fallen within the tight error bounds already known since Archimedes from comparing a circle with regular $n$-gons:

$$3.141 \simeq 3\frac{10}{71} < \pi < 3\frac{1}{7} \simeq 3.143. \tag{1.1}$$

The children's value for $\pi$ is very approximate, but improves and finally becomes exact in the limit of an infinite number of trials. This is Jacob Bernoulli's weak law of large numbers (see Subsection 1.3.2). The children also adopt a very sensible rule: they decide on the total number of throws before starting the game. The other day, in a game of "$N=4000$", they had at some point 355 hits for 452 trials—this gives a very nice approximation to the book value of $\pi$. Without hesitation, they went on until the 4000th pebble was cast. They understand that one must not stop a stochastic calculation simply because the result is just right, nor should one continue to play because the result is not close enough to what we think the answer should be.

$$\frac{355}{452} = \frac{355}{4 \times 113} = \tfrac{1}{4} \times 3.14159292\dots$$
$$\pi/4 = \tfrac{1}{4} \times 3.14159265\dots$$

### 1.1.2   Markov-chain sampling

In Monte Carlo, it is not only children who play at pebble games. We can imagine that adults, too, may play their own version at the local heliport, in the late evenings. After stowing away all their helicopters, they wander around the square-shaped landing pad (Fig. 1.2), which looks just like the area in the children's game, only bigger.

**Fig. 1.2** Adults computing the number $\pi$ at the Monte Carlo heliport.

The playing field is much wider than before. Therefore, the game must be modified. Each player starts at the clubhouse, with their expensive designer handbags filled with pebbles. With closed eyes, they throw the first little stone in a random direction, and then they walk to where this stone has landed. At that position, a new pebble is fetched from the handbag, and a new throw follows. As before, the aim of the game is to sweep out the heliport square evenly in order to compute the number $\pi$, but the distance players can cover from where they stand is much smaller than the dimensions of the field. A problem arises whenever there is a rejection, as in the case of a lady with closed eyes at a point $c$ near the boundary of the square-shaped domain, who has just thrown a pebble to a position outside the landing pad. It is not easy to understand whether she should simply move on, or climb the fence and continue until, by accident, she returns to the heliport.

What the lady should do, after a throw outside the heliport, is very surprising: where she stands, there is already a pebble on the ground. She should now ask someone to bring her the "outfielder", place it on top of the stone already on the ground, and use a new stone to try another fling. If this is again an "outfielder", she should have it fetched and increase the pile by one again, etc. Eventually, the lady moves on, visits other areas of the heliport, and also gets close to the center, which is without rejections.

**Fig. 1.3** Landing pad of the heliport at the end of the game.

The game played by the lady and her friends continues until the early morning, when the heliport has to be prepared for the day's takeoffs and landings. Before the cleaning starts, a strange pattern of pebbles on the ground may be noticed (see Fig. 1.3): far inside the square, there are only single stones, because from there, people do not throw far enough to reach the outfield. However, close to the boundaries, and especially in the corners, piles of several stones appear. This is quite mind-boggling, but does not change the fact that $\pi$ comes out as four times the ratio of hits to trials.

Those who hear this story for the first time often find it dubious. They observe that perhaps one should not pile up stones, as in Fig. 1.3, if the aim is to spread them out evenly. This objection places these modern critics in the illustrious company of prominent physicists and mathematicians who questioned the validity of this method when it was first published in 1953 (it was applied to the hard-disk system of Chapter 2). Letters were written, arguments were exchanged, and the issue was settled only after several months. Of course, at the time, helicopters and heliports were much less common than they are today.

A proof of correctness and an understanding of this method, called the Metropolis algorithm, will follow later, in Subsection 1.1.4. Here, we start by programming the adults' algorithm according to the above prescription: go from one configuration to the next by following a random throw:

$$\Delta_x \leftarrow \mathtt{ran}\left(-\delta, \delta\right),$$
$$\Delta_y \leftarrow \mathtt{ran}\left(-\delta, \delta\right)$$

(see Alg. 1.2 (`markov-pi`)). Any move that would take us outside the pad is rejected: we do not move, and count the configuration a second time (see Fig. 1.4).

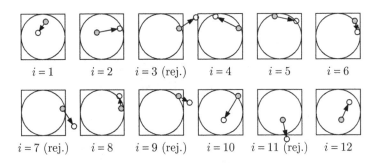

**Fig. 1.4** Simulation of Alg. 1.2 (`markov-pi`). A rejection leaves the configuration unchanged (see frames $i = 3, 7, 9, 11$).

**Table 1.2** Results of five runs of Alg. 1.2 (`markov-pi`) with $N = 4000$ and a throwing range $\delta = 0.3$

| Run | $N_{\text{hits}}$ | Estimate of $\pi$ |
|-----|------|------------|
| 1 | 3123 | 3.123 |
| 2 | 3118 | 3.118 |
| 3 | 3040 | 3.040 |
| 4 | 3066 | 3.066 |
| 5 | 3263 | 3.263 |

Table 1.2 shows the number of hits produced by Alg. 1.2 (`markov-pi`) in several runs, using each time no fewer than $N = 4000$ digital pebbles taken from the lady's bag. The results scatter around the number $\pi = 3.1415\ldots$, and we might be more inclined to admit that the idea of piling up pebbles is probably correct, even though the spread of the data, for an identical number of pebbles, is much larger than for the direct-sampling method (see Table 1.1).

In Alg. 1.2 (`markov-pi`), the throwing range $\delta$, that is to be kept fixed throughout the simulation, should not be made too small: for $\delta \gtrsim 0$, the acceptance rate is high, but the path traveled per step is small. On the other hand, if $\delta$ is too large, we also run into problems: for a large range $\delta \gg 1$, most moves would take us outside the pad. Now, the acceptance

**procedure** `markov-pi`
$N_{\text{hits}} \leftarrow 0;\ \{x, y\} \leftarrow \{1, 1\}$
**for** $i = 1, \ldots, N$ **do**
$\left\{ \begin{array}{l} \Delta_x \leftarrow \text{ran}\,(-\delta, \delta) \\ \Delta_y \leftarrow \text{ran}\,(-\delta, \delta) \\ \textbf{if } (|x + \Delta_x| < 1 \text{ and } |y + \Delta_y| < 1) \textbf{ then} \\ \quad \left\{ \begin{array}{l} x \leftarrow x + \Delta_x \\ y \leftarrow y + \Delta_y \end{array} \right. \\ \textbf{if } (x^2 + y^2 < 1)\ N_{\text{hits}} \leftarrow N_{\text{hits}} + 1 \end{array} \right.$
**output** $N_{\text{hits}}$
————

**Algorithm 1.2** `markov-pi`. Markov-chain Monte Carlo algorithm for computing $\pi$ in the adults' game.

rate is small and on average the path traveled per iteration is again small, because we almost always stay where we are. The time-honored rule of thumb consists in choosing $\delta$ neither too small, nor too large— such that the acceptance rate turns out to be of the order of $\frac{1}{2}$ (half of the attempted moves are rejected). We can experimentally check this "one-half rule" by monitoring the precision and the acceptance rate of Alg. 1.2 (`markov-pi`) at a fixed, large value of $N$.

Algorithm 1.2 (`markov-pi`) needs an initial condition. One might be tempted to use random initial conditions

$$x \leftarrow \text{ran}\,(-1, 1),$$
$$y \leftarrow \text{ran}\,(-1, 1)$$

(obtain the initial configuration through direct sampling), but this is un- realistic because Markov-chain sampling comes into play precisely when direct sampling fails. For simplicity, we stick to two standard scenar- ios: we either start from a more or less arbitrary initial condition whose only merit is that it is legal (for the heliport game, this is the club- house, at $(x, y) = (1, 1)$), or start from where a previous simulation left off. In consequence, the Markov-chain programs in this book generally omit the outer loop, and concentrate on that piece which leads from the configuration at iteration $i$ to the configuration at $i + 1$. The core heliport program then resembles Alg. 1.3 (`markov-pi(patch)`). We note that this is what defines a Markov chain: the probability of generating configuration $i + 1$ depends only on the preceding configuration, $i$, and not on earlier configurations.

The Monte Carlo games epitomize the two basic approaches to sam- pling a probability distribution $\pi(\mathbf{x})$ on a discrete or continuous space: direct sampling and Markov-chain sampling. Both approaches evaluate an observable (a function) $\mathcal{O}(\mathbf{x})$, which in our example is 1 inside the

**procedure** `markov-pi(patch)`
**input** $\{x, y\}$ (configuration $i$)
$\Delta_x \leftarrow \ldots$
$\Delta_y \leftarrow \ldots$
$\vdots$
**output** $\{x, y\}$ (configuration $i + 1$)

**Algorithm 1.3** `markov-pi(patch)`. Going from one configuration to the next, in the Markov-chain Monte Carlo algorithm.

circle and 0 elsewhere (see Fig. 1.5). In both cases, one evaluates

$$\underbrace{\frac{N_{\text{hits}}}{\text{trials}} = \frac{1}{N}\sum_{i=1}^{N}\mathcal{O}_i}_{\text{sampling}} \simeq \underbrace{\langle \mathcal{O} \rangle = \frac{\int_{-1}^{1}\mathrm{d}x\ \int_{-1}^{1}\mathrm{d}y\ \pi(x,y)\mathcal{O}(x,y)}{\int_{-1}^{1}\mathrm{d}x\ \int_{-1}^{1}\mathrm{d}y\ \pi(x,y)}}_{\text{integration}}. \qquad (1.2)$$

The probability distribution $\pi(x, y)$ no longer appears on the left: rather than being evaluated, it is sampled. This is what defines the Monte Carlo method. On the left of eqn (1.2), the multiple integrals have disappeared. This means that the Monte Carlo method allows the evaluation of high-dimensional integrals, such as appear in statistical physics and other domains, if only we can think of how to generate the samples.

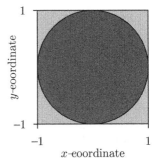

**Fig. 1.5** Probability density ($\pi = 1$ inside square, zero outside) and observable ($\mathcal{O} = 1$ inside circle, zero outside) in the Monte Carlo games.

Direct sampling, the approach inspired by the children's game, is like pure gold: a subroutine provides an independent hit at the distribution function $\pi(\mathbf{x})$, that is, it generates vectors $\mathbf{x}$ with a probability proportional to $\pi(\mathbf{x})$. Notwithstanding the randomness in the problem, direct sampling, in computation, plays a role similar to exact solutions in analytical work, and the two are closely related. In direct sampling, there is no throwing-range issue, no worrying about initial conditions (the clubhouse), and a straightforward error analysis—at least if $\pi(\mathbf{x})$ and $\mathcal{O}(\mathbf{x})$

are well behaved. Many successful Monte Carlo algorithms contain exact sampling as a key ingredient.

Markov-chain sampling, on the other hand, forces us to be much more careful with all aspects of our calculation. The critical issue here is the correlation time, during which the pebble keeps a memory of the starting configuration, the clubhouse. This time can become astronomical. In the usual applications, one is often satisfied with a handful of independent samples, obtained through week-long calculations, but it can require much thought and experience to ensure that even this modest goal is achieved. We shall continue our discussion of Markov-chain Monte Carlo methods in Subsection 1.1.4, but want to first take a brief look at the history of stochastic computing.

### 1.1.3  Historical origins

The idea of direct sampling was introduced into modern science in the late 1940s by the mathematician Ulam, not without pride, as one can find out from his autobiography *Adventures of a Mathematician* (Ulam (1991)). Much earlier, in 1777, the French naturalist Buffon (1707–1788) imagined a legendary needle-throwing experiment, and analyzed it completely. All through the eighteenth and nineteenth centuries, royal courts and learned circles were intrigued by this game, and the theory was developed further. After a basic treatment of the Buffon needle problem, we shall describe the particularly brilliant idea of Barbier (1860), which foreshadows modern techniques of variance reduction.

The Count is shown in Fig. 1.6 randomly throwing needles of length $a$ onto a wooden floor with cracks a distance $b$ apart. We introduce

**Fig. 1.6** Georges Louis Leclerc, Count of Buffon (1707–1788), performing the first recorded Monte Carlo simulation, in 1777. (Published with permission of *Le Monde*.)

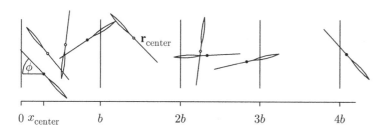

**Fig. 1.7** Variables $x_{\text{center}}$ and $\phi$ in Buffon's needle experiment. The needles are of length $a$.

coordinates $\mathbf{r}_{\text{center}}$ and $\phi$ as in Fig. 1.7, and assume that the needles' centers $\mathbf{r}_{\text{center}}$ are uniformly distributed on an infinite floor. The needles do not roll into cracks, as they do in real life, nor do they interact with each other. Furthermore, the angle $\phi$ is uniformly distributed between 0 and $2\pi$. This is the mathematical model for Buffon's experiment.

All the cracks in the floor are equivalent, and there are symmetries $x_{\text{center}} \leftrightarrow b - x_{\text{center}}$ and $\phi \leftrightarrow -\phi$. The variable $y$ is irrelevant to the problem. We may thus consider a reduced "landing pad", in which

$$0 < \phi < \frac{\pi}{2}, \tag{1.3}$$

$$0 < x_{\text{center}} < \frac{b}{2}. \tag{1.4}$$

The tip of each needle is at an $x$-coordinate $x_{\text{tip}} = x_{\text{center}} - (a/2)\cos\phi$; and every $x_{\text{tip}} < 0$ signals a hit on a crack. More precisely, the observable to be evaluated on this landing pad is (writing $x$ for $x_{\text{center}}$)

$$N_{\text{hits}}(x, \phi) = \left\{ \begin{matrix} \text{\# of hits of needle centered at } x, \\ \text{with orientation } \phi \end{matrix} \right\}$$

$$= \begin{cases} 1 & \text{for } x < a/2 \text{ and } |\phi| < \arccos\left[x/(a/2)\right] \\ 0 & \text{otherwise} \end{cases}.$$

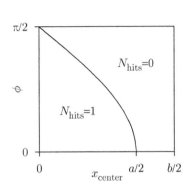

**Fig. 1.8** "Landing pad" for the Buffon needle experiment for $a < b$.

The mean number of hits of a needle of length $a$ on cracks is given, finally, by the normalized integral of the function $N_{\text{hits}}$ over the landing pad from Fig. 1.8:

$$\left\{ \begin{matrix} \text{mean number} \\ \text{of hits per needle} \end{matrix} \right\} = \langle N_{\text{hits}} \rangle = \frac{\int_0^{b/2} dx \int_0^{\pi/2} d\phi \, N_{\text{hits}}(x, \phi)}{\int_0^{b/2} dx \int_0^{\pi/2} d\phi}. \tag{1.5}$$

Integrating (over $\phi$) a function which is equal to one in a certain interval and zero elsewhere yields the length of that interval ($\arccos\left[x/(a/2)\right]$), and we find, with a suitable rescaling of $x$,

$$\langle N_{\text{hits}} \rangle = \frac{a/2}{(b/2)(\pi/2)} \int_0^1 dx \, \arccos x = \dots.$$

We might try to remember how to integrate $\arccos x$ and, in passing, marvel at how Buffon—an eighteenth-century botanist—might have done it, until it becomes apparent to us that in eqn (1.5) it is wiser to first integrate over $x$, and then over $\phi$, so that the "$\phi = \arccos x$" turns into "$x = \cos\phi$":

$$\dots = \frac{a}{b}\frac{2}{\pi} \int_0^{\pi/2} d\phi \, \cos\phi = \frac{a}{b} \cdot \frac{2}{\pi} \quad (a \le b). \tag{1.6}$$

For a needle as long as the floorboards are wide ($a = b$), the mean number of crossings is $2/\pi$. We should also realize that a needle shorter than the distance between cracks ($a \le b$) cannot hit two of them at once. The number of hits is then either 0 or 1. Therefore, the probability for a needle to hit a crack is the same as the mean number of hits:

$$\left\{ \begin{matrix} \text{probability of} \\ \text{hitting a crack} \end{matrix} \right\} = \pi(N_{\text{hits}} \ge 1),$$

$$\left\{ \begin{matrix} \text{mean number} \\ \text{of hits} \end{matrix} \right\} = \pi(N_{\text{hits}} = 1) \cdot 1 + \underbrace{\pi(N_{\text{hits}} = 2)}_{=0} \cdot 2 + \dots.$$

We can now write a program to do the Buffon experiment ourselves, by simply taking $x_{\text{center}}$ as a random number between $0$ and $b/2$ and $\phi$ as a random angle between $0$ and $\pi/2$. It remains to check whether or not the tip of the needle is on the other side of the crack (see Alg. 1.4 (`direct-needle`)).

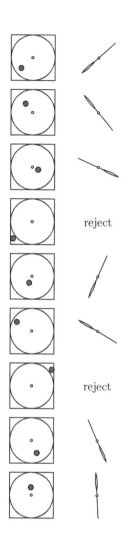

Fig. 1.9 The pebble–needle trick samples a random angle $\phi$ and allows us to compute $\sin\phi$ and $\cos\phi$.

**procedure** `direct-needle`
$N_{N_{\text{hits}}} \leftarrow 0$
**for** $i = 1, \ldots, N$ **do**
$\left\{ \begin{array}{l} x_{\text{center}} \leftarrow \mathbf{ran}\,(0, b/2) \\ \phi \leftarrow \mathbf{ran}\,(0, \pi/2) \\ x_{\text{tip}} \leftarrow x_{\text{center}} - (a/2)\mathbf{cos}\,\phi \\ \textbf{if } (x_{\text{tip}} < 0)\ N_{\text{hits}} \leftarrow N_{\text{hits}} + 1 \end{array} \right.$
**output** $N_{\text{hits}}$
———

**Algorithm 1.4** `direct-needle`. Implementing Buffon's experiment for needles of length $a$ on the reduced pad of eqn (1.4) $(a \le b)$.

On closer inspection, Alg. 1.4 (`direct-needle`) is inelegant, as it computes the number $\pi$ but also uses it as input (on line 5). There is also a call to a nontrivial cosine function, distorting the authenticity of our implementation. Because of these problems with $\pi$ and $\cos\phi$, Alg. 1.4 (`direct-needle`) is a cheat! Running it is like driving a vintage automobile (wearing a leather cap on one's head) with a computer-controlled engine just under the hood and an airbag hidden inside the wooden steering wheel. To provide a historically authentic version of Buffon's experiment, stripped down to the elementary functions, we shall adapt the children's game and replace the pebbles inside the circle by needles (see Fig. 1.9). The pebble–needle trick allows us to sample a random angle $\phi = \mathbf{ran}\,(0, 2\pi)$ in an elementary way and to compute $\sin\phi$ and $\cos\phi$ without actually calling trigonometric functions (see Alg. 1.5 (`direct-needle(patch)`)).

**procedure** `direct-needle(patch)`
$x_{\text{center}} \leftarrow \mathbf{ran}\,(0, b/2)$
$1 \quad \Delta_x \leftarrow \mathbf{ran}\,(0, 1)$
$\Delta_y \leftarrow \mathbf{ran}\,(0, 1)$
$\Upsilon \leftarrow \sqrt{\Delta_x^2 + \Delta_y^2}$
**if** $(\Upsilon > 1)$ **goto** 1
$x_{\text{tip}} \leftarrow x_{\text{center}} - (a/2)\Delta_x/\Upsilon$
$N_{\text{hits}} \leftarrow 0$
**if** $(x_{\text{tip}} < 0)\ N_{\text{hits}} \leftarrow 1$
**output** $N_{\text{hits}}$
———

**Algorithm 1.5** `direct-needle(patch)`. Historically authentic version of Buffon's experiment using the pebble–needle trick.

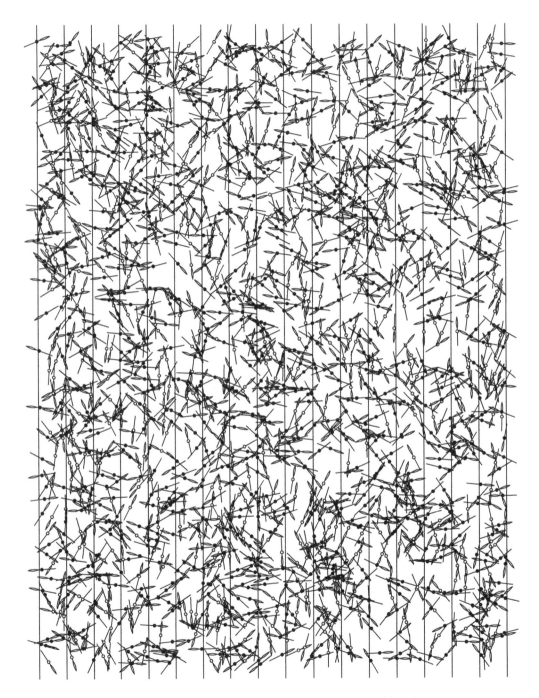

**Fig. 1.10** Buffon's experiment with 2000 needles $(a = b)$.

The pebble–needle trick is really quite clever, and we shall revisit it several times, in discussions of isotropic samplings in higher-dimensional hyperspheres, Gaussian random numbers, and the Maxwell distribution of particle velocities in a gas.

We can now follow in Count Buffon's footsteps, and perform our own needle-throwing experiments. One of these, with 2000 samples, is shown in Fig. 1.10.

Looking at this figure makes us wonder whether needles are more likely to intersect a crack at their tip, their center, or their eye. The full answer to this question, the subject of the present subsection, allows us to understand the factor $2/\pi$ in eqn (1.6) without any calculation.

Mathematically formulated, the question is the following: a needle hitting a crack does so at a certain value $l$ of its interior coordinate $0 \le l \le a$ (where, say, $l = 0$ at the tip, and $l = a$ at the end of the eye). The mean number of hits, $N_{\text{hits}}$, can be written as

$$\langle N_{\text{hits}} \rangle = \int_0^a dl \ \langle N_{\text{hits}}(l) \rangle .$$

We are thus interested in $\langle N_{\text{hits}}(l) \rangle$, the histogram of hits as a function of the interior coordinate $l$, so to speak. A probabilistic argument can be used (see Aigner and Ziegler (1992)). More transparently, we may analyze the experimental gadget shown in Fig. 1.11: two needles held together by a drop of glue.

**Fig. 1.11** Gadget No. 1: a white-centered and a black-centered needle, glued together.

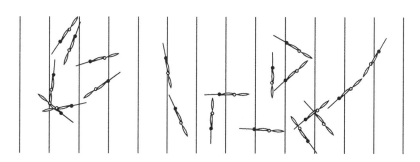

**Fig. 1.12** Buffon's experiment performed with Gadget No. 1. It is impossible to tell whether black or white needles were thrown randomly.

In Fig. 1.12, we show the result of dropping this object—with its white and dark needles—on the floor. By construction (glue!), we know that

$$\langle N_{\text{hits}}(a/2) \rangle_{\text{white needle}} = \langle N_{\text{hits}}(a) \rangle_{\text{black needle}} .$$

However, by symmetry, both needles are deposited isotropically (see Fig. 1.12). This means that

$$\langle N_{\text{hits}}(a) \rangle_{\text{black needle}} = \langle N_{\text{hits}}(a) \rangle_{\text{white needle}} ,$$

and it follows that for the white needle, $\langle N_{\text{hits}}(a/2) \rangle = \langle N_{\text{hits}}(a) \rangle$. Gluing the needles together at different positions allows us to prove analogously that $\langle N_{\text{hits}}(l) \rangle$ is independent of $l$. The argument can be carried even further: clearly, the total number of hits for the gadget in Fig. 1.12 is $3/2$ times that for a single needle, or, more generally,

$$\left\{ \begin{array}{c} \text{mean number} \\ \text{of hits} \end{array} \right\} = \Upsilon \cdot \left\{ \text{length of needle} \right\}. \tag{1.7}$$

The constant $\Upsilon$ (Upsilon) is the same for needles of any length, smaller or larger than the distance between the cracks in the floor (we have computed it already in eqn (1.6)).

Gadgets and probabilistic arguments using them are not restricted to straight needles. Let us glue a bent cobbler's (shoemaker's) needle (see Fig. 1.13) to a straight one. We see from Fig. 1.14 that the mean number of hits where the two needles touch must be the same.

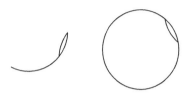

**Fig. 1.13** A cobbler's needle (*left*) and a crazy cobbler's needle (*right*).

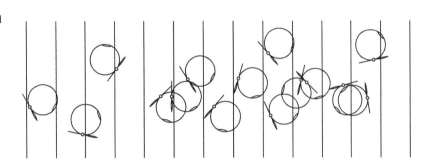

**Fig. 1.14** Buffon's experiment performed with Gadget No. 2, a straight needle and a crazy cobbler's needle glued together.

This leads immediately to a powerful generalization of eqn (1.7):

$$\left\{ \begin{array}{c} \text{mean} \\ \text{number of hits} \end{array} \right\} = \Upsilon \cdot \left\{ \begin{array}{c} \text{length of needle} \\ \text{(of any shape)} \end{array} \right\}. \tag{1.8}$$

The constant $\Upsilon$ in eqn (1.8) is the same for straight needles, cobbler's needles, and even crazy cobbler's needles, needles that are bent into full circles. Remarkably, crazy cobbler's needles of length $a = \pi b$ always have two hits (see Fig. 1.16). Trivially, the mean number of hits is equal to 2 (see Fig. 1.16). This gives $\Upsilon = 2/(\pi b)$ without any calculation, and clarifies why the number $\pi$ appears in Buffon's problem (see also Fig. 1.8). This ingenious observation goes back to Barbier (1860).

Over the last few pages, we have directly considered the mean number of hits $\langle N_{\text{hits}} \rangle$, without speaking about probabilities. We can understand this by looking at what generalizes the square in the Monte Carlo games, namely a two-dimensional rectangle with sides $b/2$ and $\pi/2$ (see Fig. 1.15). On this generalized landing pad, the observable $\mathcal{O}(x, \phi)$, the number of hits, can take on values between 0 and 4, whereas for the crazy cobbler's needles of length $\pi b$, the number of hits is always two (see Fig. 1.15). Evidently, throwing straight needles is not the same as throwing crazy cobblers' needles—the probability distributions $\pi(N_{\text{hits}})$

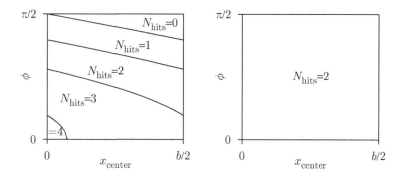

Fig. 1.15 "Landing pads" for the Buffon needle experiment with $a = \pi b$.
*Left*: straight needles. *Right*: crazy cobbler's needles.

differ, and only the mean numbers of hits (the mean of $N_{\text{hits}}$ over the whole pad) agree.

Barbier's trick is an early example of variance reduction, a powerful strategy for increasing the precision of Monte Carlo calculations. It comes in many different guises and shows that there is great freedom in finding the optimal setup for a computation.

### 1.1.4 Detailed balance

We left the lady and the heliport, in Subsection 1.1.2, without clarifying why the strange piles in Fig. 1.3 had to be built. Instead of the heliport game, let us concentrate on a simplified discrete version, the $3 \times 3$ pebble game shown in Fig. 1.17. The pebble can move in at most four directions: up, down, left, and right. In this subsection, we perform a complete analysis of Markov-chain algorithms for the pebble game, which is easily generalized to the heliport.

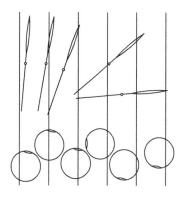

Fig. 1.16 Straight needles of length $\pi b$, with between zero and four hits, and round (crazy cobbler's) needles, which always hit twice.

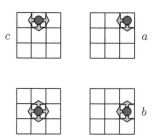

Fig. 1.17 Discrete pebble game. The corner configuration $a$ is in contact with configurations $b$ and $c$.

We seek an algorithm for moving the pebble one step at a time such that, after many iterations, it appears with the same probability in each of the fields. Anyone naive who had never watched ladies at heliports would simply chuck the pebble a few times in a random direction, i.e. one of four directions from the center, one of three directions from the edges, or one of two directions from the corners. But this natural algorithm is wrong. To understand why we must build piles, let us consider the corner configuration $a$, which is in contact with the configurations $b$ and $c$ (see Fig. 1.17). Our algorithm (yet to be found) must generate the configurations $a$, $b$, and $c$ with prescribed probabilities $\pi(a)$, $\pi(b)$, and $\pi(c)$, respectively, which we require to be equal. This means that we want to create these configurations with probabilities

$$\{\pi(a), \pi(b), \ldots\} : \left\{ \begin{array}{c} \text{stationary probability} \\ \text{for the system to be at } a, b, \text{ etc.} \end{array} \right\}, \qquad (1.9)$$

with the help of our Monte Carlo algorithm, which is nothing but a set of transition probabilities $p(a \rightarrow b)$ for moving from one configuration to the other (from $a$ to $b$),

$$\{p(a \rightarrow b), p(a \rightarrow c), \ldots\} : \left\{ \begin{array}{c} \text{probability of the algorithm} \\ \text{to move from } a \text{ to } b, \text{ etc.} \end{array} \right\}.$$

Furthermore, we enforce a normalization condition which tells us that the pebble, once at $a$, can either stay there or move on to $b$ or $c$:

$$p(a \rightarrow a) + p(a \rightarrow b) + p(a \rightarrow c) = 1. \qquad (1.10)$$

The two types of probabilities can be linked by observing that the configuration $a$ can only be generated from $b$ or $c$ or from itself:

$$\pi(a) = \pi(b)p(b \rightarrow a) + \pi(c)p(c \rightarrow a) + \pi(a)p(a \rightarrow a), \qquad (1.11)$$

which gives

$$\pi(a)[1 - p(a \rightarrow a)] = \pi(b)p(b \rightarrow a) + \pi(c)p(c \rightarrow a).$$

Writing eqn (1.10) as $1 - p(a \rightarrow a) = p(a \rightarrow b) + p(a \rightarrow c)$ and introducing it into the last equation yields

$$\pi(a)\underbrace{p(a \rightarrow b)} + \pi(a)\overbrace{p(a \rightarrow c)} = \underbrace{\pi(c)\,p(c \rightarrow a)} + \overbrace{\pi(b)\,p(b \rightarrow a)}.$$

This equation can be satisfied by equating the braced terms separately, and thus we arrive at the crucial condition of detailed balance,

$$\left\{ \begin{array}{c} \text{detailed} \\ \text{balance} \end{array} \right\} : \quad \begin{array}{c} \pi(a)p(a \rightarrow b) = \pi(b)p(b \rightarrow a) \\ \pi(a)p(a \rightarrow c) = \pi(c)p(c \rightarrow a) \end{array} \quad \text{etc.} \qquad (1.12)$$

This rate equation renders consistent the Monte Carlo algorithm (the probabilities $\{p(a \rightarrow b)\}$) and the prescribed stationary probabilities $\{\pi(a), \pi(b), \ldots\}$.

In the pebble game, detailed balance is satisfied because all probabilities for moving between neighboring sites are equal to $1/4$, and the probabilities $p(a \to b)$ and the return probabilities $p(b \to a)$ are trivially identical. Now we see why the pebbles have to pile up on the sides and in the corners: all the transition probabilities to neighbors have to be equal to 1/4. But a corner has only two neighbors, which means that half of time we can leave the site, and half of the time we must stay, building up a pile.

Of course, there are more complicated choices for the transition probabilities which also satisfy the detailed-balance condition. In addition, this condition is sufficient but not necessary for arriving at $\pi(a) = \pi(b)$ for all neighboring sites $a$ and $b$. On both counts, it is the quest for simplicity that guides our choice.

To implement the pebble game, we could simply modify Alg. 1.2 (`markov-pi`) using integer variables $\{k_x, k_y\}$, and installing the moves of Fig. 1.17 with a few lines of code. Uniform integer random numbers $\text{nran}(-1, 1)$ (that is, random integers taking values $\{-1, 0, 1\}$, see Subsection 1.2.2) would replace the real random numbers. With variables $\{k_x, k_y, k_z\}$ and another contraption to select the moves, such a program can also simulate three-dimensional pebble games.

**Table 1.3** Neighbor table for the $3 \times 3$ pebble game

| Site | Nbr(..., k) | | | |
|------|---|---|---|---|
| $k$ | 1 | 2 | 3 | 4 |
| 1 | 2 | 4 | 0 | 0 |
| 2 | 3 | 5 | 1 | 0 |
| 3 | 0 | 6 | 2 | 0 |
| 4 | 5 | 7 | 0 | 1 |
| 5 | 6 | 8 | 4 | 2 |
| 6 | 0 | 9 | 5 | 3 |
| 7 | 8 | 0 | 0 | 4 |
| 8 | 9 | 0 | 7 | 5 |
| 9 | 0 | 0 | 8 | 6 |

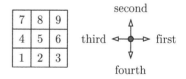

**Fig. 1.18** Numbering and neighbor scheme for the $3 \times 3$ pebble game. The first neighbor of site 5 is site 6, etc.

A smart device, the neighbor table, lets us simplify the code in a decisive fashion: this addresses each site by a number (see Fig. 1.18) rather than its Cartesian coordinates, and provides the orientation (see Table 1.3). An upward move is described, not by going from $\{k_x, k_y\}$ to $\{k_x, k_y + 1\}$, but as moving from a site $k$ to its second neighbor (Nbr(2, $k$)). All information about boundary conditions, dimensions, lattice structure, etc. can thus be outsourced into the neighbor table, whereas the core program to simulate the Markov chain remains unchanged (see Alg. 1.6 (`markov-discrete-pebble`)). This program can be written in a few moments, for neighbor relations as in Table 1.3. It visits the nine sites equally often if the run is long enough.

We have referred to $\pi(a)$ as a stationary probability. This simple concept often leads to confusion because it involves ensembles. To be completely correct, we should imagine a Monte Carlo simulation simultaneously performed on a large number of heliports, each with its own

```
procedure markov-discrete-pebble
input k (position of pebble)
n ← nran (1, 4)
if (Nbr(n, k) ≠ 0) then (see Table 1.3)
    { k ← Nbr(n, k)
output k (next position)
```

―――

**Algorithm 1.6** `markov-discrete-pebble`. Discrete Markov-chain Monte Carlo algorithm for the pebble game.

pebble-throwing lady. In this way, we can make sense of the concept of the probability of being in configuration $a$ at iteration $i$, which we have implicitly used, for example in eqn (1.11), during our derivation of the detailed-balance condition. Let us use the $3 \times 3$ pebble game to study this point in more detail. The ensemble of all transition probabilities between sites can be represented in a matrix, the system's transfer matrix $P$:

$$P = \{p(a \to b)\} = \begin{bmatrix} p(1 \to 1) & p(2 \to 1) & p(3 \to 1) & \dots \\ p(1 \to 2) & p(2 \to 2) & p(3 \to 2) & \dots \\ p(1 \to 3) & p(2 \to 3) & p(3 \to 3) & \dots \\ \vdots & \vdots & \vdots & \ddots \end{bmatrix}. \quad (1.13)$$

The normalization condition in eqn (1.10) (the pebble must go somewhere) implies that each column of the matrix in eqn (1.13) adds up to one.

With the numbering scheme of Fig. 1.18, the transfer matrix is

$$\{p(a \to b)\} = \begin{bmatrix} \boxed{\tfrac{1}{2}} & \tfrac{1}{4} & \cdot & \tfrac{1}{4} & \cdot & \cdot & \cdot & \cdot & \cdot \\ \tfrac{1}{4} & \boxed{\tfrac{1}{4}} & \tfrac{1}{4} & \cdot & \tfrac{1}{4} & \cdot & \cdot & \cdot & \cdot \\ \cdot & \tfrac{1}{4} & \boxed{\tfrac{1}{2}} & \cdot & \cdot & \tfrac{1}{4} & \cdot & \cdot & \cdot \\ \tfrac{1}{4} & \cdot & \cdot & \boxed{\tfrac{1}{4}} & \tfrac{1}{4} & \cdot & \tfrac{1}{4} & \cdot & \cdot \\ \cdot & \tfrac{1}{4} & \cdot & \tfrac{1}{4} & \boxed{0} & \tfrac{1}{4} & \cdot & \tfrac{1}{4} & \cdot \\ \cdot & \cdot & \tfrac{1}{4} & \cdot & \tfrac{1}{4} & \boxed{\tfrac{1}{4}} & \cdot & \cdot & \tfrac{1}{4} \\ \cdot & \cdot & \cdot & \tfrac{1}{4} & \cdot & \cdot & \boxed{\tfrac{1}{2}} & \tfrac{1}{4} & \cdot \\ \cdot & \cdot & \cdot & \cdot & \tfrac{1}{4} & \cdot & \tfrac{1}{4} & \boxed{\tfrac{1}{4}} & \tfrac{1}{4} \\ \cdot & \cdot & \cdot & \cdot & \cdot & \tfrac{1}{4} & \cdot & \tfrac{1}{4} & \boxed{\tfrac{1}{2}} \end{bmatrix}, \quad (1.14)$$

where the symbols "·" stand for zeros. All simulations start at the clubhouse, site 9 in our numbering scheme. For the ensemble of Monte Carlo simulations, this means that the probability vector at iteration $i = 0$ is

$$\{\pi^0(1), \dots, \pi^0(9)\} = \{0, \dots, 0, 1\}.$$

After one iteration of the Monte Carlo algorithm, the pebble is at the clubhouse with probability 1/2, and at positions 6 and 8 with probabilities 1/4. This is mirrored by the vector $\{\pi^{i=1}(1), \ldots, \pi^{i=1}(9)\}$ after one iteration, obtained by a matrix–vector multiplication

$$\pi^{i+1}(a) = \sum_{b=1}^{9} p(b \to a)\pi^i(b) \tag{1.15}$$

for $i = 0$, and $i + 1 = 1$. Equation (1.15) is easily programmed (see Alg. 1.7 (transfer-matrix); for the matrix in eqn (1.13), the eqn (1.15) corresponds to a matrix–vector multiplication, with the vector to the right). Repeated application of the transfer matrix to the initial probability vector allows us to follow explicitly the convergence of the Monte Carlo algorithm (see Table 1.4 and Fig. 1.19).

---

**procedure** transfer-matrix
**input** $\{p(a \to b)\}$ (matrix in eqn (1.14))
**input** $\{\pi^i(1), \ldots, \pi^i(9)\}$
**for** $a = 1, \ldots, 9$ **do**
$\Big\{$   $\pi^{i+1}(a) \leftarrow 0$
    **for** $b = 1, \ldots, 9$ **do**
     $\{$   $\pi^{i+1}(a) \leftarrow \pi^{i+1}(a) + p(b \to a)\pi^i(b)$
**output** $\{\pi^{i+1}(1), \ldots, \pi^{i+1}(9)\}$

---

**Algorithm 1.7** transfer-matrix. Computing pebble-game probabilities at iteration $i + 1$ from the probabilities at iteration $i$.

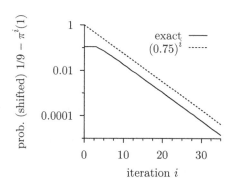

**Fig. 1.19** Pebble-game probability of site 1, shifted by 1/9 (from Alg. 1.7 (transfer-matrix); see Table 1.4).

**Table 1.4** Input/output of Alg. 1.7 (transfer-matrix), initially started at the clubhouse

| Prob. | Iteration $i$ | | | | |
|---|---|---|---|---|---|
| | 0 | 1 | 2 | ... | ∞ |
| $\pi^i(1)$ | 0 | 0 | 0 | ... | 1/9 |
| $\pi^i(2)$ | 0 | 0 | 0 | ... | 1/9 |
| $\pi^i(3)$ | 0 | 0 | 0.062 | ... | 1/9 |
| $\pi^i(4)$ | 0 | 0 | 0 | ... | 1/9 |
| $\pi^i(5)$ | 0 | 0 | 1/8 | ... | 1/9 |
| $\pi^i(6)$ | 0 | 1/4 | 0.188 | ... | 1/9 |
| $\pi^i(7)$ | 0 | 0 | 0.062 | ... | 1/9 |
| $\pi^i(8)$ | 0 | 1/4 | 0.188 | ... | 1/9 |
| $\pi^i(9)$ | 1 | 1/2 | 0.375 | ... | 1/9 |

To fully understand convergence in the pebble game, we must analyze the eigenvectors $\{\boldsymbol{\pi}^1_e, \ldots, \boldsymbol{\pi}^9_e\}$ and the eigenvalues $\{\lambda_1, \ldots, \lambda_9\}$ of the transfer matrix. The eigenvectors $\boldsymbol{\pi}^k_e$ are those vectors that essentially reproduce under the application of $P$:

$$P\boldsymbol{\pi}^k_e = \lambda_k \boldsymbol{\pi}^k_e.$$

Writing a probability vector $\boldsymbol{\pi} = \{\pi(1), \ldots, \pi(9)\}$ in terms of the eigenvectors, i.e.

$$\boldsymbol{\pi} = \alpha_1 \boldsymbol{\pi}_e^1 + \alpha_2 \boldsymbol{\pi}_e^2 + \cdots + \alpha_9 \boldsymbol{\pi}_e^9 = \sum_{k=1}^{9} \alpha_k \boldsymbol{\pi}_e^k,$$

allows us to see how it is transformed after one iteration,

$$P\boldsymbol{\pi} = \alpha_1 P\boldsymbol{\pi}_e^1 + \alpha_2 P\boldsymbol{\pi}_e^2 + \cdots + \alpha_9 P\boldsymbol{\pi}_e^9 = \sum_{k=1}^{9} \alpha_k P\boldsymbol{\pi}_e^k$$

$$= \alpha_1 \lambda_1 \boldsymbol{\pi}_e^1 + \alpha_2 \lambda_2 \boldsymbol{\pi}_e^2 + \cdots + \alpha_9 \lambda_9 \boldsymbol{\pi}_e^9 = \sum_{k=1}^{9} \alpha_k \lambda_k \boldsymbol{\pi}_e^k,$$

or after $i$ iterations,

$$P^i \boldsymbol{\pi} = \alpha_1 \lambda_1^i \boldsymbol{\pi}_e^1 + \alpha_2 \lambda_2^i \boldsymbol{\pi}_e^2 + \cdots + \alpha_9 \lambda_9^i \boldsymbol{\pi}_e^9 = \sum_{k=1}^{9} \alpha_k (\lambda_k)^i \boldsymbol{\pi}_e^k.$$

Only one eigenvector has components that are all nonnegative, so that it can be a vector of probabilities. This vector must have the largest eigenvalue $\lambda_1$ (the matrix $P$ being positive). Because of eqn (1.10), we have $\lambda_1 = 1$. Other eigenvectors and eigenvalues can be computed explicitly, at least in the $3 \times 3$ pebble game. Besides the dominant eigenvalue $\lambda_1$, there are two eigenvalues equal to 0.75, one equal to 0.5, etc. This allows us to follow the precise convergence towards the asymptotic equilibrium solution:

$$\{\pi^i(1), \ldots, \pi^i(9)\}$$
$$= \underbrace{\{\tfrac{1}{9}, \ldots, \tfrac{1}{9}\}}_{\substack{\text{first eigenvector} \\ \text{eigenvalue } \lambda_1 = 1}} + \alpha_2 \cdot (0.75)^i \underbrace{\{-0.21, \ldots, 0.21\}}_{\substack{\text{second eigenvector} \\ \text{eigenvalue } \lambda_2 = 0.75}} + \cdots.$$

In the limit $i \to \infty$, the contributions of the subdominant eigenvectors disappear and the first eigenvector, the vector of stationary probabilities in eqn (1.9), exactly reproduces under multiplication by the transfer matrix. The two are connected through the detailed-balance condition, as discussed in simpler terms at the beginning of this subsection.

The difference between $\{\pi^i(1), \ldots, \pi^i(9)\}$ and the asymptotic solution is determined by the second largest eigenvalue of the transfer matrix and is proportional to

$$(0.75)^i = e^{i \cdot \log 0.75} = \exp\left(-\frac{i}{3.476}\right). \tag{1.16}$$

The data in Fig. 1.19 clearly show the $(0.75)^i$ behavior, which is equivalent to an exponential $\propto e^{-i/\Delta_i}$ where $\Delta_i = 3.476$. $\Delta_i$ is a timescale, and allows us to define short times and long times: a short simulation has fewer than $\Delta_i$ iterations, and a long simulation has many more than that.

In conclusion, we see that transfer matrix iterations and Monte Carlo calculations reach equilibrium only after an infinite number of iterations. This is not a very serious restriction, because of the existence of a timescale for convergence, which is set by the second largest eigenvalue of the transfer matrix. To all intents and purposes, the asymptotic equilibrium solution is reached after the convergence time has passed a few times. For example, the pebble game converges to equilibrium after a few times 3.476 iterations (see eqn (1.16)). The concept of equilibrium is far-reaching, and the interest in Monte Carlo calculations is rightly strong because of this timescale, which separates fast and slow processes and leads to exponential convergence.

## 1.1.5   The Metropolis algorithm

In Subsection 1.1.4, direct inspection of the detailed-balance condition in eqn (1.12) allowed us to derive Markov-chain algorithms for simple games where the probability of each configuration was either zero or one. This is not the most general case, even for pebbles, which may be less likely to be at a position $a$ on a hilltop than at another position $b$ located in a valley (so that $\pi(a) < \pi(b)$). Moves between positions $a$ and $b$ with arbitrary probabilities $\pi(a)$ and $\pi(b)$, respecting the detailed-balance condition in eqn (1.12), are generated by the Metropolis algorithm (see Metropolis *et al.* (1953)), which accepts a move $a \to b$ with probability

$$p(a \to b) = \min\left[1, \frac{\pi(b)}{\pi(a)}\right]. \tag{1.17}$$

In the heliport game, we have unknowingly used eqn (1.17): for $a$ and $b$ both inside the square, the move was accepted without further tests ($\pi(b)/\pi(a) = 1$, $p(a \to b) = 1$). In contrast, for $a$ inside but $b$ outside the square, the move was rejected ($\pi(b)/\pi(a) = 0$, $p(a \to b) = 0$).

**Table 1.5** Metropolis algorithm represented by eqn (1.17): detailed balance holds because the second and fourth rows of this table are equal

| Case | $\pi(a) > \pi(b)$ | $\pi(b) > \pi(a)$ |
|---|---|---|
| $p(a \to b)$ | $\pi(b)/\pi(a)$ | $1$ |
| $\pi(a)p(a \to b)$ | $\pi(b)$ | $\pi(a)$ |
| $p(b \to a)$ | $1$ | $\pi(a)/\pi(b)$ |
| $\pi(b)p(b \to a)$ | $\pi(b)$ | $\pi(a)$ |

To prove eqn (1.17) for general values of $\pi(a)$ and $\pi(b)$, one has only to write down the expressions for the acceptance probabilities $p(a \to b)$ and $p(b \to a)$ from eqn (1.17) for the two cases $\pi(a) > \pi(b)$ and $\pi(b) > \pi(a)$ (see Table 1.5). For $\pi(a) > \pi(b)$, one finds that $\pi(a)p(a \to b) =$

site 0          site 1

**Fig. 1.20** Two-site problem. The probabilities to be at site 0 and site 1 are proportional to $\pi(0)$ and $\pi(1)$, respectively.

$\pi(b)p(b \rightarrow a) = \pi(b)$. In this case, and likewise for $\pi(b) > \pi(a)$, detailed balance is satisfied. This is all there is to the Metropolis algorithm.

Let us implement the Metropolis algorithm for a model with just two sites: site 0, with probability $\pi(0)$, and site 1, with $\pi(1)$, probabilities that we may choose to be arbitrary positive numbers (see Fig. 1.20). The pebble is to move between the sites such that, in the long run, the times spent on site 0 and on site 1 are proportional to $\pi(0)$ and $\pi(1)$, respectively. This is achieved by computing the ratio of statistical weights $\pi(1)/\pi(0)$ or $\pi(0)/\pi(1)$, and comparing it with a random number $\mathtt{ran}\,(0,1)$, a procedure used by almost all programs implementing the Metropolis algorithm (see Fig. 1.21 and Alg. 1.8 ($\mathtt{markov\text{-}two\text{-}site}$)).

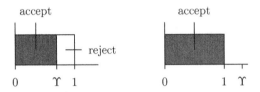

**Fig. 1.21** Accepting a move with probability $\min(1, \Upsilon)$ with the help of a random number $\mathtt{ran}\,(0,1)$.

We may run this program for a few billion iterations, using the output of iteration $i$ as the input of iteration $i+1$. While waiting for the output, we can also clean up Alg. 1.8 ($\mathtt{markov\text{-}two\text{-}site}$) a bit, noticing that if $\Upsilon > 1$, its comparison with a random number between 0 and 1 makes no sense: the move will certainly be accepted. For $\pi(l) > \pi(k)$, we should thus work around the calculation of the quotient, the generation of a random number and the comparison with that number.

> **procedure** $\mathtt{markov\text{-}two\text{-}site}$
> **input** $k$ (either 0 or 1)
> **if** $(k = 0)$ $l \leftarrow 1$
> **if** $(k = 1)$ $l \leftarrow 0$
> $\Upsilon \leftarrow \pi(l)/\pi(k)$
> **if** $(\mathtt{ran}\,(0,1) < \Upsilon)$ $k \leftarrow l$
> **output** $k$ (next site)
> ———

**Algorithm 1.8** $\mathtt{markov\text{-}two\text{-}site}$. Sampling sites 0 and 1 with stationary probabilities $\pi(0)$ and $\pi(1)$ by the Metropolis algorithm.

### 1.1.6   A priori probabilities, triangle algorithm

On the heliport, the moves $\Delta_x$ and $\Delta_y$ were restricted to a small square of edge length $2\delta$, the throwing range, centered at the present position (see Fig. 1.22(A)). This gives an example of an a priori probability distribution, denoted by $\mathcal{A}(a \rightarrow b)$, from which we sample the move $a \rightarrow b$, that is, which contains all possible moves in our Markov-chain algorithm, together with their probabilities.

The small square could be replaced by a small disk without bringing in anything new (see Fig. 1.22($B$)). A much more interesting situation arises if asymmetric a priori probabilities are allowed: in the triangle algorithm of Fig. 1.22($C$), we sample moves from an oriented equilateral triangle centered at $a$, with one edge parallel to the $x$-axis. This extravagant choice may lack motivation in the context of the adults' game, but contains a crucial ingredient of many modern Monte Carlo algorithms.

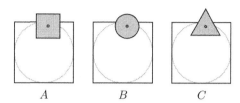

$$A \qquad\qquad B \qquad\qquad C$$

**Fig. 1.22** Throwing pattern in Alg. 1.2 (`markov-pi`) ($A$), with variants. The triangle algorithm ($C$) needs special attention.

In fact, detailed balance can be reconciled with any a priori probability $\mathcal{A}(a \to b)$, even a triangular one, by letting the probability $\mathcal{P}(a \to b)$ for moving from $a$ to $b$ be composite:

$$\mathcal{P}(a \to b) = \underbrace{\mathcal{A}(a \to b)}_{\text{consider } a \to b} \cdot \underbrace{p(a \to b)}_{\text{accept } a \to b} .$$

The probability of moving from $a$ to $b$ must satisfy $\pi(a)\mathcal{P}(a \to b) = \pi(b)\mathcal{P}(b \to a)$, so that the acceptance probabilities obey

$$\frac{p(a \to b)}{p(b \to a)} = \frac{\pi(b)}{\mathcal{A}(a \to b)} \frac{\mathcal{A}(b \to a)}{\pi(a)}.$$

This leads to a generalized Metropolis algorithm

$$p(a \to b) = \min\left[1, \frac{\pi(b)}{\mathcal{A}(a \to b)} \frac{\mathcal{A}(b \to a)}{\pi(a)}\right], \qquad (1.18)$$

also called the Metropolis–Hastings algorithm. We shall first check the new concept of an a priori probability with the familiar problem of the heliport game with the small square: as the pebble throw $a \to b$ appears with the same probability as the return throw $b \to a$, we have $\mathcal{A}(a \to b) = \mathcal{A}(b \to a)$, so that the generalized Metropolis algorithm is the same as the old one.

The triangle algorithm is more complicated: both the probability of the move $a \to b$ and that of the return move $b \to a$ must be considered in order to balance the probabilities correctly. It can happen, for example, that the probability $\mathcal{A}(a \to b)$ is finite, but that the return probability $\mathcal{A}(b \to a)$ is zero (see Fig. 1.23). In this case, the generalized Metropolis algorithm in eqn (1.18) imposes rejection of the original pebble throw

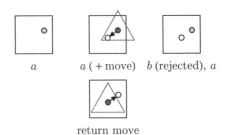

$a$          $a\,(+\text{move})$    $b\,(\text{rejected}),\,a$

return move

**Fig. 1.23** Rejected move $a \to b$ in the triangle algorithm.

from $a$ to $b$. (Alg. 7.3 (`direct-triangle`) allows us to sample a random point inside an arbitrary triangle).

The triangle algorithm can be generalized to an arbitrary a priori probability $\mathcal{A}(a \to b)$, and the generalized Metropolis algorithm (eqn (1.18)) will ensure that the detailed-balance condition remains satisfied. However, only good choices for $\mathcal{A}(a \to b)$ have an appreciable acceptance rate (the acceptance probability of each move averaged over all moves) and actually move the chain forward. As a simple example, we can think of a configuration $a$ with a high probability ($\pi(a)$ large), close to configurations $b$ with $\pi(b)$ small. The original Metropolis algorithm leads to many rejections in this situation, slowing down the simulation. Introducing a priori probabilities to propose configurations $b$ less frequently wastes less computer time with rejections. Numerous examples in later chapters illustrate this point.

A case worthy of special attention is $\mathcal{A}(a \to b) = \pi(b)$ and $\mathcal{A}(b \to a) = \pi(a)$, for which the acceptance rate in eqn (1.18) of the generalized Metropolis algorithm is equal to unity: we are back to direct sampling, which we abandoned because we did not know how to put it into place. However, no circular argument is involved. A priori probabilities are crucial when we can almost do direct sampling, or when we can almost directly sample a subsystem. A priori probabilities then present the computational analogue of perturbation theory in theoretical physics.

### 1.1.7   Perfect sampling with Markov chains

The difference between the ideal world of children (direct sampling) and that of adults (Markov-chain sampling) is clear-cut: in the former, direct access to the probability distribution $\pi(\mathbf{x})$ is possible, but in the latter, convergence towards $\pi(\mathbf{x})$ is reached only in the long-time limit. Controlling the error from within the simulation poses serious difficulties: we may have the impression that we have decorrelated from the clubhouse, without suspecting that it is—figuratively speaking—still around the corner. It has taken half a century to notice that this difficulty can sometimes be resolved, within the framework of Markov chains, by producing perfect chain samples, which are equivalent to the children's throws and guaranteed to be totally decorrelated from the initial condition.

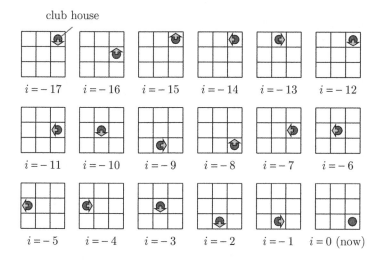

Fig. 1.24 A 3 × 3 pebble game starting at the clubhouse at iteration $i = -17$, arriving at the present configuration at $i = 0$ (now).

For concreteness, we discuss perfect sampling in the context of the 3 × 3 pebble game. In Fig. 1.24, the stone has moved in 17 steps from the clubhouse to the lower right corner. As the first subtle change in the setting, we let the simulation start at time $i = -17$, and lead up to the present time $i = 0$. Because we started at the clubhouse, the probability of being in the lower right corner at $i = 0$ is slightly smaller than 1/9. This correlation goes to zero exponentially in the limit of long running times, as we have seen (see Fig. 1.19).

The second small change is to consider random maps rather than random moves (see Fig. 1.25: from the upper right corner, the pebble must move down; from the upper left corner, it must move right; etc.). At each iteration $i$, a new random map is drawn. Random maps give a consistent, alternative definition of the Markov-chain Monte Carlo method, and for any given trajectory it is impossible to tell whether it was produced by random maps or by a regular Monte Carlo algorithm (in Fig. 1.26, the trajectory obtained using random maps is the same as in Fig. 1.24).

In the random-map Markov chain of Fig. 1.26, it can, furthermore, be verified explicitly that any pebble position at time $i = -17$ leads to the lower right corner at iteration $i = 0$. In addition, we can imagine that $i = -17$ is not really the initial time, but that the simulation has been going on since $i = -\infty$. There have been random maps all along the way, and Fig. 1.26 shows only the last stretch. The pebble position at $i = 0$ is the same for any configuration at $i = -17$: it is also the outcome of an infinite simulation, with an initial position at $i = -\infty$, from which it has

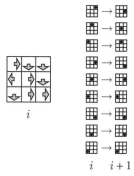

Fig. 1.25 A random map at iteration $i$ and its action on all possible pebble positions.

decorrelated. The $i = 0$ pebble position in the lower right corner is thus a direct sample—obtained by a Markov-chain Monte Carlo method.

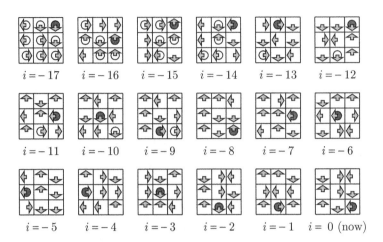

$i = -17$  $i = -16$  $i = -15$  $i = -14$  $i = -13$  $i = -12$

$i = -11$  $i = -10$  $i = -9$  $i = -8$  $i = -7$  $i = -6$

$i = -5$  $i = -4$  $i = -3$  $i = -2$  $i = -1$  $i = 0$ (now)

**Fig. 1.26** Monte Carlo dynamics using time-dependent random maps. All positions at $i = -17$ give an identical output at $i = 0$.

The idea of perfect sampling is due to Propp and Wilson (1996). It mixes a minimal conceptual extension of the Monte Carlo method (random maps) with the insight that a finite amount of backtracking (called coupling from the past) may be sufficient to figure out the present state of a Markov chain that has been running forever (see Fig. 1.27).

$i = -\infty$

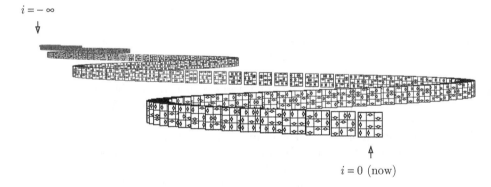

$i = 0$ (now)

**Fig. 1.27** A random-map Markov chain that has been running since $i = -\infty$.

Producing direct samples for a $3 \times 3$ pebble game by Markov chains is a conceptual breakthrough, but not yet a great technical achievement. Later on, in Chapter 5, we shall construct direct samples (using Markov chains) with $2^{100} = 1\,267\,650\,600\,228\,229\,401\,496\,703\,205\,376$ configurations. Going through all of them to see whether they have merged is out of the question, but we shall see that it is sometimes possible to squeeze all configurations in between two extremal ones: if those two configurations have come together, all others have merged, too.

Understanding and running a coupling-from-the-past program is the ultimate in Monte Carlo style—much more elegant than walking around a heliport, well dressed and with a fancy handbag over one's shoulder, waiting for the memory of the clubhouse to more or less fade away.

## 1.2 Basic sampling

On several occasions already, we have informally used uniform random numbers $x$ generated through a call $x \leftarrow \mathtt{ran}\,(a,b)$. We now discuss the two principal aspects of random numbers. First we must understand how random numbers enter a computer, a fundamentally deterministic machine. In this first part, we need little operational understanding, as we shall always use routines written by experts in the field. We merely have to be aware of what can go wrong with those routines. Second, we shall learn how to reshape the basic building block of randomness—$\mathtt{ran}\,(0,1)$—into various distributions of random integers and real numbers, permutations and combinations, $N$-dimensional random vectors, random coordinate systems, etc. Later chapters will take this program much further: $\mathtt{ran}\,(0,1)$ will be remodeled into random configurations of liquids and solids, boson condensates, and mixtures, among other things.

### 1.2.1 Real random numbers

Random number generators (more precisely, pseudorandom number generators), the subroutines which produce $\mathtt{ran}\,(0,1)$, are intricate deterministic algorithms that condense into a few dozen lines a lot of clever number theory and probabilities, all rigorously tested. The output of these algorithms looks random, but is not: when run a second time, under exactly the same initial conditions, they always produce an identical output. Generators that run in the same way on different computers are called "portable". They are generally to be preferred. Random numbers have many uses besides Monte Carlo calculations, and rely on a solid theoretical and empirical basis. Routines are widely available, and their writing is a mature branch of science. Progress has been fostered by essential commercial applications in coding and cryptography. We certainly do not have to conceive such algorithms ourselves and, in essence, only need to understand how to test them for our specific applications.

Every modern, good $\mathtt{ran}\,(0,1)$ routine has a flat probability distribution. It has passed a battery of standard statistical tests which would have detected unusual correlations between certain values $\{x_i,\ldots,x_{i+k}\}$ and other values $\{x_j,\ldots,x_{j+k'}\}$ later down the sequence. Last but not least, the standard routines have been successfully used by many people before us. However, all the meticulous care taken and all the endorsement by others do not insure us against the small risk that the particular random number generator we are using may in fact fail in our particular problem. To truly convince ourselves of the quality of a complicated calculation that uses a given random number generator, it remains for us (as end users) to replace the random number generator *in the very simulation program we are using* by a second, different algorithm. By the definition of what constitutes randomness, this change of routine should have no influence on the results (inside the error bars). Therefore, if changing the random number generator in our simulation program leads to no systematic variations, then the two generators are almost certainly OK for our application. There is nothing more we can do and nothing less we should do to calm our anxiety about this crucial ingredient of Monte Carlo programs.

Algorithm 1.9 (naive-ran) is a simple example—useful for study, but unsuited for research—of linear congruential[3] random number generators, which are widely installed in computers, pocket calculators, and other digital devices. Very often, such generators are the building blocks of good algorithms.

**procedure** naive-ran
$m \leftarrow 134456$
$n \leftarrow 8121$
$k \leftarrow 28411$
**input** idum
$\text{idum} \leftarrow \text{mod}(\text{idum} \cdot n + k, m)$
$\text{ran} \leftarrow \text{idum}/\text{real}(m)$
**output** idum, ran

Table 1.6 Repeated calls to Alg. 1.9 (naive-ran). Initially, the seed was set to idum ← 89053.

| # | idum | ran |
|---|---|---|
| 1 | 123456 | 0.91819 |
| 2 | 110651 | 0.82295 |
| 3 | 55734 | 0.41451 |
| 4 | 65329 | 0.48588 |
| 5 | 1844 | 0.01371 |
| 6 | 78919 | 0.58695 |
| ... | ... | ... |
| 134457 | 123456 | ... |
| 134458 | 110651 | ... |
| ... | ... | ... |

**Algorithm 1.9** naive-ran. Low-quality portable random number generator, naive-ran$(0, 1)$, using a linear congruential method.

In Alg. 1.9 (naive-ran), the parameters $\{m, n, k\}$ have been carefully adjusted, whereas the variable {idum}, called the seed, is set at the beginning, but never touched again from the outside during a run of the program. Once started, the sequence of pseudorandom numbers unravels. Just like the sequence of any other generator, even the high-quality ones, it is periodic. In Alg. 1.9 (naive-ran), the periodicity is $134\,456$ (see Table 1.6); in good generators, the periodicity is much larger than we shall ever be able to observe.

Let us denote real random numbers, uniformly distributed between values $a$ and $b$, by the abstract symbol ran$(a, b)$, without regard for initialization and the choice of algorithm (we suppose it to be perfect). In the printed routines, repeated calls to ran$(a, b)$, such as

$$x \leftarrow \text{ran}(-1, 1),$$
$$y \leftarrow \text{ran}(-1, 1), \qquad (1.19)$$

generate statistically independent random values for $x$ and $y$. Later, we shall often use a concise vector notation in our programs. The two

---

[3]Two numbers are *congruent* if they agree with each other, i.e. if their difference is divisible by a given modulus: 12 is congruent to 2 (modulo 5), since $12 - 2 = 2 \times 5$.

variables $\{x, y\}$ in eqn (1.19), for example, may be part of a vector $\mathbf{x}$, and we may assign independent random values to the components of this vector by the call

$$\mathbf{x} \leftarrow \{\mathtt{ran}\,(-1, 1)\,, \mathtt{ran}\,(-1, 1)\}.$$

(For a discussion of possible conflicts between vectors and random numbers, see Subsection 1.2.6.)

Depending on the context, random numbers may need a little care. For example, the logarithm of a random number between 0 and 1, $x \leftarrow \log \mathtt{ran}\,(0, 1)$, may have to be replaced by

$$
\begin{aligned}
&1 \quad \Upsilon \leftarrow \mathtt{ran}\,(0, 1) \\
&\quad\quad \mathbf{if}\ (\Upsilon = 0)\ \mathbf{goto}\ 1\ \text{\small(reject number)} \\
&\quad\quad x \leftarrow \log \Upsilon
\end{aligned}
$$

to avoid overflow ($\Upsilon = 0$, $x = -\infty$) and a crash of the program after a few hours of running. To avoid this problem, we might define the random number $\mathtt{ran}\,(0, 1)$ to be always larger than 0 and smaller than 1. However this does not get us out of trouble: a well-implemented random number generator between 0 and 1, always satisfying

$$0 < \mathtt{ran}\,(0, 1) < 1,$$

might be used to implement a routine $\mathtt{ran}\,(1, 2)$. The errors of finite-precision arithmetic could lead to an inconsistent implementation where, owing to rounding, $1 + \mathtt{ran}\,(0, 1)$ could turn out to be exactly equal to one, even though we want $\mathtt{ran}\,(1, 2)$ to satisfy

$$1 < \mathtt{ran}\,(1, 2) < 2.$$

Clearly, great care is called for, in this special case but also in general: Monte Carlo programs, notwithstanding their random nature, are extremely sensitive to small bugs and irregularities. They have to be meticulously written: under no circumstance should we accept routines that need an occasional manual restart after a crash, or that sometimes produce data which has to be eliminated by hand. Rare problems, for example logarithms of zero or random numbers that are equal to 1 but should be strictly larger, quickly get out of control, lead to a loss of trust in the output, and, in short, leave us with a big mess ....

## 1.2.2   Random integers, permutations, and combinations

Random variables in a Monte Carlo calculation are not necessarily real-valued. Very often, we need uniformly distributed random integers $m$, between (and including) $k$ and $l$. In this book, such a random integer is generated by the call $m \leftarrow \mathtt{nran}\,(k, l)$. In the implementation in Alg. 1.10 (**nran**), the **if ( )** statement (on line 4) provides extra protection against rounding problems in the underlying $\mathtt{ran}\,(k, l + 1)$ routine.

```
         procedure nran
         input {k, l}
    1    m ← int(ran (k, l + 1))
         if (m > l) goto 1
         output m
```

---

**Algorithm 1.10 nran.** Uniform random integer nran $(k, l)$ between (and including) $k$ and $l$.


The next more complicated objects, after integers, are permutations of $K$ distinct objects, which we may take to be the integers $\{1, \ldots, K\}$. A permutation $P$ can be written as a two-row matrix[4]

$$P = \begin{pmatrix} P_1 & P_2 & P_3 & P_4 & P_5 \\ 1 & 2 & 3 & 4 & 5 \end{pmatrix}. \tag{1.20}$$

We can think of the permutation in eqn (1.20) as balls labeled $\{1, \ldots, 5\}$ in the order $\{P_1, \ldots, P_5\}$, on a shelf (ball $P_k$ is at position $k$). The order of the columns is without importance in eqn (1.20), and $P$ can also be written as $P = \begin{pmatrix} P_1 & P_3 & P_4 & P_2 & P_5 \\ 1 & 3 & 4 & 2 & 5 \end{pmatrix}$: information about the placing of balls is not lost, and we still know that ball $P_k$ is in the $k$th position on the shelf. Two permutations $P$ and $Q$ can be multiplied, as shown in the example below, where the columns of $Q$ are first rearranged such that the lower row of $Q$ agrees with the upper row of $P$. The product $PQ$ consists of the lower row of $P$ and the upper row of the rearranged $Q$:

$$\overbrace{\begin{pmatrix} 1 & 4 & 3 & 2 & 5 \\ 1 & 2 & 3 & 4 & 5 \end{pmatrix}}^{P} \overbrace{\begin{pmatrix} 1 & 3 & 2 & 4 & 5 \\ 1 & 2 & 3 & 4 & 5 \end{pmatrix}}^{Q} = \overbrace{\begin{pmatrix} 1 & 4 & 3 & 2 & 5 \\ 1 & 2 & 3 & 4 & 5 \end{pmatrix}}^{P} \overbrace{\begin{pmatrix} 1 & 4 & 2 & 3 & 5 \\ 1 & 4 & 3 & 2 & 5 \end{pmatrix}}^{Q \text{ (rearranged)}} = \overbrace{\begin{pmatrix} 1 & 4 & 2 & 3 & 5 \\ 1 & 2 & 3 & 4 & 5 \end{pmatrix}}^{PQ}. \tag{1.21}$$

On the shelf, with balls arranged in the order $\{P_1, \ldots, P_5\}$, the multiplication of $P$ from the right by another permutation $Q = \begin{pmatrix} Q_1 & Q_2 & Q_3 & Q_4 & Q_5 \\ 1 & 2 & 3 & 4 & 5 \end{pmatrix}$ replaces the ball $k$ with $Q_k$ (or, equivalently, the ball $P_k$ by $Q_{P_k}$).

The identity $\begin{pmatrix} 1 & 2 & \cdots & K \\ 1 & 2 & \cdots & K \end{pmatrix}$ is a special permutation. Transpositions are the same as the identity permutation, except for two elements which are interchanged ($P_k = l$ and $P_l = k$). The second factor in the product in eqn (1.21) is a transposition. Any permutation of $K$ elements can be built up from at most $K - 1$ transpositions.

Permutations can also be arranged into disjoint cycles

$$P = \begin{pmatrix} P_2 & P_3 & P_4 & P_1 & P_6 & P_7 & P_8 & P_9 & P_5 & \cdot & \cdot & \cdot & \cdot & \cdot \\ P_1 & P_2 & P_3 & P_4 & P_5 & P_6 & P_7 & P_8 & P_9 & \cdot & \cdot & \cdot & \cdot & \cdot \end{pmatrix},$$

$$\underbrace{\phantom{P_2 \; P_3 \; P_4 \; P_1}}_{\text{first cycle}} \underbrace{\phantom{P_6 \; P_7 \; P_8 \; P_9 \; P_5}}_{\text{second cycle}} \underbrace{\phantom{\cdot \; \cdot \; \cdot}}_{\text{other cycles}}$$

$$\tag{1.22}$$

which can be written in a cycle representation as

$$P = (P_1, P_2, P_3, P_4)(P_5, \ldots, P_9)(\ldots)(\ldots). \tag{1.23}$$

---

[4]Anticipating later applications in quantum physics, we write permutations "bottom-up" as $\begin{pmatrix} P_1 & \cdots & P_K \\ 1 & \cdots & K \end{pmatrix}$ rather than "top-down" $\begin{pmatrix} 1 & \cdots & K \\ P_1 & \cdots & P_K \end{pmatrix}$, as is more common.

In this representation, we simply record in one pair of parentheses that $P_1$ is followed by $P_2$, which is in turn followed by $P_3$, etc., until we come back to $P_1$. The order of writing the cycles is without importance. In addition, each cycle of length $k$ has $k$ equivalent representations. We could, for example, write the permutation $P$ of eqn (1.23) as

$$P = (P_5, \ldots, P_9)(P_4, P_1, P_2, P_3)(\ldots)(\ldots).$$

Cycle representations will be of central importance in later chapters; in particular, the fact that every cycle of $k$ elements can be reached from the identity by means of $k-1$ transpositions. As an example, we can see that multiplying the identity permutation $\left(\begin{smallmatrix}1&2&3&4\\1&2&3&4\end{smallmatrix}\right)$ by the three transpositions of $(1, 2)$, $(1, 3)$, and $(1, 4)$ gives

$$\underbrace{\left(\begin{smallmatrix}1&2&3&4\\1&2&3&4\end{smallmatrix}\right)}_{\text{identity}}\underbrace{\left(\begin{smallmatrix}2&1&3&4\\1&2&3&4\end{smallmatrix}\right)}_{1\leftrightarrow2}\underbrace{\left(\begin{smallmatrix}2&3&1&4\\2&1&3&4\end{smallmatrix}\right)}_{1\leftrightarrow3}\underbrace{\left(\begin{smallmatrix}2&3&4&1\\2&3&1&4\end{smallmatrix}\right)}_{1\leftrightarrow4} = \left(\begin{smallmatrix}2&3&4&1\\1&2&3&4\end{smallmatrix}\right) = (1, 2, 3, 4),$$

a cycle of four elements. More generally, we can now consider a permutation $P$ of $K$ elements containing $n$ cycles, with $\{k_1, \ldots, k_n\}$ elements. The first cycle, which has $k_1$ elements, is generated from the identity by $k_1 - 1$ transpositions, the second cycle by $k_2 - 1$, etc. The total number of transpositions needed to reach $P$ is $(k_1 - 1) + \cdots + (k_n - 1)$, but since $K = k_1 + \cdots + k_n$, we can see that the number of transpositions is $K - n$. The sign of a permutation is positive if the number of transpositions from the identity is even, and odd otherwise (we then speak of even and odd permutations). We see that

$$\text{sign } P = (-1)^{K-n} = (-1)^{K+n} = (-1)^{\# \text{ of transpositions}}. \qquad (1.24)$$

We can always add extra transpositions and undo them later, but the sign of the permutation remains the same. Let us illustrate the crucial relation between the number of cycles and the sign of a permutation by an example:

$$P = \left(\begin{smallmatrix}4&2&5&7&6&3&8&1\\1&2&3&4&5&6&7&8\end{smallmatrix}\right) = (1, 4, 7, 8)(2)(3, 5, 6).$$

This is a permutation of $K = 8$ elements with $n = 3$ cycles. It must be odd, because of eqn (1.24). To see this, we rearrange the columns of $P$ to make the elements of the same cycle come next to each other:

$$P = \begin{pmatrix} 4 & 7 & 8 & 1 & 2 & 5 & 6 & 3 \\ \underbrace{1 \quad 4 \quad 7 \quad 8}_{\text{first cycle}} & 2 & \underbrace{3 \quad 5 \quad 6}_{\text{third cycle}} \end{pmatrix}. \qquad (1.25)$$

In this representation, it is easy to see that the first cycle is generated in three transpositions from $\left(\begin{smallmatrix}1&4&7&8\\1&4&7&8\end{smallmatrix}\right)$; the second cycle, consisting of the element 2, needs no transposition; and the third cycle is generated in two transpositions from $\left(\begin{smallmatrix}3&5&6\\3&5&6\end{smallmatrix}\right)$. The total number of transpositions is five, and the permutation is indeed odd.

Let us now sample random permutations, i.e. generate one of the $K!$ permutations with equal probability. In our picture of permutations as balls on a shelf, it is clear that a random permutation can be created by placing all $K$ balls into a bucket, and randomly picking out one after the other and putting them on the shelf, in their order of appearance. Remarkably, one can implement this procedure with a single vector of length $K$, which serves both as the bucket and as the shelf (see Alg. 1.11 (ran-perm); for an example, see Table 1.7). We may instead stop the

**procedure ran-perm**
$$\{P_1, \ldots, P_K\} \leftarrow \{1, \ldots, K\}$$
**for** $k = 1, \ldots, K - 1$ **do**
$$\begin{cases} l \leftarrow \mathtt{nran}(k, K) \\ P_l \leftrightarrow P_k \end{cases}$$
**output** $\{P_1, \ldots, P_K\}$

**Table 1.7** Example run of Alg. 1.11 (ran-perm). In each step $k$, the numbers $k$ and $l$ are underlined.

| #  | $P_1$ | $P_2$ | $P_3$ | $P_4$ | $P_5$ |
|----|-------|-------|-------|-------|-------|
| 1  | $\underline{1}$ | 2 | 3 | $\underline{4}$ | 5 |
| 2  | 4 | $\underline{2}$ | $\underline{3}$ | 1 | 5 |
| 3  | 4 | 3 | $\underline{\underline{2}}$ | 1 | 5 |
| 4  | 4 | 3 | 2 | $\underline{1}$ | $\underline{5}$ |
|    | 4 | 3 | 2 | 5 | 1 |

**Algorithm 1.11 ran-perm**. Generating a uniformly distributed random permutation of $K$ elements.

process after $M$ steps, rather than $K$ ($M < K$), to sample a random combination (see Alg. 1.12 (ran-combination)).

**procedure ran-combination**
$$\{P_1, \ldots, P_K\} \leftarrow \{1, \ldots, K\}$$
**for** $k = 1, \ldots, M$ **do**
$$\begin{cases} l \leftarrow \mathtt{nran}(k, K) \\ P_l \leftrightarrow P_k \end{cases}$$
**output** $\{P_1, \ldots, P_M\}$

**Algorithm 1.12 ran-combination**. Generating a uniformly distributed random combination of $M$ elements from $K$.

These two programs are the most basic examples of random combinatorics, a mature branch of mathematics, where natural, easy-to-prove algorithms are right next to tough ones. For illustration, we consider a closely related problem of (a few) black dots and many white dots

$$\left( \bullet \; \bullet \; \bullet \; \circ \; \circ \; \circ \; \circ \; \circ \; \circ \; \circ \; \circ \; \circ \; \circ \; \circ \; \circ \; \circ \; \circ \; \circ \right),$$

which we want to mix, as in

$$\left( \circ \; \bullet \; \circ \; \circ \; \circ \; \circ \; \circ \; \circ \; \circ \; \bullet \; \circ \; \circ \; \circ \; \circ \; \bullet \; \circ \; \circ \; \circ \right).$$

The fastest algorithm for mixing the three black dots with the white ones is a slightly adapted version of Alg. 1.12 (ran-combination): first swap the black dot at position 1 with the element at position $j = \mathtt{nran}(1, 18)$, then exchange whatever is at position 2 with the element at $\mathtt{nran}(2, 18)$, and finally swap the contents of the positions 3 and $\mathtt{nran}(3, 18)$. The resulting configuration of dots is indeed a random mixture, but it takes a bit of thought to prove this.

### 1.2.3 Finite distributions

The sampling of nonuniform finite distributions has an archetypal example in the Saturday night problem. We imagine $K$ possible evening activities that we do not feel equally enthusiastic about: study ($k = 1$, probability $\pi_1 \gtrsim 0$), chores ($k = 2$, probability $\pi_2 \ll 1$), cinema, book writing, etc. The probabilities $\pi_k$ are all known, but we may still have trouble deciding what to do. This means that we have trouble in sampling the distribution $\{\pi_1, \ldots, \pi_K\}$. Two methods allow us to solve the Saturday night problem: a basic rejection algorithm, and tower sampling, a rejection-free approach.

**procedure** `reject-finite`
$\pi_{\max} \leftarrow \max_{k=1}^{K} \pi_k$
1  $k \leftarrow \mathtt{nran}\,(1, K)$
$\Upsilon \leftarrow \mathtt{ran}\,(0, \pi_{\max})$
**if** $(\Upsilon > \pi_k)$ **goto** 1
**output** $k$

**Algorithm 1.13** `reject-finite`. Sampling a finite distribution $\{\pi_1, \ldots, \pi_K\}$ with a rejection algorithm.

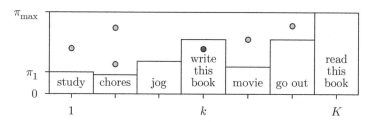

**Fig. 1.28** Saturday night problem solved by Alg. 1.13 (`reject-finite`).

In the rejection method (see Alg. 1.13 (`reject-finite`)), pebbles are randomly thrown into a big frame containing boxes for all the activities, whose sizes represent their probabilities. Eventually, one pebble falls into one of them and makes our choice. Clearly, the acceptance rate of the algorithm is given by the ratio of the sum of the volumes of all boxes to the volume of the big frame and is equal to $\langle \pi \rangle / \pi_{\max}$, where the mean probability is $\langle \pi \rangle = \sum_k \pi_k / K$. This implies that on average we have to throw $\pi_{\max} / \langle \pi \rangle$ pebbles before we have a hit. This number can easily become so large that the rejection algorithm is not really an option.

Tower sampling is a vastly more elegant solution to the Saturday night problem. Instead of placing the boxes next to each other, as in Fig. 1.28, we pile them up (see Fig. 1.29). Algorithm 1.14 (`tower-sample`) keeps track of the numbers $\Pi_1 = \pi_1, \Pi_2 = \pi_1 + \pi_2$, etc. With a single random number $\mathtt{ran}\,(0, \Pi_K)$, an activity $k$ is then chosen. There is no rejection.

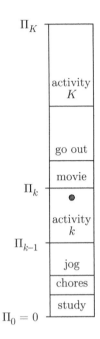

$\Pi_K$

activity
$K$

go out

movie

$\Pi_k$  ●

activity
$k$

$\Pi_{k-1}$

jog

chores

study

$\Pi_0 = 0$

**Fig. 1.29** Saturday night problem solved by tower sampling.

**procedure** `tower-sample`
**input** $\{\pi_1, \ldots, \pi_K\}$
$\Pi_0 \leftarrow 0$
**for** $l = 1, \ldots, K$ **do** $\Pi_l \leftarrow \Pi_{l-1} + \pi_l$
$\Upsilon \leftarrow \mathrm{ran}\,(0, \Pi_K)$
\*     find $k$ with $\Pi_{k-1} < \Upsilon < \Pi_k$
**output** $k$

———

**Algorithm 1.14** `tower-sample`. Tower sampling of a finite distribution $\{\pi_1, \ldots, \pi_K\}$ without rejections.

Tower sampling can be applied to discrete distributions with a total number $K$ in the hundreds, thousands, or even millions. It often works when the naive rejection method of Fig. 1.28 fails because of too many rejections. Tower sampling becomes impracticable only when the probabilities $\{\pi_1, \ldots, \pi_K\}$ can no longer be listed.

In Alg. 1.14 (`tower-sample`), we must clarify how we actually find the element $k$, i.e. how we implement the line marked by an asterisk. For small $K$, we may go through the ordered table $\{\Pi_0, \ldots, \Pi_K\}$ one by one, until the element $k$, with $\Pi_{k-1} < \Upsilon \le \Pi_k$, is encountered. For large $K$, a bisection method should be implemented if we are making heavy use of tower sampling: we first check whether $\Upsilon$ is smaller or larger than $\Pi_{K/2}$, and then cut the possible range of indices $k$ in half on each subsequent iteration. Algorithm 1.15 (`bisection-search`) terminates in about $\log_2 K$ steps.

**procedure** `bisection-search`
**input** $\Upsilon, \{\Pi_0, \Pi_1, \ldots, \Pi_K\}$ (ordered table with $\Pi_k \ge \Pi_{k-1}$)
$k_{\min} \leftarrow 0$
$k_{\max} \leftarrow K + 1$
**for** $i = 1, 2, \ldots$ **do**
$\left\{ \begin{array}{l} k \leftarrow (k_{\min} + k_{\max})/2 \text{ (integer arithmetic)} \\ \textbf{if } (\Pi_k < \Upsilon) \textbf{ then} \\ \quad \{\ k_{\min} \leftarrow k \\ \textbf{else if } (\Pi_{k-1} > \Upsilon) \textbf{ then} \\ \quad \{\ k_{\max} \leftarrow k \\ \textbf{else} \\ \quad \left\{ \begin{array}{l} \textbf{output } k \\ \textbf{exit} \end{array} \right. \end{array} \right.$

———

**Algorithm 1.15** `bisection-search`. Locating the element $k$ with $\Pi_{k-1} < \Upsilon < \Pi_k$ in an ordered table $\{\Pi_0, \ldots, \Pi_K\}$.

### 1.2.4 Continuous distributions and sample transformation

The two discrete methods of Subsection 1.2.3 remain meaningful in the continuum limit. For the rejection method, the arrangement of boxes in Fig. 1.28 simply becomes a continuous curve $\pi(x)$ in some range $x_{min} < x < x_{max}$ (see Alg. 1.16 (`reject-continuous`)). We shall often use a refinement of this simple scheme, where the function $\pi(x)$, which we want to sample, is compared not with a constant function $\pi_{max}$ but with another function $\tilde{\pi}(x)$ that we know how to sample, and is everywhere larger than $\pi(x)$ (see Subsection 2.3.4).

**procedure** `reject-continuous`
1    $x \leftarrow \mathbf{ran}\,(x_{min}, x_{max})$
     $\Upsilon \leftarrow \mathbf{ran}\,(0, \pi_{max})$
     **if** $(\Upsilon > \pi(x))$ **goto** 1 (reject sample)
     **output** $x$
——

**Algorithm 1.16** `reject-continuous`. Sampling a value $x$ with probability $\pi(x) < \pi_{max}$ in the interval $[x_{min}, x_{max}]$ with the rejection method.

For the continuum limit of tower sampling, we change the discrete index $k$ in Alg. 1.14 (`tower-sample`) into a real variable $x$:

$$\{k, \pi_k\} \longrightarrow \{x, \pi(x)\}$$

(see Fig. 1.30). This gives us the transformation method: the loop in the third line of Alg. 1.14 (`tower-sample`) turns into an integral formula:

$$\overbrace{\Pi_k \leftarrow \Pi_{k-1} + \pi_k}^{\text{in Alg. 1.14 (tower-sample)}} \longrightarrow \Pi(x) = \overbrace{\int_{-\infty}^{x} dx'\, \pi(x')}^{\Pi(x)=\Pi(x-dx)+\pi(x)dx}. \qquad (1.26)$$

Likewise, the line marked by an asterisk in Alg. 1.14 (`tower-sample`) has an explicit solution:

$$\overbrace{\text{find } k \text{ with } \Pi_{k-1} < \Upsilon < \Pi_k}^{\text{in Alg. 1.14 (tower-sample)}} \longrightarrow \overbrace{\text{find } x \text{ with } \Pi(x) = \Upsilon}^{\text{i.e. } x=\Pi^{-1}(\Upsilon)}, \qquad (1.27)$$

where $\Pi^{-1}$ is the inverse function of $\Pi$.

As an example, let us sample random numbers $0 < x < 1$ distributed according to an algebraic function $\pi(x) \propto x^\gamma$ (with $\gamma > -1$) (see Fig. 1.30, which shows the case $\gamma = -\frac{1}{2}$). We find

$$\pi(x) = (\gamma + 1)x^\gamma \text{ for } 0 < x < 1,$$
$$\Pi(x) = \int_0^x dx\, \pi(x') = x^{\gamma+1} = \mathbf{ran}\,(0,1),$$
$$x = \mathbf{ran}\,(0,1)^{1/(\gamma+1)}. \qquad (1.28)$$

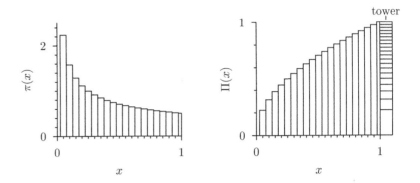

**Fig. 1.30** Transformation method as the continuum limit of tower sampling.

The transformation method as laid out in eqns (1.26) and (1.27) can be interpreted as a sample transformation, stressing the unity between integration and sampling: any change of variables in an integral can be done directly with random numbers, i.e. with samples. Indeed, in the above example of an algebraic function, we can transform the integral over a flat distribution into the integral of the target distribution:

$$\int_0^1 d\Upsilon \xrightarrow[\text{transform}]{\text{integral}} \text{const} \int_0^1 dx \; x^\gamma.$$

The same transformation works for samples:

$$\left\{ \begin{array}{c} \text{sample } \Upsilon \\ \Upsilon = \mathtt{ran}\,(0,1) \end{array} \right\} \xrightarrow[\text{transform}]{\text{sample}} \left\{ \begin{array}{c} \text{sample } x \\ \text{with } \pi(x) \propto x^\gamma \end{array} \right\}.$$

We now seek the transformation between $x$ and $\Upsilon$:

$$d\Upsilon = \text{const} \cdot dx \; x^\gamma.$$

The sample $\Upsilon = \mathtt{ran}\,(0,1)$ is thus transformed as follows:

$$\mathtt{ran}\,(0,1) = \Upsilon = \text{const}' \cdot x^{\gamma+1} + \text{const}''.$$

Finally (checking that the bounds of $\mathtt{ran}\,(0,1)$ correspond to $x = 0$ and $x = 1$), this results in

$$x = \mathtt{ran}\,(0,1)^{1/(\gamma+1)}, \tag{1.29}$$

in agreement with eqn (1.28). (In Subsection 1.4.2, we shall consider algebraic distributions for $x$ between 1 and $\infty$.)

As a second example of sample transformation, we consider random numbers that are exponentially distributed, so that $\pi(x) \propto e^{-\lambda x}$ for $x \geq 0$. As before, we write

$$\int_0^1 d\Upsilon = \text{const} \int_0^\infty dx \; e^{-\lambda x} \tag{1.30}$$

and seek a transformation of a flat distribution of $\Upsilon$ in the interval $[0, 1]$ into the target distribution of $x$:

$$d\Upsilon = \text{const} \cdot dx \, e^{-\lambda x},$$
$$\text{ran}\,(0, 1) = \Upsilon = \text{const}' \cdot e^{-\lambda x} + \text{const}''.$$

Checking the bounds $x = 0$ and $x = \infty$, this leads to

$$-\frac{1}{\lambda} \log \text{ran}\,(0, 1) = x. \tag{1.31}$$

In this book, we shall often transform samples under the integral sign, in the way we have seen in the two examples of the present subsection.

## 1.2.5 Gaussians

In many applications, we need Gaussian random numbers $y$ distributed with a probability

$$\pi(y) = \frac{1}{\sqrt{2\pi}\sigma} \exp\left[-\frac{(y - \langle y \rangle)^2}{2\sigma^2}\right].$$

(The parameter $\langle y \rangle$ is the mean value of the distribution, and $\sigma$ is the standard deviation.) One can always change variables using $x = (y - \langle y \rangle)/\sigma$ to obtain normally distributed variables $x$ with a distribution

$$\pi(x) = \frac{1}{\sqrt{2\pi}} \exp\left(-x^2/2\right). \tag{1.32}$$

Inversely, the normally distributed variables $x$ can be rescaled into $y = \sigma x + \langle y \rangle$.

Naively, to sample Gaussians such as those described in eqn (1.32), one can compute the sum of a handful of independent random variables, essentially $\text{ran}\,(-1, 1)$, and rescale them properly (see Alg. 1.17 (naive-gauss), the factor $1/12$ will be derived in eqn (1.55)). This program illustrates the power of the central limit theorem, which will be discussed further in Subsection 1.3.3. Even for $K = 3$, the distribution takes on the characteristic Gaussian bell shape (see Fig. 1.31).

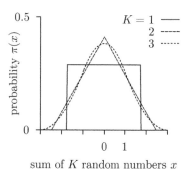

Fig. 1.31 Probability for the sum of $K$ random numbers $\text{ran}\left(-\frac{1}{2}, \frac{1}{2}\right)$ for small $K$, rescaled as in Alg. 1.17 (naive-gauss).

**procedure naive-gauss**
$\sigma \leftarrow \sqrt{K/12}$
$\Sigma \leftarrow 0$
**for** $k = 1, \ldots, K$ **do**
$\quad\left\{ \; \Sigma \leftarrow \Sigma + \text{ran}\left(-\frac{1}{2}, \frac{1}{2}\right)\right.$
$x \leftarrow \Sigma/\sigma$
**output** $x$

**Algorithm 1.17** naive-gauss. An approximately Gaussian random number obtained from the rescaled sum of $K$ uniform random numbers.

With Alg. 1.17 (naive-gauss), we shall always worry whether its parameter $K$ is large enough. For practical calculations, it is preferable to sample Gaussian random numbers without approximations. To do so, we recall the trick used to evaluate the error integral

$$\int_{-\infty}^{\infty} \frac{dx}{\sqrt{2\pi}} e^{-x^2/2} = 1. \tag{1.33}$$

We square eqn (1.33)

$$\left[\int_{-\infty}^{\infty} \frac{dx}{\sqrt{2\pi}} \exp\left(-x^2/2\right)\right]^2 = \int_{-\infty}^{\infty} \frac{dx}{\sqrt{2\pi}} e^{-x^2/2} \int_{-\infty}^{\infty} \frac{dy}{\sqrt{2\pi}} e^{-y^2/2} \tag{1.34}$$

$$= \int_{-\infty}^{\infty} \frac{dx\,dy}{2\pi} \exp\left[-(x^2+y^2)/2\right], \tag{1.35}$$

introduce polar coordinates $(dx\,dy = r\,dr\,d\phi)$,

$$\ldots = \int_0^{2\pi} \frac{d\phi}{2\pi} \int_0^{\infty} r\,dr\,\exp\left(-r^2/2\right),$$

and finally substitute $r^2/2 = \Upsilon$ $(r\,dr = d\Upsilon)$

$$\ldots = \underbrace{\int_0^{2\pi} \frac{d\phi}{2\pi}}_{1} \underbrace{\int_0^{\infty} d\Upsilon\, e^{-\Upsilon}}_{1}. \tag{1.36}$$

Equation (1.36) proves the integral formula in eqn (1.33). In addition, the two independent integrals in eqn (1.36) are easily sampled: $\phi$ is uniformly distributed between 0 and $2\pi$, and $\Upsilon$ is exponentially distributed and can be sampled through the negative logarithm of $\mathbf{ran}\,(0,1)$ (see eqn (1.31), with $\lambda = 1$). After generating $\Upsilon = -\log \mathbf{ran}\,(0,1)$ and $\phi = \mathbf{ran}\,(0,2\pi)$,

**procedure gauss**
**input** $\sigma$
$\phi \leftarrow \mathbf{ran}\,(0,2\pi)$
$\Upsilon \leftarrow -\mathbf{log}\,\mathbf{ran}\,(0,1)$
$r \leftarrow \sigma\sqrt{2\Upsilon}$
$x \leftarrow r\mathbf{cos}\,\phi$
$y \leftarrow r\mathbf{sin}\,\phi$
**output** $\{x,y\}$

**Algorithm 1.18 gauss.** Two independent Gaussian random numbers obtained by sample transformation. See Alg. 1.19 (gauss(patch)).

we transform the sample, as discussed in Subsection 1.2.4: a crab's walk leads from eqn (1.36) back to eqn (1.34) $(r = \sqrt{2\Upsilon}, x = r\cos\phi, y = r\sin\phi)$. We finally get two normally distributed Gaussians, in one of the nicest applications of multidimensional sample transformation (see Alg. 1.18 (gauss)).

Algorithm 1.18 (**gauss**) can be simplified further. As discussed in Subsection 1.1.3, we can generate uniform random angles by throwing pebbles into the children's square and retaining those inside the circle (see Fig. 1.9). This pebble–needle trick yields a random angle $\phi$, but allows us also to determine $\sin\phi$ and $\cos\phi$ without explicit use of trigonometric tables (see Alg. 1.19 (**gauss**(patch))). In the printed routine, $x/\sqrt{\Upsilon'} = \cos\phi$, etc. Moreover, the variable $\Upsilon' = x^2 + y^2$, for $\{x,y\}$ inside the circle, is itself uniformly distributed, so that $\Upsilon'$ in Alg. 1.19 (**gauss**(patch)) can replace the **ran**$(0,1)$ in the original Alg. 1.18 (**gauss**). This Box–Muller algorithm is statistically equivalent to but marginally faster than Alg. 1.18 (**gauss**).

> **procedure gauss**(patch)
> **input** $\sigma$
> 1  $x \leftarrow$ **ran**$(-1,1)$
> $y \leftarrow$ **ran**$(-1,1)$
> $\Upsilon' \leftarrow x^2 + y^2$
> **if** $(\Upsilon' > 1$ **or** $\Upsilon' = 0)$ **goto** 1 (reject sample)
> $\Upsilon \leftarrow -\log\Upsilon'$
> $\Upsilon'' \leftarrow \sigma\sqrt{2\Upsilon/\Upsilon'}$
> $x \leftarrow \Upsilon''x$
> $y \leftarrow \Upsilon''y$
> **output** $\{x,y\}$
> ——

**Algorithm 1.19** **gauss**(patch). Gaussian random numbers, as in Alg. 1.18 (**gauss**), but without calls to trigonometric functions.

## 1.2.6 Random points in/on a sphere

The pebble–needle trick, in its recycled version in Subsection 1.2.5, shows that any random point inside the unit disk of Fig. 1.1 can be transformed into two independent Gaussian random numbers:

$$\left\{\begin{array}{c} 2\text{ Gaussian} \\ \text{samples} \end{array}\right\} \Longleftarrow \left\{\begin{array}{c} \text{random pebble in} \\ \text{unit disk} \end{array}\right\}.$$

This remarkable relation between Gaussians and pebbles is unknown to most players of games in Monaco, short or tall. The unit disk is the same as the two-dimensional unit sphere, and we may wonder whether a random point inside a $d$-dimensional unit sphere could be helpful for generating $d$ Gaussians at a time. Well, the truth is exactly the other way around: sampling $d$ Gaussians provides a unique technique for obtaining a random point in a $d$-dimensional unit sphere:

$$\left\{\begin{array}{c} d\text{ Gaussian} \\ \text{samples} \end{array}\right\} \Longrightarrow \left\{\begin{array}{c} \text{random point in} \\ d\text{-dimensional unit sphere} \end{array}\right\}. \tag{1.37}$$

This relationship is closely linked to the Maxwell distribution of velocities in a gas, which we shall study in Chapter 2.

In higher dimensions, the sampling of random points inside the unit sphere cannot really be achieved by first sampling points in the cube surrounding the sphere and rejecting all those positions which have a radius in excess of one (see Alg. 1.20 (`naive-sphere`)). Such a modified children's algorithm has a very high rejection rate, as we shall show now before returning to eqn (1.37).

**procedure** `naive-sphere`
1   $\Sigma \leftarrow 0$
 **for** $k = 1, \ldots, d$ **do**
  $\Big\{$ $x_k \leftarrow \mathbf{ran}\,(-1, 1)$
   $\Sigma \leftarrow \Sigma + x_k^2$
   **if** $(\Sigma > 1)$ **goto** 1 (reject sample)
 **output** $\{x_1, \ldots, x_d\}$

——

**Algorithm 1.20** `naive-sphere`. Sampling a uniform random vector inside the $d$-dimensional unit sphere.

The acceptance rate of Alg. 1.20 (`naive-sphere`) is related to the ratio of volumes of the unit hypersphere and a hypercube of side length 2:

$$\left\{\begin{matrix}\text{volume}\\\text{of unit sphere}\\\text{in } d \text{ dim.}\end{matrix}\right\} = \left\{\begin{matrix}\text{volume of}\\d\text{-dim. cube}\\\text{of length 2}\end{matrix}\right\}\left\{\begin{matrix}\text{acceptance rate of}\\\text{Alg. 1.20}\end{matrix}\right\}. \quad (1.38)$$

The volume $V_d(R)$ of a $d$-dimensional sphere of radius $R$ is

$$V_d(R) = \int_{x_1^2 + \cdots + x_d^2 \leq R^2} dx_1 \ldots dx_d = \underbrace{\left[\frac{\pi^{d/2}}{\Gamma(d/2 + 1)}\right]}_{V_d(1)} R^d, \quad (1.39)$$

where $\Gamma(x) = \int_0^\infty dt\, t^{x-1} e^{-t}$, the gamma function, generalizes the factorial. This function satisfies $\Gamma(x+1) = x\Gamma(x)$. For an integer argument $n$, $\Gamma(n) = (n-1)!$. The value $\Gamma(1/2) = \sqrt{\pi}$ allows us to construct the gamma function for all half-integer $x$.

In this book, we have a mission to derive all equations explicitly, and thus need to prove eqn (1.39). Clearly, the volume of a $d$-dimensional sphere of radius $R$ is proportional to $R^d$ (the area of a disk goes as the radius squared, the volume of a three-dimensional sphere goes as the cube, etc.), and we only have to determine $V_d(1)$, the volume of the unit sphere. Splitting the $d$-dimensional vector $\mathbf{R} = \{x_1, \ldots, x_d\}$ as

$$\mathbf{R} = \{\underbrace{x_1, x_2}_{\mathbf{r}}, \underbrace{x_3, \ldots, x_d}_{\mathbf{u}}\}, \quad (1.40)$$

where $R^2 = r^2 + u^2$, and writing $\mathbf{r}$ in two-dimensional polar coordinates

$\mathbf{r} = \{r, \phi\}$, we find

$$V_d(1) = \int_0^1 r \, dr \int_0^{2\pi} d\phi \overbrace{\int_{x_3^2 + \cdots + x_d^2 \le 1 - r^2}}^{(d-2)\text{-dim. sphere of radius } \sqrt{1-r^2}} dx_3 \ldots dx_d \qquad (1.41)$$

$$= 2\pi \int_0^1 dr \, r V_{d-2}(\sqrt{1 - r^2})$$

$$= 2\pi V_{d-2}(1) \int_0^1 dr \, r \left(\sqrt{1 - r^2}\right)^{d-2}$$

$$= \pi V_{d-2}(1) \int_0^1 du \, u^{d/2 - 1} = \frac{\pi}{d/2} V_{d-2}(1). \qquad (1.42)$$

The relation in eqns (1.41) and (1.42) sets up a recursion for the volume of the $d$-dimensional unit sphere that starts with the elementary values $V_1(1) = 2$ (line) and $V_2(1) = \pi$ (disk), and immediately leads to eqn (1.39).

It follows from eqn (1.39) that, for large $d$,

$$\left\{ \begin{array}{c} \text{volume of} \\ \text{unit hypersphere} \end{array} \right\} \ll \left\{ \begin{array}{c} \text{volume of} \\ \text{hypercube of side 2} \end{array} \right\},$$

and this in turn implies (from eqn (1.38)) that the acceptance rate of Alg. 1.20 (naive-sphere) is close to zero for $d \gg 1$ (see Table 1.8). The naive algorithm thus cannot be used for sampling points inside the $d$-dimensional sphere for large $d$.

Going back to eqn (1.37), we now consider $d$ independent Gaussian random numbers $\{x_1, \ldots, x_d\}$. Raising the error integral in eqn (1.33) to the $d$th power, we obtain

$$1 = \int \ldots \int dx_1 \ldots dx_d \left(\frac{1}{\sqrt{2\pi}}\right)^d \exp\left[-\frac{1}{2}(x_1^2 + \cdots + x_d^2)\right], \qquad (1.43)$$

an integral which can be sampled by $d$ independent Gaussians in the same way that the integral in eqn (1.34) is sampled by two of them.

The integrand in eqn (1.43) depends only on $r^2 = (x_1^2 + \cdots + x_d^2)$, i.e. on the square of the radius of the $d$-dimensional sphere. It is therefore appropriate to write it in polar coordinates with the $d - 1$ angular variables; the solid angle $\Omega$ is taken all over the unit sphere, the radial variable is $r$, and the volume element is

$$dV = dx_1 \ldots dx_d = r^{d-1} \, dr \, d\Omega.$$

Equation (1.43), the integral sampled by $d$ Gaussians, becomes

$$1 = \left(\frac{1}{\sqrt{2\pi}}\right)^d \int_0^\infty dr \, r^{d-1} \exp\left(-r^2/2\right) \int d\Omega. \qquad (1.44)$$

This is almost what we want, because our goal is to sample random points in the unit sphere, that is, to sample, in polar coordinates, the integral

$$V_d(1) = \int_0^1 dr \, r^{d-1} \int d\Omega \quad \text{(unit sphere)}. \qquad (1.45)$$

Table 1.8 Volume $V_d(1)$ and acceptance rate in $d$ dimensions for Alg. 1.20 (naive-sphere)

| $d$ | $V_d(1)$ | Acceptance rate |
|---|---|---|
| 2 | $\pi$ | $\pi/4 = 0.785$ |
| 4 | 4.9348 | 0.308 |
| 10 | 2.55016 | 0.002 |
| 20 | 0.026 | $2.46 \times 10^{-8}$ |
| 60 | $3.1 \times 10^{-18}$ | $2.7 \times 10^{-36}$ |

```
procedure direct-sphere
  Σ ← 0
  for k = 1, ..., d do
    {  x_k ← gauss(σ)
       Σ ← Σ + x_k^2
    Υ ← ran(0,1)^{1/d}
  for k = 1, ..., d do
    {  x_k ← Υx_k/√Σ
  output {x_1, ..., x_d}
  ─────
```

**Algorithm 1.21** direct-sphere. Uniform random vector inside the $d$-dimensional unit sphere. The output is independent of $\sigma$.

The angular parts of the integrals in eqns (1.44) and (1.45) are identical, and this means that the $d$ Gaussians sample angles isotropically in $d$ dimensions, and only get the radius wrong. This radius should be sampled from a distribution $\pi(r) \propto r^{d-1}$. From eqn (1.29), we obtain the direct distribution of $r$ by taking the $d$th root of a random number $\text{ran}(0,1)$. A simple rescaling thus yields the essentially rejection-free Alg. 1.21 (direct-sphere), one of the crown jewels of the principality of Monte Carlo.

```
procedure direct-surface
  σ ← 1/√d
  Σ ← 0
  for k = 1, ..., d do
    {  x_k ← gauss(σ)
       Σ ← Σ + x_k^2
  for k = 1, ..., d do
    {  x_k ← x_k/√Σ
  output {x_1, ..., x_d}
  ─────
```

**Algorithm 1.22** direct-surface. Random vector on the surface of the $d$-dimensional unit sphere. For large $d$, $\Sigma$ approaches one (see Fig. 1.32).

The above program illustrates the profound relationship between sampling and integration, and we can take the argument leading to Alg. 1.21 (direct-sphere) a little further to rederive the volume of the unit sphere. In eqns (1.45) and (1.44), the multiple integrals over the angular variables $d\Omega$ are the same, and we may divide them out:

The integral in the denominator of eqn (1.46), where $r^2/2 = u$ and $r\,dr = du$, becomes

$$2^{d/2-1} \underbrace{\int_0^\infty du\ u^{d/2-1} e^{-u}}_{\Gamma(d/2)}.$$

$$V_d(1) = \frac{\text{eqn (1.45)}}{\text{eqn (1.44)}} = \frac{\left(\sqrt{2\pi}\right)^d \int_0^1 dr\ r^{d-1}}{\int_0^\infty dr\ r^{d-1} \exp\left(-r^2/2\right)} = \frac{\pi^{d/2}}{(d/2)\Gamma(d/2)}. \quad (1.46)$$

This agrees with the earlier expression in eqn (1.39). The derivation in eqn (1.46) of the volume of the unit sphere in $d$ dimensions is lightning-

fast, a little abstract, and as elegant as Alg. 1.21 (`direct-sphere`) itself. After sampling the Gaussians $\{x_1, \ldots, x_d\}$, we may also rescale the
vector to unit length. This amounts to sampling a random point on
the surface of the $d$-dimensional unit sphere, not inside its volume (see
Alg. 1.22 (`direct-surface`)).

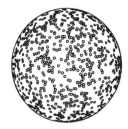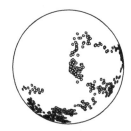

**Fig. 1.32** Random pebbles on the surface of a sphere (from Alg. 1.22
(direct sampling, *left*) and Alg. 1.24 (Markov-chain sampling, *right*)).

The sampling algorithm for pebbles on the surface of a $d$-dimensional
sphere can be translated into many computer languages similarly to the
way Alg. 1.22 (`direct-surface`) is written. However, vector notation
is more concise, not only in formulas but also in programs. This means
that we collect the $d$ components of a vector into a single symbol (as
already used before)

$$\mathbf{x} = \{x_1, \ldots, x_d\}.$$

The Cartesian scalar product of two vectors is

$$(\mathbf{x} \cdot \mathbf{x}') = x_1 x_1' + \cdots + x_d x_d',$$

and the square root of the scalar product of a vector with itself gives the
vector's norm,

$$|\mathbf{x}| = \sqrt{(\mathbf{x} \cdot \mathbf{x})} = \sqrt{x_1^2 + \cdots + x_d^2}.$$

**procedure** direct-surface(vector notation)
$\sigma \leftarrow 1/\sqrt{d}$
$\mathbf{x} \leftarrow \{\text{gauss}\,(\sigma), \ldots, \text{gauss}\,(\sigma)\}$
$\mathbf{x} \leftarrow \mathbf{x}/|\mathbf{x}|$
**output** $\mathbf{x}$
———

**Algorithm 1.23** direct-surface(vector notation). Same program as
Alg. 1.22, in vector notation, with a $d$-dimensional vector $\mathbf{x}$.

The difference between Alg. 1.22 and Alg. 1.23 is purely notational.
The actual implementation in a computer language will depends on how
vectors can be addressed. In addition, it depends on the random number
generator whether a line such as $\mathbf{x} \leftarrow \{\text{gauss}\,(\sigma), \ldots, \text{gauss}\,(\sigma)\}$ can

be implemented as a vector instruction or must be broken up into a sequential loop because of possible conflicts with the seed-and-update mechanism of the random number generator.

For later use, we also present a Markov-chain algorithm on the surface of a $d$-dimensional unit sphere, written directly in vector notation (see Alg. 1.24 (`markov-surface`)). The algorithm constructs a random vector $\epsilon$ from Gaussians, orthogonalizes it, and normalizes it with respect to the input vector $\mathbf{x}$, the current position of the Markov chain. The step taken is in the direction of this reworked $\epsilon$, a random unit vector in the hyperplane orthogonal to $\mathbf{x}$.

**procedure** `markov-surface`
**input** $\mathbf{x}$ (unit vector $|\mathbf{x}| = 1$)
$\epsilon \leftarrow \{\mathtt{gauss}\,(\sigma), \dots, \mathtt{gauss}\,(\sigma)\}$ ($d$ independent Gaussians)
$\Sigma \leftarrow (\epsilon \cdot \mathbf{x})$
$\epsilon \leftarrow \epsilon - \Sigma \mathbf{x}$
$\epsilon \leftarrow \epsilon/|\epsilon|$
$\Upsilon \leftarrow \mathtt{ran}\,(-\delta, \delta)$ ($\delta$: step size)
$\mathbf{x} \leftarrow \mathbf{x} + \Upsilon \epsilon$
$\mathbf{x} \leftarrow \mathbf{x}/|\mathbf{x}|$
**output** $\mathbf{x}$

**Algorithm 1.24** `markov-surface`. Markov-chain Monte Carlo algorithm for random vectors on the surface of a $d$-dimensional unit sphere.

## 1.3   Statistical data analysis

In the first sections of this book, we familiarized ourselves with sampling as an ingenious method for evaluating integrals which are unsolvable by other methods. However, we stayed on the sunny side of the subject, avoiding the dark areas: statistical errors and the difficulties related to their evaluation. In the present section, we concentrate on the description and estimation of errors, both for independent variables and for Markov chains. We first discuss the fundamental mechanism which connects the probability distribution of a single random variable to that of a sum of independent variables, then discuss the moments of distributions, especially the variance, and finally review the basic facts of probability theory—Chebyshev's inequality, the law of large numbers, and the central limit theorem—that are relevant to statistical physics and Monte Carlo calculations. We then study the problem of estimating mean values for independent random variables (direct sampling) and for those coming from Markov-chain simulations.

### 1.3.1   Sum of random variables, convolution

In this subsection, we discuss random variables and sums of random variables in more general terms than before. The simplest example of a random variable, which we shall call $\xi_i$, is the outcome of the $i$th trial in the children's game. This is called a Bernoulli variable. It takes the value one with probability $\theta$ and the value zero with probability $(1 - \theta)$. $\xi_i$ can also stand for the $i$th call to a random number generator $\mathtt{ran}\,(0, 1)$ or $\mathtt{gauss}\,(\sigma)$, etc., or more complicated quantities. We note that in our algorithms, the index $i$ of the random variable $\xi_i$ is hidden in the seed-and-update mechanism of the random number generator.

The number of hits in the children's game is itself a random variable, and we denote it by a symbol similar to the others, $\xi$:

$$\xi = \xi_1 + \cdots + \xi_N.$$

$\xi_1$ takes values $k_1$, $\xi_2$ takes values $k_2$, etc., and the probability of obtaining $\{k_1, \ldots, k_N\}$ is, for independent variables,

$$\pi(\{k_1, \ldots, k_N\}) = \pi(k_1) \cdot \cdots \cdot \pi(k_N).$$

The sum random variable $\xi$ takes the values $\{0, \ldots, N\}$ with probabilities $\{\pi_0, \ldots, \pi_N\}$, which we shall now calculate. Clearly, only sets $\{k_1, \ldots, k_N\}$ that produce $k$ hits contribute to $\pi_k$:

$$\underbrace{\pi_k}_{\substack{N \text{ trials} \\ k \text{ hits}}} = \sum_{\substack{k_1=0,1,\ldots,k_N=0,1 \\ k_1+\cdots+k_N=k}} \overbrace{\pi(k_1, \ldots, k_N)}^{\pi(k_1)\pi(k_2)\cdots\pi(k_N)}.$$

The $k$ hits and $N - k$ nonhits among the values $\{k_1, \ldots, k_N\}$ occur with probabilities $\theta$ and $(1 - \theta)$, respectively. In addition, we have to take into account the number of ways of distributing $k$ hits amongst $N$ trials by multiplying by the combinatorial factor $\binom{N}{k}$. The random variable $\xi$, the number of hits, thus takes values $k$ with a probability distributed according to the binomial distribution

$$\pi_k = \binom{N}{k} \theta^k (1 - \theta)^{N-k} \quad (0 \leq k \leq N). \tag{1.47}$$

In this equation, the binomial coefficients are given by

$$\binom{N}{k} = \frac{N!}{k!(N-k)!} = \frac{(N-k+1)(N-k+2)\cdots N}{1 \times 2 \times \cdots \times k}. \tag{1.48}$$

We note that Alg. 1.1 ($\mathtt{direct\text{-}pi}$)—if we look only at the number of hits—merely samples this binomial distribution for $\theta = \pi/4$.

The explicit formula for the binomial distribution (eqn (1.47)) is inconvenient for practical evaluation because of the large numbers entering the coefficients in eqn (1.48). It is better to relate the probability distribution $\pi'$ for $N + 1$ trials to that for $N$ trials: the $N + 1$th trial is independent of what happened before, so that the probability $\pi'_k$ for $k$ hits with $N + 1$ trials can be put together from the independent probabilities $\pi_{k-1}$ for $k - 1$ hits and $\pi_k$ for $k$ hits with $N$ trials, and the Bernoulli distribution for a single trial:

$$\underbrace{\pi'_k}_{\substack{N+1 \text{ trials} \\ k \text{ hits}}} = \underbrace{\pi_k}_{\substack{N \text{ trials} \\ k \text{ hits}}} \cdot \underbrace{(1 - \theta)}_{\text{no hit}} + \underbrace{\pi_{k-1}}_{\substack{N \text{ trials} \\ k - 1 \text{ hits}}} \cdot \underbrace{\theta}_{\text{hit}}, \tag{1.49}$$

(see Alg. 1.25 (`binomial-convolution`)).

**procedure** `binomial-convolution`
**input** $\{\pi_0, \ldots, \pi_N\}$ ($N$ trials)
$\pi'_0 \leftarrow (1 - \theta)\pi_0$ ($\theta$: probability of hit)
**for** $k = 1, \ldots, N$ **do**
$\quad \{\ \pi'_k \leftarrow \theta\pi_{k-1} + (1 - \theta)\pi_k$
$\pi'_{N+1} \leftarrow \theta\pi_N$
**output** $\{\pi'_0, \ldots, \pi'_{N+1}\}$ ($N + 1$ trials)
———

**Table 1.9** Probabilities $\{\pi_0, \ldots, \pi_N\}$ in Alg. 1.25 (`binomial-convolution`) for small values of $N$, with $\theta = \pi/4$

| $N$ | $\pi_0$ | $\pi_1$ | $\pi_2$ | $\pi_3$ |
|-----|---------|---------|---------|---------|
| 1 | 0.215 | 0.785 | . | . |
| 2 | 0.046 | 0.337 | 0.617 | . |
| 3 | 0.010 | 0.109 | 0.397 | 0.484 |

**Algorithm 1.25** `binomial-convolution`. Probabilities of numbers of hits for $N + 1$ trials obtained from those for $N$ trials.

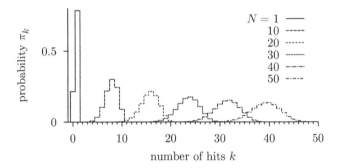

**Fig. 1.33** Probabilities for the number of hits in the children's game with $N$ trials (from Alg. 1.25 (`binomial-convolution`)), using $\theta = \pi/4$.

In evaluating the binomial distribution using eqn (1.49), we implicitly piece together the binomial coefficients in eqn (1.47) through the relations

$$\binom{N}{0} = \binom{N}{N} = 1, \quad \binom{N}{k-1} + \binom{N}{k} = \binom{N+1}{k} \quad (1 \le k \le N).$$

Replacing $\theta$ and $(1 - \theta)$ by 1 in Alg. 1.25 (`binomial-convolution`) generates Pascal's triangle of binomial coefficients.

Algorithm 1.25 (`binomial-convolution`) implements a convolution[5] of the distribution function for $N$ trials with the Bernoulli distribution of one single trial, and gives the binomial distribution (see Fig. 1.33 and Table 1.9).

Convolutions are a powerful tool, as we show now for the case where $\xi_i$ is uniformly distributed in the interval $[0, 1]$. As before, we are interested in the distribution $\pi(x)$ of the sum variable

$$\xi = \xi_1 + \cdots + \xi_N,$$

---

[5]From the Latin *convolvere*: to roll together, fold together, or intertwine.

which takes values $x$ in the interval $[0, N]$. Again thinking recursively, we can obtain the probability distribution $\pi'(x)$ for the sum of $N + 1$ random numbers by convoluting $\pi(y)$, the distribution for $N$ variables, with the uniform distribution $\pi^1(y - x)$ in the interval $[0, 1]$ for a single variable ($\pi^1(x) = 1$ for $0 < x < 1$). The arguments $y$ and $x - y$ have been chosen so that their sum is equal to $x$. To complete the convolution, the product of the probabilities must be integrated over all possible values of $y$:

$$\underbrace{\pi'(x)}_{\substack{x:\text{ sum} \\ \text{of } N+1}} = \int_{x-1}^{x} dy \underbrace{\pi(y)}_{\substack{y:\text{ sum} \\ \text{of } N}} \underbrace{\pi^1(x-y)}_{\substack{x-y: \\ \text{one term}}}. \tag{1.50}$$

Here, the integration generalizes the sum over two terms used in the binomial distribution. The integration limits implement the condition that $\pi^1(x-y)$ is different from zero only for $0 < x - y < 1$. For numerical calculation, the convolution in eqn (1.50) is most easily discretized by cutting the interval $[0, 1]$ into $l$ equal segments $\{x_0 = 0, x_1, \ldots, x_{l-1}, x_l = 1\}$ so that $x_k \equiv k/l$. In addition, weights $\{\pi_0, \ldots, \pi_l\}$ are assigned as shown in Alg. 1.26 (**ran01-convolution**). The program allows us to compute the probability distribution for the sum of $N$ random numbers, as sampled by Alg. 1.17 (**naive-gauss**) (see Fig. 1.34). The distribution of the sum of $N$ uniform random numbers is also analytically known, but its expression is very cumbersome for large $N$.

**procedure ran01-convolution**
$$\{\pi_0^1, \ldots, \pi_l^1\} \leftarrow \{\tfrac{1}{2l}, \tfrac{1}{l}, \ldots, \tfrac{1}{l}, \tfrac{1}{2l}\}$$
**input** $\{\pi_0, \ldots, \pi_{Nl}\}$ (probabilities for sum of $N$ variables)
**for** $k = 0, \ldots, Nl + l$ **do**
$$\left\{\ \pi_k' \leftarrow \sum_{m=\max(0, k-Nl)}^{\min(l, k)} \pi_{k-m} \pi_m^1 \right.$$
**output** $\{\pi_0', \ldots, \pi_{Nl+l}'\}$ (probabilities for $N + 1$ variables)

———

**Algorithm 1.26 ran01-convolution.** Probabilities for the sum of $N+1$ random numbers obtained from the probabilities for $N$ numbers.

Equations (1.49) and (1.50) are special cases of general convolutions for the sum $\xi + \eta$ of two independent random variables $\xi$ and $\eta$, taking values $x$ and $y$ with probabilities $\pi_\xi(x)$ and $\pi_\eta(y)$, respectively. The sum variable $\xi + \eta$ takes values $x$ with probability

$$\pi_{\xi+\eta}(x) = \int_{-\infty}^{\infty} dy\ \pi_\xi(y)\pi_\eta(x - y).$$

Again, the arguments of the two independent variables, $y$ and $x - y$, have been chosen such that their sum is equal to $x$. The value of $y$ is arbitrary, and must be integrated (summed) over. In Subsection 1.4.4, we shall revisit convolutions, and generalize Alg. 1.26 (**ran01-convolution**) to distributions which stretch out to infinity in such a way that $\pi(x)$ cannot be cut off at large arguments $x$.

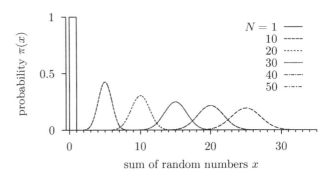

**Fig. 1.34** Probability distribution for the sum of $N$ random numbers $\mathtt{ran}\,(0,1)$ (from Alg. 1.26 ($\mathtt{ran01\text{-}convolution}$), with $l = 100$).

### 1.3.2   Mean value and variance

In this subsection, we discuss the moments of probability distributions, and the connection between the moments of a single variable and those of a sum of variables. The variance—the second moment—stands out: it is additive and sets a scale for an interval containing a large fraction of the events (through the Chebyshev inequality). These two properties will be discussed in detail. The variance also gives a necessary and sufficient condition for the convergence of a sum of identically distributed random variables to a Gaussian, as we shall discuss in Subsection 1.3.3.

The mean value (also called the expectation value) of a distribution for a discrete variable $\xi$ is given by

$$\langle \xi \rangle = \sum_k k\pi_k,$$

where the sum runs over all possible values of $k$. For a continuous variable, the mean value is given, analogously, by

$$\langle \xi \rangle = \int \mathrm{d}x\; x\pi(x). \tag{1.51}$$

The mean value of the Bernoulli distribution is $\theta$, and the mean of the random number $\xi = \mathtt{ran}\,(0,1)$ is obviously $1/2$.

The mean value of a sum of $N$ random variables is equal to the sum of the means of the variables. In fact, for two variables $\xi_1$ and $\xi_2$, and for their joint probability distribution $\pi(x_1, x_2)$, we may still define the probabilities of $x_1$ and $x_2$ by integrating over the other variable:

$$\pi_1(x_1) = \int \mathrm{d}x_2\; \pi(x_1, x_2), \quad \pi_2(x_2) = \int \mathrm{d}x_1\; \pi(x_1, x_2),$$

with $\langle \xi_1 \rangle = \int \mathrm{d}x_1\; x_1\pi_1(x_1)$, etc. This gives

$$\underline{\langle \xi_1 + \xi_2 \rangle} = \int \mathrm{d}x_1\; \mathrm{d}x_2\; (x_1 + x_2)\pi(x_1, x_2) = \underline{\langle \xi_1 \rangle + \langle \xi_2 \rangle}.$$

The additivity of the mean value holds for variables that are not independent (i.e. do not satisfy $\pi(x_1, x_2) = \pi_1(x_1)\pi_2(x_2)$). This was naively assumed in the heliport game: we strove to make the stationary distribution function $\pi(x_1, x_2)$ uniform, so that the probability of falling into the circle was equal to $\theta$. It was natural to assume that the mean value of the sum of $N$ trials would be the same as $N$ times the mean value of a single trial.

Among the higher moments of a probability distribution, the variance

$$\text{Var}\,(\xi) = \left\{\begin{matrix}\text{average squared distance}\\\text{from the mean value}\end{matrix}\right\} = \langle(\xi - \langle\xi\rangle)^2\rangle \qquad (1.54)$$

is quintessential. It can be determined, as indicated in eqn (1.54), from the squared deviations from the mean value. For the Bernoulli distribution, the mean value is $\theta$, so that

$$\text{Var}\,(\xi) = \underbrace{\theta^2}_{(0-\langle\xi\rangle)^2}\;\underbrace{(1-\theta)}_{\pi(0)} + \underbrace{(1-\theta)^2}_{(1-\langle\xi\rangle)^2}\;\underbrace{\theta}_{\pi(1)} = \theta(1-\theta).$$

It is usually simpler to compute the variance from another formula, obtained from eqn (1.54) by writing out the square:

$$\underline{\text{Var}\,(\xi)} = \langle(\xi - \langle\xi\rangle)^2\rangle = \langle\xi^2\rangle - 2\underbrace{\langle\xi\cdot\langle\xi\rangle\rangle}_{\langle\xi\rangle\langle\xi\rangle} + \langle\xi\rangle^2 = \underline{\langle\xi^2\rangle - \langle\xi\rangle^2}.$$

This gives the following for the Bernoulli distribution:

$$\text{Var}\,(\xi) = \underbrace{\langle\xi^2\rangle}_{0^2\cdot(1-\theta)+1^2\cdot\theta} - \underbrace{\langle\xi\rangle^2}_{\theta^2} = \theta(1-\theta).$$

The variance of the uniform random number $\xi = \texttt{ran}\,(0, 1)$ is

$$\text{Var}\,(\xi) = \int_0^1 \mathrm{d}x\,\underbrace{\pi(x)}_{=1}\,x^2 - \left[\int_0^1 \mathrm{d}x\,\underbrace{\pi(x)}_{=1}\,x\right]^2 = \frac{1}{12}, \qquad (1.55)$$

which explains the factor $1/12$ in Alg. 1.17 (**naive-gauss**). The variance of a Gaussian random number $\texttt{gauss}(\sigma)$ is

$$\int_{-\infty}^{\infty} \frac{\mathrm{d}x}{\sqrt{2\pi}\sigma}x^2\exp\left(-\frac{x^2}{2\sigma^2}\right) = \sigma^2.$$

The variance (the mean square deviation) has the dimensions of a squared length, and it is useful to consider its square root, the root mean square deviation (also called the standard deviation):

$$\left\{\begin{matrix}\text{root mean square}\\\text{(standard) deviation}\end{matrix}\right\} : \sqrt{\langle(\xi - \langle\xi\rangle)^2\rangle} = \sqrt{\text{Var}\,(\xi)} = \sigma.$$

This should be compared with the mean absolute deviation $\langle|\xi - \langle\xi\rangle|\rangle$, which measures how far on average a random variable is away from

The fundamental relations

$$\langle a\xi + b\rangle = a\,\langle\xi\rangle + b, \qquad (1.52)$$

$$\text{Var}\,(a\xi + b) = a^2\text{Var}\,(\xi), \qquad (1.53)$$

are direct consequences of the definitions in eqns (1.51) and (1.54).

its mean value. For the Gaussian distribution, for example, the mean absolute deviation is

$$\left\{ \begin{matrix} \text{mean absolute} \\ \text{deviation} \end{matrix} \right\} : \quad \int_{-\infty}^{\infty} \frac{dx}{\sqrt{2\pi}\sigma} |x| \exp\left(-\frac{x^2}{2\sigma^2}\right) = \underbrace{\sqrt{\frac{2}{\pi}}}_{0.798} \sigma, \quad (1.56)$$

which is clearly different from the standard deviation.

The numerical difference between the root mean square deviation and the mean absolute deviation is good to keep in mind, but represents little more than a subsidiary detail, which depends on the distribution. The great importance of the variance and the reason why absolute deviations play no role reside in the way they can be generalized from a single variable to a sum of variables. To see this, we must first consider the correlation function of two independent variables $\xi_i$ and $\xi_j$, taking values $x_i$ and $x_j$ (where $i \neq j$):

$$\int dx_i \int dx_j \; \pi(x_i)\pi(x_j)x_i x_j = \left[\int dx_i \; \pi(x_i)x_i\right]\left[\int dx_j \; \pi(x_j)x_j\right],$$

which is better written—and remembered—as

$$\langle \xi_i \xi_j \rangle = \begin{cases} \langle \xi_i \rangle \langle \xi_j \rangle & \text{for } i \neq j \\ \langle \xi_i^2 \rangle & \text{for } i = j \end{cases} \quad \left\{ \begin{matrix} \text{independent} \\ \text{variables} \end{matrix} \right\}.$$

This factorization of correlation functions implies that the variance of a sum of independent random variables is additive. In view of eqn (1.53), it suffices to show this for random variables of zero mean, where we find

$$\underline{\text{Var}\,(\xi_1 + \cdots + \xi_N)} = \langle (\underbrace{\xi_1 + \cdots + \xi_N}_{N})^2 \rangle = \left\langle \left(\sum_i \xi_i\right)\left(\sum_j \xi_j\right)\right\rangle$$

$$= \sum_i \langle \xi_i^2 \rangle + \underbrace{\sum_{i \neq j} \langle \xi_i \rangle \langle \xi_j \rangle}_{=0} = \underline{\text{Var}\,(\xi_1) + \cdots + \text{Var}\,(\xi_N)}. \quad (1.57)$$

The additivity of variances for the sum of independent random variables is of great importance. No analogous relation exists for the mean absolute deviation.

Independent random variables $\xi_i$ with the same finite variance satisfy, from eqns (1.52) and (1.53):

$$\text{Var}\,(\xi_1 + \cdots + \xi_N) = N\,\text{Var}\,(\xi_i),$$

$$\text{Var}\left(\frac{\xi_1 + \cdots + \xi_N}{N}\right) = \frac{1}{N}\,\text{Var}\,(\xi_i).$$

For concreteness, we shall now apply these two formulas to the children's game, with $N_{\text{hits}} = \xi_1 + \cdots + \xi_N$:

$$\text{Var}\,(N_{\text{hits}}) = \left\langle \left(N_{\text{hits}} - \frac{\pi}{4}N\right)^2\right\rangle = N\,\text{Var}\,(\xi_i) = N\,\overbrace{\theta(1-\theta)}^{0.169},$$

$$\mathrm{Var}\left(\frac{N_{\mathrm{hits}}}{N}\right) = \left\langle\left(\frac{N_{\mathrm{hits}}}{N} - \frac{\pi}{4}\right)^2\right\rangle = \frac{1}{N}\mathrm{Var}\left(\xi_i\right) = \overbrace{\frac{\theta(1-\theta)}{N}}^{0.169}. \quad (1.58)$$

In the initial simulation in this book, the number of hits was usually of the order of 20 counts away from 3141 (for $N = 4000$ trials), because the variance is $\mathrm{Var}\left(N_{\mathrm{hits}}\right) = N \cdot 0.169 = 674.2$, so that the mean square deviation comes out as $\sqrt{674} = 26$. This quantity corresponds to the square root of the average of the last column in Table 1.10 (which analyzes the first table in this chapter). The mean absolute distance $\langle|\Delta N_{\mathrm{hits}}|\rangle$, equal to 20, is smaller than the root mean square difference by a factor $\sqrt{2/\pi}$, as in eqn (1.56), because the binomial distribution is virtually Gaussian when $N$ is large and $\theta$ is of order unity.

The variance not only governs average properties, such as the mean square deviation of a random variable, but also allows us to perform interval estimates. Jacob Bernoulli's weak law of large numbers is of this type (for sums of Bernoulli-distributed variables, as in the children's game). It states that for any interval $[\pi/4 - \epsilon, \pi/4 + \epsilon]$, the probability for $N_{\mathrm{hits}}/N$ to fall within this interval goes to one as $N \to \infty$. This law is best discussed in the general context of the Chebyshev inequality, which we need to understand only for distributions with zero mean:

$$\mathrm{Var}\left(\xi\right) = \int_{-\infty}^{\infty} dx\, x^2 \pi(x) \geq \int_{|x|>\epsilon} dx\, x^2 \pi(x) \geq \epsilon^2 \underbrace{\int_{|x|>\epsilon} dx\, \pi(x)}_{\substack{\text{prob. that} \\ |x - \langle x\rangle| > \epsilon}}.$$

Table 1.10 Reanalysis of Table 1.1 using eqn (1.58) ($N = 4000, \theta = \pi/4$)

| # | $N_{\mathrm{hits}}$ | $|\Delta N_{\mathrm{hits}}|$ | $(\Delta N_{\mathrm{hits}})^2$ |
|---|---|---|---|
| 1 | 3156 | 14.4 | 207.6 |
| 2 | 3150 | 8.4 | 70.7 |
| 3 | 3127 | 14.6 | 212.9 |
| 4 | 3171 | 29.4 | 864.8 |
| 5 | 3148 | 6.4 | 41.1 |
| ... | ... | ... | |
| $\langle\rangle$ | 3141. | 20.7 | 674.2 |

This gives

$$\left\{\begin{array}{l}\text{Chebyshev} \\ \text{inequality}\end{array}\right\} : \left\{\begin{array}{l}\text{probability that} \\ |x - \langle x\rangle| > \epsilon\end{array}\right\} < \frac{\mathrm{Var}\left(\xi\right)}{\epsilon^2}. \quad (1.59)$$

In the children's game, the variance of the number of hits, $\mathrm{Var}\left(N_{\mathrm{hits}}/N\right)$, is smaller than $1/(4N)$ because, for $\theta \in [0, 1]$, $\theta(1-\theta) \leq 1/4$. This allows us to write

$$\left\{\begin{array}{l}\text{weak law of} \\ \text{large numbers}\end{array}\right\} : \left\{\begin{array}{l}\text{probability that} \\ |N_{\mathrm{hits}}/N - \pi/4| < \epsilon\end{array}\right\} > 1 - \frac{1}{4\epsilon^2 N}.$$

In this equation, we can keep the interval parameter $\epsilon$ fixed. The probability inside the interval approaches 1 as $N \to \infty$. We can also bound the interval containing, say, 99% of the probability, as a function of $N$. We thus enter 0.99 into the above inequality, and find

$$\left\{\begin{array}{l}\text{size of interval containing} \\ \text{99\% of probability}\end{array}\right\} : \epsilon < \frac{5}{\sqrt{N}}.$$

Chebyshev's inequality (1.59) shows that a (finite) variance plays the role of a scale delimiting an interval of probable values of $x$: whatever the distribution, it is improbable that a sample will be more than a few standard deviations away from the mean value. This basic regularity property of distributions with a finite variance must be kept in mind in practical calculations. In particular, we must keep this property separate from the central limit theorem, which involves an $N \to \infty$ limit. Applied to the sum of independent variables, Chebyshev's inequality turns into the weak law of large numbers and is the simplest way to understand why $N$-sample averages must converge (in probability) towards the mean value, i.e. why the width of the interval containing any fixed amount of probability goes to zero as $\propto 1/\sqrt{N}$.

### 1.3.3   The central limit theorem

We have discussed the crucial role played by the mean value of a distribution and by its variance. If the variance is finite, we can shift any random variable by the mean value, and then rescale it by the standard deviation, the square root of the variance, such that it has zero mean and unity variance. We suppose a random variable $\xi$ taking values $y$ with probability $\pi(y)$. The rescaling is done so that $\xi_{\text{resc}}$ takes values $x = (y - \langle \xi \rangle)/\sigma$ with probability

$$\pi_{\text{resc}}(x) = \sigma \pi (\underbrace{\sigma x + \langle \xi \rangle}_{y}) \tag{1.60}$$

and has zero mean and unit variance. As an example, we have plotted in Fig. 1.35 the probability distribution of a random variable $\xi$ corresponding to the sum of 50 random numbers $\mathtt{ran}\,(0,1)$. This distribution is characterized by a standard deviation $\sigma = \sqrt{50} \times \sqrt{1/12} = 2.04$, and mean $\langle \xi \rangle = 25$. As an example, $y = 25$, with $\pi(y) = 0.193$, is rescaled to $x = 0$, with $\pi_{\text{resc}}(0) = 2.04 \times 0.193 = 0.39$ (see Fig. 1.35 for the rescaled distributions for the children's game and for the sum of $N$ random numbers $\mathtt{ran}\,(0,1)$).

The central limit theorem[6] states that this rescaled random variable is Gaussian in the limit $N \to \infty$ if $\xi$ itself is a sum of independent random variables $\xi = \xi_1 + \cdots + \xi_N$ of finite variance.

In this subsection, we prove this fundamental theorem of probability theory under the simplifying assumption that all moments of the distribution are finite:

$$\begin{array}{ccccc} \langle \xi_i \rangle & \langle \xi_i^2 \rangle & \langle \xi_i^3 \rangle & \cdots & \langle \xi_i^k \rangle \\ \| & \| & \| & \cdots & \| \\ 0 & 1 & \multicolumn{3}{c}{\text{all moments finite}} \end{array} \tag{1.61}$$

(Odd moments may differ from zero for asymmetric distributions.) For independent random variables with identical distributions, the finite variance by itself constitutes a necessary and sufficient condition for convergence towards a Gaussian (see Gnedenko and Kolmogorov (1954)). The finiteness of the moments higher than the second renders the proof

---

[6]To be pronounced as *central* limit *theorem* rather than *central limit* theorem. The expression was coined by mathematician G. Polya in 1920 in a paper written in German, where it is clear that the theorem is central, not the limit.

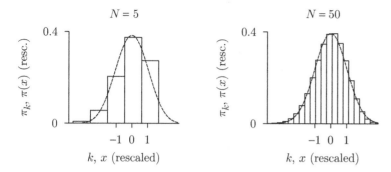

**Fig. 1.35** Distribution functions corresponding to Figs 1.33 and 1.34, rescaled as in eqn (1.60).

**Table 1.11** The 40 choices of indices $\{i, j, k, l\}$ for $N = 4$ (see eqn (1.63)) for which $\langle \xi_i \xi_j \xi_k \xi_l \rangle$ is different from zero

| # | $i$ | $j$ | $k$ | $l$ |
|---|---|---|---|---|
| 1 | 1 | 1 | 1 | 1 |
| 2 | 1 | 1 | 2 | 2 |
| 3 | 1 | 1 | 3 | 3 |
| 4 | 1 | 1 | 4 | 4 |
| 5 | 1 | 2 | 1 | 2 |
| 6 | 1 | 2 | 2 | 1 |
| 7 | 1 | 3 | 1 | 3 |
| 8 | 1 | 3 | 3 | 1 |
| 9 | 1 | 4 | 1 | 4 |
| 10 | 1 | 4 | 4 | 1 |
| 11 | 2 | 1 | 1 | 2 |
| 12 | 2 | 1 | 2 | 1 |
| 13 | 2 | 2 | 1 | 1 |
| 14 | 2 | 2 | 2 | 2 |
| 15 | 2 | 2 | 3 | 3 |
| 16 | 2 | 2 | 4 | 4 |
| 17 | 2 | 3 | 2 | 3 |
| 18 | 2 | 3 | 3 | 2 |
| 19 | 2 | 4 | 2 | 4 |
| 20 | 2 | 4 | 4 | 2 |
| 21 | 3 | 1 | 1 | 3 |
| 22 | 3 | 1 | 3 | 1 |
| 23 | 3 | 2 | 2 | 3 |
| 24 | 3 | 2 | 3 | 2 |
| 25 | 3 | 3 | 1 | 1 |
| 26 | 3 | 3 | 2 | 2 |
| 27 | 3 | 3 | 3 | 3 |
| 28 | 3 | 3 | 4 | 4 |
| 29 | 3 | 4 | 3 | 4 |
| 30 | 3 | 4 | 4 | 3 |
| 31 | 4 | 1 | 1 | 4 |
| 32 | 4 | 1 | 4 | 1 |
| 33 | 4 | 2 | 2 | 4 |
| 34 | 4 | 2 | 4 | 2 |
| 35 | 4 | 3 | 3 | 4 |
| 36 | 4 | 3 | 4 | 3 |
| 37 | 4 | 4 | 1 | 1 |
| 38 | 4 | 4 | 2 | 2 |
| 39 | 4 | 4 | 3 | 3 |
| 40 | 4 | 4 | 4 | 4 |

trivial: in this case, two distributions are identical if all their moments agree. We thus need only compare the moments of the sum random variable, rescaled as

$$\xi = \frac{1}{\sqrt{N}}(\xi_1 + \cdots + \xi_N),\qquad(1.62)$$

with the moments of the Gaussian to verify that they are the same. Let us start with the third moment:

$$\left\langle \left( \frac{\xi_1 + \cdots + \xi_N}{\sqrt{N}} \right)^3 \right\rangle = \frac{1}{N^{3/2}} \sum_{ijk=1}^{N} \langle \xi_i \xi_j \xi_k \rangle$$

$$= \frac{1}{N^{3/2}} (\underbrace{\langle \xi_1 \xi_1 \xi_1 \rangle}_{\langle \xi_1^3 \rangle} + \underbrace{\langle \xi_1 \xi_1 \xi_2 \rangle}_{\langle \xi_1^2 \rangle \langle \xi_2 \rangle = 0} + \cdots + \underbrace{\langle \xi_2 \xi_2 \xi_2 \rangle}_{\langle \xi_2^3 \rangle} + \cdots).$$

In this sum, only $N$ terms are nonzero, and they remain finite because of our simplifying assumption in eqn (1.61). We must divide by $N^{3/2}$; therefore $\langle \xi^3 \rangle \to 0$ for $N \to \infty$. In the same manner, we scrutinize the fourth moment:

$$\left\langle \left( \frac{\xi_1 + \cdots + \xi_N}{\sqrt{N}} \right)^4 \right\rangle = \frac{1}{N^2} \sum_{ijkl=1}^{N} \langle \xi_i \xi_j \xi_k \xi_l \rangle$$

$$= \frac{1}{N^2} [\underbrace{\langle \xi_1 \xi_1 \xi_1 \xi_1 \rangle}_{\langle \xi_1^4 \rangle} + \underbrace{\langle \xi_1 \xi_1 \xi_1 \xi_2 \rangle}_{\langle \xi_1^3 \rangle \langle \xi_2 \rangle = 0} + \cdots + \underbrace{\langle \xi_1 \xi_1 \xi_2 \xi_2 \rangle}_{\langle \xi_1^2 \rangle \langle \xi_2^2 \rangle} + \cdots]. \quad(1.63)$$

Only terms not containing a solitary index $i$ can be different from zero, because $\langle \xi_i \rangle = 0$: for $N = 4$, there are 40 such terms (see Table 1.11). Of these, 36 are of type "$iijj$" (and permutations) and four are of type "$iiii$". For general $N$, the total correlation function is

$$\langle \xi^4 \rangle = \frac{1}{N^2} \left[ 3N(N-1) \langle \xi_i^2 \rangle^2 + N \langle \xi_i^4 \rangle \right].$$

In the limit of large $N$, where $N \simeq N - 1$, we have $\langle \xi^4 \rangle = 3$.

In the same way, we can compute higher moments of $\xi$. Odd moments approach zero in the large-$N$ limit. For example, the fifth moment is put together from const$\cdot N^2$ terms of type "$iijjj$", in addition to the $N$ terms of type "$iiiii$". Dividing by $N^{5/2}$ indeed gives zero. The even moments are finite, and the dominant contributions for large $N$ come, for the sixth moment, from terms of type "$iijjkk$" and their permutations. The total number of these terms, for large $N$ is $\simeq N^3 \cdot 6!/3!/2^3 = N^3 \times 1 \times 3 \times 5$. For arbitrary $k$, the number of ways is

$$\left\{ \begin{matrix} \text{moments} \\ \text{in eqn (1.62)} \end{matrix} \right\} \left\{ \begin{matrix} \langle \xi^{2k} \rangle \to 1 \times 3 \times \cdots \times (2k-1) \\ \langle x^{2k-1} \rangle \to 0 \end{matrix} \right. \quad \text{for } N \to \infty. \quad (1.64)$$

We now check that the Gaussian distribution also has the moments given by eqn (1.64). This follows trivially for the odd moments, as the Gaussian is a symmetric function. The even moments can be computed from

$$\underbrace{\int_{-\infty}^{\infty} \frac{\mathrm{d}x}{\sqrt{2\pi}} \exp\left( -\frac{x^2}{2} + xh \right) = \exp\left( \frac{h^2}{2} \right)}_{\text{because } \int \frac{\mathrm{d}x}{\sqrt{2\pi}} \exp\left[ -(x-h)^2/2 \right] = 1} = 1 + \frac{h^2}{2} + \frac{(h^2/2)^2}{2!} + \cdots .$$

We may differentiate this equation $2k$ times (under the integral on the left, as it is absolutely convergent) and then set $h = 0$. On the left, this gives the $(2k)$th moment of the Gaussian distribution. On the right, we obtain an explicit expression:

$$\begin{aligned} \langle x^{2k} \rangle &= \frac{\partial^{2k}}{\partial h^{2k}} \int_{-\infty}^{\infty} \frac{\mathrm{d}x}{\sqrt{2\pi}} \exp\left( -\frac{x^2}{2} + xh \right) \Bigg|_{h=0} \\ &= \frac{\partial^{2k}}{\partial h^{2k}} \left[ 1 + \frac{h^2}{2} + \frac{(h^2/2)^2}{2!} + \frac{(h^2/2)^3}{3!} + \cdots \right]_{h=0} \\ &= \frac{\partial^{2k}}{\partial h^{2k}} \left( \frac{h^{2k}}{k!2^k} \right) = \frac{(2k)!}{k!2^k} = 1 \times 3 \times \cdots \times (2k-1). \end{aligned}$$

This is the same as eqn (1.64), and the distribution function of the sum of $N$ random variables, in the limit $N \to \infty$, has no choice but to converge to a Gaussian, so that we have proven the central limit theorem (for the simplified case that all the moments of the distribution are finite). The Gaussian has 68% of its weight concentrated within the interval $[-\sigma, \sigma]$ and 95% of its weight within the interval $[-2\sigma, 2\sigma]$ (see Fig. 1.36). These numbers are of prime importance for statistical calculations in general, and for Monte Carlo error analysis in particular. The probability of being more than a few standard deviations away from the mean value drops precipitously (see Table 1.12, where the error function is $\mathrm{erf}(x) = (2/\sqrt{\pi}) \int_0^t \mathrm{d}t \, \exp\left(-t^2\right)$).

In practical calculations, we are sometimes confronted with a few exceptional sample averages that are many (estimated) standard deviations away from our (estimated) mean value. We may, for example,

**Table 1.12** Probability of being outside a certain interval for any distribution (from the Chebyshev inequality) and for a Gaussian distribution.

| Excluded interval | Probability of being $\notin$ interval | |
|---|---|---|
| | Chebyshev | Gaussian |
| $[-\sigma, \sigma]$ | Less than 100% | 32% |
| $[-2\sigma, 2\sigma]$ | Less than 25% | 5% |
| $[-3\sigma, 3\sigma]$ | Less than 11% | 0.3% |
| $[-4\sigma, 4\sigma]$ | Less than 6% | 0.006% |
| $[-k\sigma, k\sigma]$ | Less than $\frac{1}{k^2}$ | $1 - \text{erf}(k/\sqrt{2})$ |

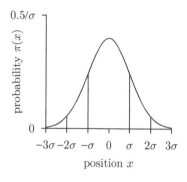

**Fig. 1.36** The Gaussian distribution. The probability of being outside the interval $[-\sigma, \sigma]$ is 0.32, etc. (see Table 1.12)

compute an observable (say a magnetization) as a function of an external parameter (say the temperature). A number of values lie nicely on our favorite theoretical curve, but the exceptional ones are 10 or 15 standard deviations off. With the central limit theorem, such an outcome is extremely unlikely. In this situation, it is a natural tendency to think that the number of samples might be too small for the central limit theorem to apply. This reasoning is usually erroneous because, first of all, this limit is reached extremely quickly if only the distribution function of the random variables $\xi_i$ is well behaved. Secondly, the Chebyshev inequality applies to arbitrary distributions with finite variance. It also limits deviations from the mean value, even though less strictly (see Table 1.12). In the above case of data points which are very far off a theoretical curve, it is likely that the estimated main characteristics of the distribution function are not at all what we estimate them to be, or (in a Markov-chain Monte Carlo calculation) that the number of independent data sets has been severely overestimated. Another possibility is that our favorite theoretical curve is simply wrong.

## 1.3.4  Data analysis for independent variables

Experiments serve no purpose without data analysis, and Monte Carlo calculations are useful only if we can come up with an estimate of the observables, such as an approximate value for the mathematical constant $\pi$, or an estimate of the quark masses, a value for a condensate fraction, an approximate equation of state, etc. Our very first simulation in this book generated 3156 hits for 4000 trials (see Table 1.1). We shall now see what this result tells us about $\pi$, at the most fundamental level of understanding. Hits and nonhits were generated by the Bernoulli distribution:

$$\xi_i = \begin{cases} 1 & \text{with probability } \theta \\ 0 & \text{with probability } (1 - \theta) \end{cases}, \qquad (1.65)$$

but the value of $\pi/4 = \theta = \langle \xi_i \rangle$ is supposed unknown. Instead of the original variables $\xi_i$, we consider random variables $\eta_i$ shifted by this unknown mean value:

$$\eta_i = \xi_i - \theta.$$

The shifted random variables $\eta_i$ now have zero mean and the same variance as the original variables $\xi_i$ (see eqns (1.52) and (1.53)):

$$\langle \eta_i \rangle = 0, \quad \mathrm{Var}\,(\eta_i) = \mathrm{Var}\,(\xi_i) = \theta(1-\theta) \leq \frac{1}{4}.$$

Without invoking the limit $N \to \infty$, we can use the Chebyshev inequality eqn (1.59) to obtain an interval around zero containing at least 68% of the probability:

In our example, the realizations of the $\eta_i$ satisfy

$$\frac{1}{N}\sum_{i=1}^{N}\eta_i = \underbrace{\frac{3156}{4000} - \frac{\pi}{4}}_{0.789}. \qquad (1.66)$$

$$\left\{ \begin{array}{c} \text{with 68\%} \\ \text{probability} \end{array} \right\} : \quad \underbrace{\left| \frac{1}{N}\sum_{i=1}^{N}\eta_i \right|}_{\text{see eqn (1.66)}} < \frac{1.77\sigma}{\sqrt{N}} < \frac{1.77}{2\sqrt{4000}} = 0.014.$$

This has implications for the difference between our experimental result, 0.789, and the mathematical constant $\pi$. The difference between the two, with more than 68% chance, is smaller than 0.014:

$$\frac{\pi}{4} = 0.789 \pm 0.014 \Leftrightarrow \pi = 3.156 \pm 0.056, \qquad (1.67)$$

where the value 0.056 is an upper bound for the 68% confidence interval that in physics is called an error bar. The quite abstract reasoning leading from eqn (1.65) to eqn (1.67)—in other words from the experimental result 3156 to the estimate of $\pi$ with an error bar—is extremely powerful, and not always well understood. To derive the error bar, we did not use the central limit theorem, but the more general Chebyshev inequality. We also used an upper bound for the variance. With these precautions we arrived at the following result. We know with certainty that among an infinite number of beach parties, at which participants would play the same game of 4000 as we described in Subsection 1.1.1 and which would yield Monte Carlo results analogous to ours, more than 68% would hold the mathematical value of $\pi$ inside their error bars. In arriving at this result, we did not treat the number $\pi$ as a random variable—that would be nonsense, because $\pi$ is a mathematical constant.

We must now relax our standards. Instead of reasoning in a way that holds for all $N$, we suppose the validity of the central limit theorem. The above 68% confidence limit for the error bar now leads to the following estimate of the mean value:

$$\langle \xi \rangle = \frac{1}{N}\sum_{i=1}^{N}\xi_i \pm \frac{\sigma}{\sqrt{N}}$$

(see Table 1.12; the 68% confidence interval corresponds to one standard deviation of the Gaussian distribution).

In addition, we must also give up the bound for the variance in favor of an estimate of the variance, through the expression

$$\mathrm{Var}\,(\xi_i) = \mathrm{Var}\,(\eta_i) \simeq \frac{1}{N-1}\left[ \sum_{j=1}^{N}\left(\xi_j - \frac{1}{N}\sum_{i=1}^{N}\xi_i\right)^2 \right]. \qquad (1.68)$$

The mean value on the right side of eqn (1.68) is equal to the variance for all $N$ (and one therefore speaks of an unbiased estimator). With the replacement $\xi_j \to \eta_j$, we obtain

$$
\left\langle \frac{1}{N-1} \left[ \sum_{j=1}^{N} \left( \eta_j - \frac{1}{N} \sum_{i=1}^{N} \eta_i \right)^2 \right] \right\rangle
$$

$$
= \frac{1}{N-1} \left[ \sum_{jk} \underbrace{\langle \eta_j \eta_k \rangle}_{\mathrm{Var}(\eta_j)\delta_{jk}} - \frac{2}{N} \sum_{ij=1}^{N} \langle \eta_i \eta_j \rangle + \frac{1}{N^2} \sum_{ijk} \langle \eta_i \eta_k \rangle \right] = \mathrm{Var}\,(\eta_i)
$$

(a more detailed analysis is possible under the conditions of the central limit theorem). In practice, we always replace the denominator $\frac{1}{N-1} \to \frac{1}{N}$ in eqn (1.68) (the slight difference plays no role at all). We arrive, with eqn (1.69), at the standard formulas for the data analysis of independent random variables:

$$
\langle \xi \rangle = \frac{1}{N} \sum_{i=1}^{N} \xi_i \pm \text{error},
$$

where

$$
\text{error} = \frac{1}{\sqrt{N}} \sqrt{\frac{1}{N} \sum_i \xi_i^2 - \left( \frac{1}{N} \sum_i \xi_i \right)^2}.
$$

We note that

$$
\sum_{j=1}^{N} \left( \xi_j - \frac{1}{N} \sum_{i=1}^{N} \xi_i \right)^2 =
$$

$$
\sum_{j=1}^{N} \xi_j^2 - \frac{2}{N} \left( \sum_{j=1}^{N} \xi_j \right)\left( \sum_{i=1}^{N} \xi_i \right) + \frac{1}{N} \left( \sum_{i=1}^{N} \xi_i \right)^2
$$

$$
= \sum_j \xi_j^2 - N \left( \frac{1}{N} \sum_j \xi_j \right)^2. \qquad (1.69)
$$

The error has a prefactor $1/\sqrt{N}$ and not $1/N$, because it goes as the standard deviation, itself the square root of the variance. It should be noted that the replacement of the variance by an estimate is far from innocent. This replacement can lead to very serious problems, as we shall discuss in an example, namely the $\gamma$-integral of Section 1.4. In Markov-chain Monte Carlo applications, another problem arises because the number of samples, $N$, must be replaced by an effective number of independent samples, which must itself be estimated with care. This will be taken up in Subsection 1.3.5.

We now briefly discuss the Bayesian approach to statistics, not because we need it for computing error bars, but because of its close connections with statistical physics (see Section 2.2). We stay in the imagery of the children's game. Bayesian statistics attempts to determine the probability that our experimental result (3156 hits) was the result of a certain value $\pi_{\text{test}}$. If $\pi_{\text{test}}$ was very large ($\pi_{\text{test}} \lesssim 4$, so that $\theta \lesssim 1$), and also if $\pi_{\text{test}} \gtrsim 0$, the experimental result, 3156 hits for 4000 trials, would be very unlikely. Let us make this argument more quantitative and suppose that the test values of $\pi$ are drawn from an a priori probability distribution (not to be confused with the a priori probability of the generalized Metropolis algorithm). The test values give hits and nonhits with the appropriate Bernoulli distribution and values of $N_{\text{hits}}$ with their binomial distribution (with parameters $\theta = \pi_{\text{test}}/4$ and $N = 4000$), but there is a great difference between those test values that give 3156 hits, and those that do not. To illustrate this point, let us sample the values of $\pi_{\text{test}}$ that yield 3156 hits with a program, Alg. 1.27 (naive-bayes-pi). Like all other naive algorithms in this book, it is conceptually correct, but unconcerned with computational efficiency. This program picks a random value of $\pi_{\text{test}}$ (from the a priori probability distribution), then samples this test value's binomial distribution. The "good" test values are kept in a table (see Table 1.13). They have a probability distribution (the a posteriori probability distribution), and it is reasonable to say that this distribution contains information on the mathematical constant $\pi$.

**procedure** `naive-bayes-pi`

1   $\pi_{\text{test}} \leftarrow \mathbf{ran}\,(0,4)$   (sampled with the a priori probability)

$N_{\text{hits}} \leftarrow 0$

**for** $i = 1, \ldots, 4000$ **do**

$\left\{ \begin{array}{l} \mathbf{if}\ (\mathbf{ran}\,(0,1) < \pi_{\text{test}}/4)\ \mathbf{then} \\ \quad \{\ N_{\text{hits}} \leftarrow N_{\text{hits}} + 1 \end{array} \right.$

**if** $(N_{\text{hits}} \neq 3156)$ **goto** 1   (reject $\pi_{\text{test}}$)

**output** $\pi_{\text{test}}$   (output with the a posteriori probability)

**Table 1.13** Values of $\pi_{\text{test}}$ that lead to 3156 hits for 4000 trials (from Alg. 1.27 (`naive-bayes-pi`))

| Run | $\pi_{\text{test}}$ |
|-----|--------------------|
| 1 | 3.16816878 |
| 2 | 3.17387056 |
| 3 | 3.16035151 |
| 4 | 3.13338971 |
| 5 | 3.16499329 |
| $\vdots$ | $\vdots$ |

**Algorithm 1.27** `naive-bayes-pi`. Generating a test value $\pi_{\text{test}}$, which leads to $N_{\text{hits}} = 3156$ for $N = 4000$ trials.

In the Bayesian approach, the choice of the a priori probability in Alg. 1.27 (`naive-bayes-pi`) influences the outcome in Table 1.13. We could use nonuniform a priori probability distributions, as for example $\pi_{\text{test}}^2 \leftarrow \mathbf{ran}\,(0,16)$, or $\sqrt{\pi_{\text{test}}} \leftarrow \mathbf{ran}\,(0,2)$. Since we know Archimedes' result, it would be an even better idea to use $\pi_{\text{test}} \leftarrow \mathbf{ran}\,(3\frac{10}{71}, 3\frac{1}{7})$ (which does not contain the point 3.156...). Some choices are better than others. However, there is no best choice for the a priori probability, and the final outcome, the a posteriori probability distribution, will (for all finite $N$) carry the mark of this input distribution.

Let us rephrase the above discussion in terms of probability distributions, rather than in terms of samples, for random $\pi_{\text{test}}$ between 0 and 4:

$$\underbrace{\left\{ \begin{array}{c} \text{probability of having} \\ \pi_{\text{test}} \text{ with 3156 hits} \end{array} \right\}}_{\substack{\text{a posteriori probability for } \pi}} = \int_0^4 \underbrace{\mathrm{d}\pi_{\text{test}}}_{\substack{\text{a priori} \\ \text{probability}}} \underbrace{\left\{ \begin{array}{c} \text{probability that } \pi_{\text{test}} \\ \text{yields 3156 hits} \end{array} \right\}}_{\substack{\text{binomial probability of obtaining} \\ \text{3156 hits, given } \pi_{\text{test}}}}.$$

This integral is easily written down and evaluated analytically. For the binomial distribution, it leads to essentially equivalent expressions as the error analysis from the beginning of this subsection.

Other choices for the a priori probability are given by

$$\int_0^4 \mathrm{d}\pi_{\text{test}}\,, \quad \int_0^{16} \mathrm{d}(\pi_{\text{test}}^2)\,, \quad \int_0^2 \mathrm{d}\sqrt{\pi_{\text{test}}}\,, \ldots \underbrace{\int_{3\frac{10}{71}}^{3\frac{1}{7}} \mathrm{d}\pi_{\text{test}}}_{\substack{\text{Archimedes,} \\ \text{see eqn (1.1)}}}\,, \text{ etc.} \quad (1.70)$$

Some a priori probabilities are clearly preferable to others, and they all give different values a posteriori probability distributions, even though the differences are rarely as striking as the ones shown in Fig. 1.37. In the limit $N \to \infty$, all choices become equivalent.

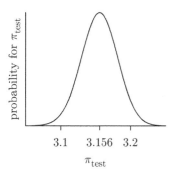 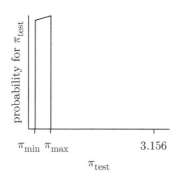

**Fig. 1.37** A posteriori probability for a priori choices $\pi_{\text{test}} \leftarrow \mathbf{ran}\,(0, 4)$ (*left*) and $\pi_{\text{test}} \leftarrow \mathbf{ran}\left(3\frac{10}{71}, 3\frac{1}{7}\right)$ (*right*).

Bayesian statistics is a field of great importance for complicated classification and recognition problems, which are all beyond the scope of this book. We saw how easy it is to incorporate Archimedes' bias $3\frac{10}{71} < \pi < 3\frac{1}{7}$. Within Bayesian statistics, it is nevertheless impossible to decide which choice would least influence data analysis (be an unbiased choice), for example among the first three a priori probabilities in eqn (1.70). We stress that no such ambiguity affected the derivation of the error bar of eqn (1.67) (using the shifted variables $\eta_i$). The concept of an unbiased choice will come up again in the discussion of entropy and equiprobability in statistical mechanics (see Section 2.2).

We have seen in the first part of the present subsection that error analysis stands on an extremely solid basis. The children's Monte Carlo calculation of $\pi$ could have been done a long time before Archimedes obtained a good analytical bound for it. In the same way, we may obtain, using Monte Carlo methods, numerical results for a model in statistical mechanics or other sciences, years before the model is analytically solved. Our numerical results should agree closely, and we should claim credit for our solution. If, on the other hand, our numerical results turn out to be wrong, we were most probably sloppy in generating, or in analyzing, our results. Under no circumstances can we excuse ourselves by saying that it was "just a simulation".

## 1.3.5 Error estimates for Markov chains

We have so far discussed the statistical errors of independent data, as produced by Alg. 1.1 (`direct-pi`) and other direct-sampling algorithms. We must now develop an equally firm grip on the analysis of correlated data. Let us look again at Alg. 1.2 (`markov-pi`). This archetypal Markov-chain program has left 20 000 pebbles lying on the heliport (from five

**Table 1.14** Naive reanalysis of the Markov-chain data from Table 1.2, treating $N_\text{hits}$ for each run as independent

| Run | $N_\text{hits}$ | Estimate of $\pi$ |
|-----|-----------------|-------------------|
| 1   | 3123            |                   |
| 2   | 3118            | 3.122             |
| 3   | 3040            | $\pm$             |
| 4   | 3066            | 0.04              |
| 5   | 3263            |                   |

runs with 4000 pebbles each), which we cannot treat as independent. The pebbles on the ground are distributed with a probability $\pi(x, y)$, but they tend to be grouped, and even lie in piles on top of each other. What we can learn about $\pi(x, y)$ by sampling is less detailed than what is contained in the same number of independent pebbles. As a simple consequence, the spread of the distribution of run averages is wider than before.

It is more sensible to treat not the $5 \times 4000$ pebbles but the five run averages for $4N_\text{hits}/N$ (that is, the values $\{3.123, \ldots, 3.263\}$) as independent, approximately Gaussian variables (see Table 1.14). We may then compute an error estimate from the means and mean square deviations of these five numbers. The result of this naive reanalysis of the heliport data is shown in Table 1.14.

**Fig. 1.38** Markov chains on the heliport. *Left*: all chains start at the clubhouse. *Right*: one chain starts where the previous one stops.

Analyzing a few very long runs is a surefooted, fundamentally sound strategy for obtaining a first error estimate, especially when the influence of initial conditions is negligible. In cases such as in the left frame of Fig. 1.38, however, the choice of starting point clearly biases the estimation of $\pi$, and we want each individual run to be as long as possible. On the other hand, we also need a large number of runs in order to minimize the uncertainty in the error estimate itself. This objective favors short runs. With naive data analysis, it is not evident how to find the best compromise between the length of each run and the number of runs, for a given budget of computer time.

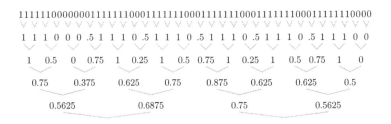

**Fig. 1.39** Four iterations of Alg. 1.28 (**data-bunch**) applied to the correlated output of Alg. 1.2 (**markov-pi**).

In the case of the heliport, and also in general, there is an easy way out of this dilemma: rather than have all Markov chains start at the clubhouse, we should let one chain start where the last one left off (see the right frame of Fig. 1.38). This gives a single, much longer Markov chain. In this case, the cutting-up into five bunches is arbitrary. We could equally well produce bunches of size $\{2, 4, 8, \ldots\}$, especially if the number of data points is a power of two. The bunching into sets of increasing length can be done iteratively, by repeatedly replacing two adjacent samples by their average value (see Fig. 1.39 and Alg. 1.28 (data-bunch)). At each iteration, we compute the apparent error, as if the data were independent. The average value remains unchanged.

Bunching makes the data become increasingly independent, and makes the apparent error approach the true error of the Markov chain. We want to understand why the bunched data are less correlated, even though this can already be seen from Fig. 1.38. In the $k$th iteration, bunches of size $2^k$ are generated: let us suppose that the samples are correlated on a length scale $\xi \lesssim 2^k$, but that original samples distant by more than $2^k$ are fully independent. It follows that, at level $k$, these correlations still affect neighboring bunches, but not next-nearest ones (see Fig. 1.39): the correlation length of the data decreases from length $2^k$ to a length $\lesssim 2$. In practice, we may do the error analysis on all bunches, rather than on every other one.

**procedure** data-bunch
**input** $\{x_1, \ldots, x_{2N}\}$ (Markov-chain data)
$\Sigma \leftarrow 0$
$\Sigma' \leftarrow 0$
**for** $i = 1, \ldots, N$ **do**
$\begin{cases} \Sigma \leftarrow \Sigma + x_{2i-1} + x_{2i} \\ \Sigma' \leftarrow \Sigma' + x_{2i-1}^2 + x_{2i}^2 \\ x_i' \leftarrow (x_{2i-1} + x_{2i})/2 \end{cases}$
error $\leftarrow \sqrt{\Sigma'/(2N) - (\Sigma/(2N))^2}/\sqrt{2N}$
**output** $\Sigma/(2N)$, error, $\{x_1', \ldots, x_N'\}$
———

**Algorithm 1.28** data-bunch. Computing the apparent error (treating data as independent) for $2N$ data points and bunching them into pairs.

It is interesting to apply Alg. 1.28 (data-bunch) repeatedly to the data generated by a long simulation of the heliport (see Fig. 1.40). In this figure, we can identify three regimes. For bunching intervals smaller than the correlation time (here $\simeq 64$), the error is underestimated. For larger bunching intervals, a characteristic plateau indicates the true error of our simulation. This is because bunching of uncorrelated data does not change the expected variance of the data. Finally, for a very small number of intervals, the data remain uncorrelated, but the error estimate itself becomes noisy.

Algorithm 1.28 (data-bunch), part of the folklore of Monte Carlo computation, provides an unbeaten analysis of Markov-chain data, and

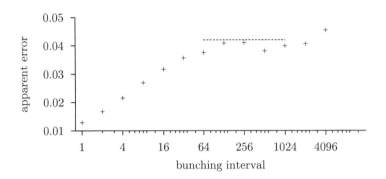

**Fig. 1.40** Repeated bunching of Markov-chain data (from Alg. 1.2 (`markov-pi`) ($N = 2^{14}$, $\delta = 0.3$), analysis by Alg. 1.28 (`data-bunch`)).

is the only technique needed in this book. Data bunching is not fail-safe, as we shall discuss in Section 1.4, but it is the best we can do. What is missing to convince us of the pertinence of our numerical results must be made up for by critical judgment, rigorous programming, comparison with other methods, consideration of test cases, etc.

## 1.4   Computing

Since the beginning of this chapter, we have illustrated the theory of Monte Carlo calculation in simple, unproblematic settings. It is time to become more adventurous, and to advance into areas where things can go wrong. This will acquaint us with the limitations and pitfalls of the Monte Carlo method. Two distinct issues will be addressed. One is the violation of ergodicity—the possibility that a Markov chain never visits all possible configurations. The second limiting issue of the Monte Carlo method shows up when fluctuations become so important that averages of many random variables can no longer be understood as a dominant mean value with a small admixture of noise.

### 1.4.1   Ergodicity

The most characteristic limitation of the Monte Carlo method is the slow convergence of Markov chains, already partly discussed in Subsection 1.1.2: millions of chain links in a simulation most often correspond to only a handful of independent configurations. Such a simulation may resemble our random walk on the surface of a sphere (see Fig. 1.32): many samples were generated, but we have not even gone around the sphere once.

It routinely happens that a computer program has trouble decorrelating from the initial configuration and settling into the stationary probability distribution. In the worst case, independent samples are not even created in the limit of infinite computer time. One then speaks of a nonergodic algorithm. However, we should stress that a practically nonergodic algorithm that runs for a month without decorrelating from the clubhouse is just as useless and just as common.[7]

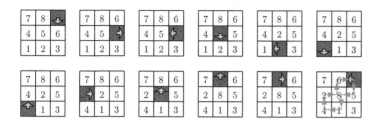

**Fig. 1.41** Local Monte Carlo moves applied to the sliding puzzle, a variant of the pebble game.

In later chapters, we shall be closely concerned with slow (practically nonergodic) algorithms, and the concept of ergodicity will be much discussed. At the present stage, however, let us first look at a nonergodic Monte Carlo algorithm in a familiar setting that is easy to analyze: a variant of our pebble game, namely the sliding puzzle, a popular toy (see Fig. 1.41).

The configurations of the sliding puzzle correspond to permutations of the numbers $\{0, \ldots, 8\}$, the zero square being empty. Let us see why not all configurations can be reached by repeated sliding moves in the local Monte Carlo algorithm, whereby a successful move of the empty square corresponds to its transposition with one of its neighbors. We saw in Subsection 1.2.2 that all permutations can be constructed from transpositions of two elements. But taking the zero square around a closed loop makes us go left and right the same number of times, and we also perform equal numbers of successful up and down moves (a loop is shown in the last frame of Fig. 1.41). Therefore, on a square lattice, the zero square can return to its starting position only after an even number of moves, and on completion of an even number of transpositions. Any accessible configuration with this square in the upper right corner thus corresponds to an even permutation (compare with Subsection 1.2.2), and odd permutations with the zero square in the upper right corner, as in Fig. 1.42, can never be reached: the local Monte Carlo algorithm for the sliding puzzle is nonergodic.

## 1.4.2 Importance sampling

Common sense tells us that nonergodic or practically nonergodic algorithms are of little help in solving computational problems. It is less evident that even ergodic sampling methods can leave us with a tricky job.

**Fig. 1.42** A puzzle configuration that cannot be reached from the configurations in Fig. 1.41.

---

[7]This definition of ergodicity, for a random process, is slightly different from that for a physical system.

Difficulties appear whenever there are rare configurations—configurations with a low probability—which contribute in an important way to the value of our integral. If these rare yet important configurations are hardly ever seen, say in a week-long Monte Carlo simulation, we clearly face a problem. This problem, though, has nothing to do with Markov chains and shows up whether we use direct sampling as in the children's algorithm or walk around like adults on the heliport. It is linked only to the properties of the distribution that we are trying to sample.

In this context, there are two archetypal examples that we should familiarize ourselves with. The first is the Monte Carlo golf course. We imagine trying to win the Monaco golf trophy not by the usual playing technique but by randomly dropping golf balls (replacing pebbles) onto the greens (direct sampling) or by randomly driving balls across the golf course, with our eyes closed (Markov-chain sampling). Either way, random sampling would take a lot of trials to hit even a single hole out there on the course, let alone bring home a little prize money, or achieve championship fame. The golf course problem and its cousin, that of a needle in a haystack, are examples of genuinely hard sampling problems which cannot really be simplified.

The second archetypal example containing rare yet important configurations belongs to a class of models which have more structure than the golf course and the needle in a haystack. For these problems, a concept called importance sampling allows us to reweight probability densities and observables and to overcome the basic difficulty. This example is defined by the integral

$$I(\gamma) = \int_0^1 \mathrm{d}x\ x^\gamma = \frac{1}{\gamma + 1} \text{ for } \gamma > -1. \tag{1.71}$$

We refer to this as the $\gamma$-integral, and shall go through the computational and mathematical aspects of its evaluation by sampling methods in the remainder of this section. The $\gamma$-integral has rare yet important configurations because, for $\gamma < 0$, the integrand all of a sudden becomes very large close to $x = 0$. Sampling the $\gamma$-integral is simple only in appearance: what happens just below the surface of this problem was cleared up only recently (on the timescale of mathematics) by prominent mathematicians, most notably P. Lévy. His analysis from the 1930s will directly influence the understanding of our naive Monte Carlo programs.

To evaluate the $\gamma$-integral—at least initially—one generates uniform random numbers $x_i$ between 0 and 1, and averages the observable $\mathcal{O}_i = x_i^\gamma$ (see Alg. 1.29 (direct-gamma)). The output of this program should approximate $I(\gamma)$.

To obtain the error, a few lines must be added to the program, in addition to the mean

$$\frac{\Sigma}{N} = \frac{1}{N} \sum_{i=1}^N \mathcal{O}_i \simeq \langle \mathcal{O} \rangle,$$

**procedure** direct-gamma
$\Sigma \leftarrow 0$
**for** $i = 1, \dots, N$ **do**
$\left\{ \begin{array}{l} x_i \leftarrow \mathtt{ran}\,(0,1) \\ \Sigma \leftarrow \Sigma + x_i^\gamma \ \text{(running average: } \Sigma/i) \end{array} \right.$
**output** $\Sigma/N$

——

**Algorithm 1.29** direct-gamma. Computing the $\gamma$-integral in eqn (1.71) by direct sampling.

which is already programmed. We also need the average of the squared observables,

$$\frac{1}{N} \sum_{i=1}^{N} x_i^{2\gamma} = \frac{1}{N} \sum_{i=1}^{N} \mathcal{O}_i^2 \simeq \langle \mathcal{O}^2 \rangle.$$

This allows us to estimate the variance (see Subsection 1.3.5):

$$\text{error} = \frac{\sqrt{\langle \mathcal{O}^2 \rangle - \langle \mathcal{O} \rangle^2}}{\sqrt{N}}. \tag{1.72}$$

With direct sampling, there are no correlation times to worry about. The output for various values of $\gamma$ is shown in Table 1.15.

Most of the results in this table agree with the analytical results to within error bars, and we should congratulate ourselves on this nice success! In passing, we should look at the way the precision in our calculation increases with computer time, i.e. with the number $N$ of iterations. The calculation recorded in Table 1.15 (with $N = 10\,000$) for $\gamma = 2$, on a year 2005 laptop computer, takes less than $1/100$ s and, as we see, reaches a precision of 0.003 (two digits OK). Using 100 times more samples ($N = 10^6$, 0.26 s), we obtain the result 0.33298 (not shown, three significant digits). One hundred million samples are obtained in 25 seconds, and the result obtained is 0.3333049 (not shown, four digits correct). The precision increases as the square root of the number of samples, and gains one significant digit for each hundred-fold increase of computer time. This slow but sure convergence is a hallmark of Monte Carlo integration and can be found in any program that runs without bugs and irregularities and has a flawless random number generator.

However, the calculation for $\gamma = -0.8$ in Table 1.15 is in trouble: the average of the $\mathcal{O}_i$'s is much further off $I(\gamma)$ than the internally computed error indicates. What should we do about this? There is no rescue in sarcasm[8] as it is nonsense, because of Chebyshev's inequality, to think that one could be 10 standard deviations off the mean value. Furthermore, Alg. 1.29 (direct-gamma) is too simple to have bugs, and even our analytical calculation in eqn (1.71) is watertight: $I(-0.8)$ is indeed 5 and

**Table 1.15** Output of Alg. 1.29 (direct-gamma) for various values of $\gamma$ ($N = 10\,000$). The computation for $\gamma = -0.8$ is in trouble.

| $\gamma$ | $\Sigma/N \pm$ Error | $1/(\gamma+1)$ |
|---|---|---|
| 2.0 | $0.334 \pm 0.003$ | $0.333\dots$ |
| 1.0 | $0.501 \pm 0.003$ | 0.5 |
| 0.0 | $1.000 \pm 0.000$ | 1 |
| -0.2 | $1.249 \pm 0.003$ | 1.25 |
| -0.4 | $1.682 \pm 0.014$ | $1.666\dots$ |
| -0.8 | $3.959 \pm 0.110$ | 5.0 |

---

[8]An example of sarcasm: there are three kinds of lies: lies, damned lies, and statistics.

not 4.... To make progress, we monitor our simulation, and output running averages, as indicated in Alg. 1.29 (direct-gamma). In Fig. 1.43, this average calmly approaches a wrong mean value. Then chance lets the program hit a very small value $x_i$, with an incredibly large $\mathcal{O}_i$ (we remember that $\gamma < 0$ so that small values of $x$ give large observables). Thus one sample, in a simulation of a million trials, single-handedly hikes up the running average.

Figure 1.43 contains a nightmare scenario: a Monte Carlo simulation converges nicely (with an average of about 4.75, well established for a computational eternity), until a seemingly pathological sample changes the course of the simulation. In this situation, real insight and strength of human character are called for: *we must not doctor the data and suppress even a single sample* (which would let us continue with an average near 4.75). Sooner or later, someone would find out that our data were botched!

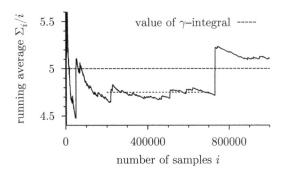

**Fig. 1.43** Running average of Alg. 1.29 (direct-gamma) for $\gamma = -0.8$.

What happens in Table 1.15 is that the error formula in eqn (1.72) involves an average over the squared observable:

$$\underbrace{\frac{1}{N}\sum_{i=1}^{N}\mathcal{O}_i^2}_{\substack{\text{estimate}\\ \text{(always finite)}}} \simeq \underbrace{\int_0^1 dx\ x^{2\gamma}}_{\substack{\text{variance}\\ \text{(infinite for } \gamma<-\frac{1}{2}\text{)}}} \quad . \tag{1.73}$$

We see that the algorithm to evaluate the $\gamma$-integral, in the range $-1 < \gamma < -\frac{1}{2}$, where it is still finite, has a problem in estimating the error, because it uses a finite sum to approximate an integral with an infinite variance. Clearly, the situation is difficult to analyze in the absence of an analytic solution.

We can salvage the Monte Carlo calculation of the $\gamma$-integral by preferentially visiting regions of space where the expression $\mathcal{O}(x)\pi(x)$ is large. In our example, we split the integrand of the $\gamma$-integral appropriately into a new probability density $\pi(x) = x^\varsigma$ and a new observable

$\mathcal{O}(x) = x^{\gamma - \zeta}$ (see Fig. 1.44). For negative $\zeta$ ($\gamma < \zeta < 0$), small values of $x$, with a large integrand, are visited more often and the variance of the observable is reduced, while the product of the probability density and the observable remains unchanged. This crucial technique is called importance sampling.

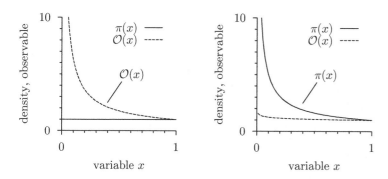

**Fig. 1.44** Splits of the $\gamma$-integrand into a density and an observable ($\gamma = -0.8$). Left: $\pi(x) = 1$; $\mathcal{O}(x) = x^{-0.8}$. Right: $\pi(x) = x^{-0.7}$; $\mathcal{O}(x) = x^{-0.1}$.

**procedure** direct-gamma-zeta
$\Sigma \leftarrow 0$
**for** $i = 1, \ldots, N$ **do**
$\left\{ \begin{array}{l} x_i \leftarrow \mathbf{ran}\,(0,1)^{1/(\zeta+1)} \quad (\pi(x_i) \propto x_i^\zeta, \text{ see eqn (1.29)}) \\ \Sigma \leftarrow \Sigma + x_i^{\gamma - \zeta} \end{array} \right.$
**output** $\Sigma/N$
————

**Algorithm 1.30** direct-gamma-zeta. Using importance sampling to compute the $\gamma$-integral (see eqn (1.74)).

**Table 1.16** Output of Alg. 1.30 (direct-gamma-zeta) with $N = 10\,000$. All pairs $\{\gamma, \zeta\}$ satisfy $2\gamma - \zeta > -1$ so that $\langle \mathcal{O}^2 \rangle < \infty$.

| $\gamma$ | $\zeta$ | $\Sigma/N$ | $\frac{\zeta+1}{\gamma+1}$ |
|---|---|---|---|
| $-0.4$ | $0.0$ | $1.685 \pm 0.017$ | $1.66$ |
| $-0.6$ | $-0.4$ | $1.495 \pm 0.008$ | $1.5$ |
| $-0.7$ | $-0.6$ | $1.331 \pm 0.004$ | $1.33$ |
| $-0.8$ | $-0.7$ | $1.508 \pm 0.008$ | $1.5$ |

The idea is implemented in Alg. 1.30 (direct-gamma-zeta), our first application of importance sampling. The output of the program, $\Sigma/N$, corresponds to the ratio of two $\gamma$-integrals:

$$\Sigma/N = \frac{1}{N} \sum_{i=1}^{N} \mathcal{O}_i \simeq \langle \mathcal{O} \rangle = \frac{\int_0^1 \mathrm{d}x\, \pi(x)\mathcal{O}(x)}{\int_0^1 \mathrm{d}x\, \pi(x)}$$

$$= \frac{\int_0^1 \mathrm{d}x\, x^\zeta x^{\gamma-\zeta}}{\int_0^1 \mathrm{d}x\, x^\zeta} = \frac{\int_0^1 \mathrm{d}x\, x^\gamma}{\int_0^1 \mathrm{d}x\, x^\zeta} = \frac{I(\gamma)}{I(\zeta)} = \frac{\zeta+1}{\gamma+1}. \quad (1.74)$$

This is because Monte Carlo simulations compute $\int \mathrm{d}x\, \pi(x)\mathcal{O}(x)$ only if the density function $\pi$ is normalized. Otherwise, we have to normalize with $\int \mathrm{d}x\, \pi(x)$. A glimpse at Table 1.16 shows that the calculation comes out just right.

We must understand how to compute the error, and why reweighting is useful. The error formula eqn (1.72) remains valid for the observable $\mathcal{O}(x) = x^{\gamma - \zeta}$ (a few lines of code have to be added to the program). The variance of $\mathcal{O}$ is finite under the following condition:

$$\frac{1}{N} \sum_{i=1}^{N} \mathcal{O}_i^2 \simeq \langle \mathcal{O}^2 \rangle = \frac{\int_0^1 dx \, \pi(x) \mathcal{O}^2(x)}{\int_0^1 dx \, \pi(x)}$$

$$= \int_0^1 dx \, x^\zeta x^{2\gamma - 2\zeta} < \infty \Leftrightarrow \gamma > -\frac{1}{2} + \zeta/2. \quad (1.75)$$

This gives a much larger range of possible $\gamma$ values than does eqn (1.73) for negative values of $\zeta$. All pairs $\{\gamma, \zeta\}$ in Table 1.16 satisfy the inequality (1.75). Together, the five different pairs in the table give a product of all the ratios equal to

$$\underbrace{\frac{\int_0^1 dx \, x^{-0.8}}{\int_0^1 dx \, x^{-0.7}}}_{\gamma = -0.8, \zeta = -0.7} \frac{\int_0^1 dx \, x^{-0.7}}{\int_0^1 dx \, x^{-0.6}} \frac{\int_0^1 dx \, x^{-0.6}}{\int_0^1 dx \, x^{-0.4}} \frac{\int_0^1 dx \, x^{-0.4}}{\int_0^1 dx \, x^{0.0}} = I(-0.8),$$

with the numerical result

$$\langle \mathcal{O} \rangle \simeq 1 \times 1.685 \times 1.495 \times 1.331 \times 1.508 = 5.057.$$

Using the rules of Gaussian error propagation, the variance is

$$\text{Var} \, (\mathcal{O}) = \left[ \left( \frac{0.017}{1.685} \right)^2 + \left( \frac{0.008}{1.495} \right)^2 + \left( \frac{0.004}{1.331} \right)^2 + \left( \frac{0.008}{1.508} \right)^2 \right] \times 5.057^2.$$

From the above expression, the final result for the $\gamma$-integral is

$$I(\gamma = -0.8) = 5.057 \pm 0.06,$$

obtained entirely by a controlled Monte Carlo calculation.

### 1.4.3   Monte Carlo quality control

In Subsection 1.4.2, the Monte Carlo evaluation of the $\gamma$-integral proceeded on quite shaky ground, because the output was sometimes correct and sometimes wrong. Distressingly, neither the run nor the data analysis produced any warning signals of coming trouble. We relied on an analytical quality control provided by the inequalities

$$\text{if } \gamma > \begin{cases} -1 & : \text{integral exists} \\ -\frac{1}{2} & : \text{variance exists} \\ -\frac{1}{2} + \zeta/2 & : \text{variance exists (importance sampling)} \end{cases} \quad . \quad (1.76)$$

To have analytical control over a calculation is very nice, but this cannot always be achieved. Often, we simply do not completely understand the structure of high-dimensional integrals. We are then forced to replace the above analytical "back office" with a numerical procedure that warns us, with a clear and intelligible voice, about a diverging integral or an infinite variance. Remarkably, the Markov-chain Monte Carlo algorithm can serve this purpose: we must simply reformulate the integration problem at hand such that any nonintegrable singularity shows up as an infinite statistical weight which attracts Markov chains without ever allowing them to get away again.

For concreteness, we discuss the $\gamma$-integral of Subsection 1.4.2, which we shall now try to evaluate using a Markov-chain Monte Carlo algorithm (see Alg. 1.31 (markov-zeta)). To see whether the $\gamma$-integral for $\gamma = -0.8$ exists and whether its variance is finite, we run Alg. 1.31 (markov-zeta) twice, once with statistical weights $\pi'(x) = |\mathcal{O}(x)\pi(x)| = x^{-0.8}$ and then with $\pi''(x) = |\mathcal{O}^2(x)\pi(x)| = x^{-1.6}$. The first case corresponds to putting $\zeta = -0.8$ in the algorithm, and the second to $\zeta = -1.6$.

**procedure** markov-zeta
**input** $x$
$\tilde{x} \leftarrow x + \mathrm{ran}\,(-\delta, \delta)$
**if** $(0 < \tilde{x} < 1)$ **then**
$\quad\begin{cases} p_{\mathrm{accept}} \leftarrow (\tilde{x}/x)^{\zeta} \\ \text{if } (\mathrm{ran}\,(0,1) < p_{\mathrm{accept}})\ x \leftarrow \tilde{x} \end{cases}$
**output** $x$
——

**Algorithm 1.31** markov-zeta. Markov-chain Monte Carlo algorithm for a point $x$ on the interval $[0,1]$ with $\pi(x) \propto x^{\zeta}$.

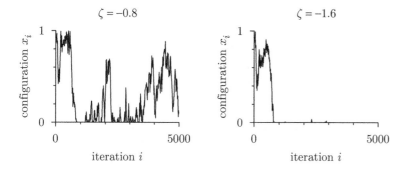

**Fig. 1.45** Runs of Alg. 1.31 (markov-zeta). *Left*: exponent $\zeta = -0.8$: integrable singularity at $x = 0$. *Right*: $\zeta = -1.6$: nonintegrable singularity.

By simply monitoring the position $x_i$ as a function of the iteration number $i$, we are able to decide whether the densities $\pi'(x)$ and $\pi''(x)$ are integrable. In the second case above, there will be a region around a point $x'$ where $\int_{x-\epsilon}^{x+\epsilon} \mathrm{d}x'\, \pi(x') = \infty$. The simulation will be unable to escape from the vicinity of this point. On the other hand, an integrable singularity at $x$ ($\pi(x) \to \infty$ but $\int_{x-\epsilon}^{x+\epsilon} dx' \, \pi(x')$ is finite) does not trap the simulation. Data for the two cases, $\pi'(x)$ and $\pi''(x)$, are shown in Fig. 1.45. We can correctly infer from the numerical evidence that the $\gamma$-integral exists for $\gamma = -0.8$ ($\langle \mathcal{O} \rangle$ is finite) but not for $\gamma = -1.6$ ($\langle \mathcal{O}^2 \rangle$ is infinite). In the present context, the Metropolis algorithm is purely qualitative, as no observables are computed. This qualitative method allows us to learn about the analytic properties of high-dimensional integrals when analytical information analogous to eqn (1.76) is unavailable (see Krauth and Staudacher (1999)).

### 1.4.4   Stable distributions

Many problems in the natural and social sciences, economics, engineering, etc. involve probability distributions which are governed by rare yet important events. For a geologist, the running-average plot in Fig. 1.43 might represent seismic activity over time. Likewise, a financial analyst might have to deal with similar data (with inverted $y$-axis) in a record of stock-exchange prices: much small-scale activity, an occasional Black Monday, and a cataclysmic crash on Wall Street. Neither the geologist nor the financial analyst can choose the easy way out provided by importance sampling, because earthquakes and stock crashes cannot be made to go away through a clever change of variables. It must be understood how often accidents like the ones shown in the running average in Fig. 1.43 happen. Both the geologist and the financial analyst must study the probability distribution of running averages outside the regime $\gamma > -\frac{1}{2}$, that is, when the variance is infinite and Monte Carlo calculations are impossible to do without importance sampling. The subject of such distributions with infinite variances was pioneered by Lévy in the 1930s. He showed that highly erratic sums of $N$ random variables such as that in Fig. 1.43 tend towards universal distributions, analogously to the way that sums of random variables with a finite variance tend towards the Gaussian, as dictated by the central limit theorem. The limit distributions depend on the power (the parameter $\gamma$) and on the precise asymptotic behavior for $x \to \pm\infty$.

We shall first generate these limit distributions from rescaled outputs of Alg. 1.29 (`direct-gamma`), similarly to what was done for uniform bounded random numbers in Alg. 1.17 (`naive-gauss`). Secondly, we shall numerically convolute distributions in much the same way as we did for the bounded uniform distribution in Alg. 1.26 (`ran01-convolution`). Finally, we shall explicitly construct the limiting distributions using characteristic functions, i.e. the Fourier transforms of the distribution functions.

We run Alg. 1.29 (`direct-gamma`), not once, as for the initial running-average plot of Fig. 1.43, but a few million times, in order to produce histograms of the sample average $\Sigma/N$ at fixed $N$ (see Fig. 1.46, for $\gamma = -0.8$). The mean value of all the histograms is equal to 5, the value of the $\gamma$-integral, but this is not a very probable outcome of a single run: from

the histogram, a single run with $N = 1000$ samples is much more likely
to give a sample average around 3.7, and for $N = 10\,000$, we are most
likely to obtain around 4.2. This is consistent with our very first one-shot
simulation recorded in Table 1.15, where we obtained $\Sigma/N = 3.95$. We
note that the peak position of the distribution approaches $\langle x_i \rangle = 5$ very
slowly as $N \to \infty$ (see Table 1.17).

The averages generated by Alg. 1.29 (`direct-gamma`) must be rescaled
in order to fall onto a unique curve. The correct rescaling,

$$\Upsilon = \frac{\Sigma/N - \langle x_i \rangle}{N^{-1-\gamma}}, \tag{1.77}$$

will be taken for granted for the moment, and derived later, in eqn (1.80).
Rescaling the output of Alg. 1.29 (`direct-gamma`), for our value $\gamma = -0.8$, consists in subtracting the mean value, 5, from each average of
$N$ terms and then dividing by $N^{-0.2}$ (see Fig. 1.46). These histograms
of rescaled averages illustrate the fact that the distribution function
of a sum of random variables, which we computed for the $\gamma$-integral,
converges to a limit distribution for large values of $N$. We recall from
Subsection 1.3.3 that for random variables with a finite variance, $\Sigma/\sqrt{N}$
gives a unique curve. This case is reached for $\gamma \to -\frac{1}{2}$.

Table 1.17 Peak positions of the histogram of sample averages $\Sigma/N$ (from Alg. 1.29 (`direct-gamma`), with $\gamma = -0.8$)

| $N$ | Peak position of $\Sigma/N$ |
|---|---|
| 10 | 1.85 |
| 100 | 3.01 |
| 1000 | 3.74 |
| 10 000 | 4.21 |
| 100 000 | 4.50 |
| 1 000 000 | 4.68 |
| 10 000 000 | 4.80 |
| 100 000 000 | 4.87 |

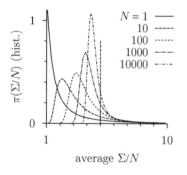 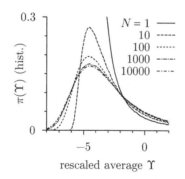

**Fig. 1.46** Histograms of averages (*left*) and rescaled averages (*right*)
(from Alg. 1.29 (`direct-gamma`), with $\gamma = -0.8$ and $\Upsilon = (\Sigma/N - 5)/N^{-0.2}$).

Up to now, we have formulated the $\gamma$-integral in terms of a uniform
random variable $x = \mathtt{ran}\,(0, 1)$ and an observable $\mathcal{O}(x) = x^\gamma$. In what
follows, it is better to incorporate the observable into the random variable $x = \mathtt{ran}\,(0, 1)^\gamma$. That is, we consider random variables $\xi_i$ taking
values $x$ with probability

$$\pi(x) = \begin{cases} -\frac{1}{\gamma} \cdot 1/x^{1-1/\gamma} & \text{for } 1 < x < \infty \\ 0 & \text{otherwise} \end{cases} \tag{1.78}$$

(see eqn (1.29)). We shall also use $\alpha = -1/\gamma$, as is standard notation in
this problem since the time of Lévy, who considered the sum of random

variables $\xi_i$ taking values $x$ with probability $\pi(x)$ characterized by a zero mean and by the asymptotic behavior

$$\pi(x) \simeq \begin{cases} A_+/x^{1+\alpha} & \text{for } x \to \infty \\ A_-/|x|^{1+\alpha} & \text{for } x \to -\infty \end{cases},$$

where $1 < \alpha < 2$. When rescaled as in eqn (1.77), the probability distribution keeps the same asymptotic behavior and eventually converges to a stable distribution which depends only on the three parameters $\{\alpha, A_+, A_-\}$. For the $\gamma$-integral with $\gamma = -0.8$, these parameters are $\alpha = -1/\gamma = 1.25$, $A_+ = 1.25$ and $A_- = 0$ (see eqn (1.78)). We shall later derive this behavior from an analysis of characteristic functions. As the first step, it is instructive to again assume this rescaling, and to empirically generate the universal function for the $\gamma$-integral (where $\alpha = 1.25, A_+ = 1.25$, and $A_- = 0$) from repeated convolutions of a starting distribution with itself. We may recycle Alg. 1.26 (`ran01-convolution`), but cannot cut off the distribution at large $|x|$. This would make the distribution acquire a finite variance, and drive the convolution towards a Gaussian. We must instead pad the function at large arguments, as shown in Fig. 1.47, and as implemented in Alg. 1.32 (`levy-convolution`) for $A_- = 0$. After a number of iterations, the grid of points $x_k$, between the fixed values $x_{\min}$ and $x_{\max}$, becomes very fine, and will eventually have to be decimated. Furthermore, during the iterations, the function $\pi(x)$ may lose its normalization and cease to be of zero mean. This is repaired by computing the norm of $\pi$—partly continuous, partly discrete—as follows:

$$\int dx\ \pi(x) = \underbrace{\int_{-\infty}^{x_0} dx\ \frac{A_-}{|x|^{1+\alpha}}}_{A_- \cdot \frac{1}{\alpha} \frac{1}{|x_0|^{\alpha}}} + \Delta \left( \frac{\pi_0 + \pi_K}{2} + \sum_{k=1}^{K-1} \pi_k \right) + \underbrace{\int_{x_K}^{\infty} dx\ \frac{A_+}{|x|^{1+\alpha}}}_{A_+ \cdot \frac{1}{\alpha} \frac{1}{x_K^{\alpha}}}.$$

The mean value can be kept at zero in the same way.

---

**procedure** `levy-convolution`
**input** $\{\{\pi_0, x_0\}, \ldots, \{\pi_K, x_K\}\}$ (see Fig. 1.47)
**for** $k = K+1, K+2 \ldots$ **do** (padding)
$\quad \begin{cases} x_k \leftarrow x_0 + k\Delta \\ \pi_k \leftarrow A_+/x_k^{1+\alpha} \end{cases}$
**for** $k = 0, \ldots, 2K$ **do** (convolution)
$\quad \begin{cases} x_k' \leftarrow (x_0 + x_k)/2^{1/\alpha} \\ \pi_k' \leftarrow (\Delta \sum_{l=0}^{k} \pi_l \pi_{k-l}) \cdot 2^{1/\alpha} \end{cases}$
**output** $\{\{\pi_m', x_m'\}, \ldots, \{\pi_n', x_n'\}\}$ (all $x'$ in interval $[x_{\min}, x_{\max}]$)

---

**Algorithm 1.32** `levy-convolution`. Convolution of $\pi(x)$ with itself. $\pi(x)$ is padded as in Fig. 1.47, with $A_- = 0$.

We now discuss stable distributions in terms of their characteristic functions. Lévy first understood that a non-Gaussian stable law (of zero

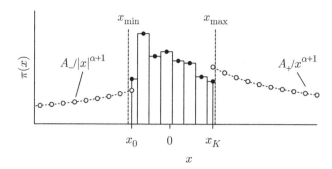

**Fig. 1.47** Padding of a discrete function $\pi(x_i)$ with continuous functions $A_+/x^{1+\alpha}$ (for $x > x_{\max} \gg 0$) and $A_-/|x|^{1+\alpha}$ (for $x < x_{\min} \ll 0$).

mean) can only have a characteristic function

$$\phi(t) = \int_{-\infty}^{\infty} dx\ \pi(x) e^{itx} = \exp\left[-\left(c_0 + ic_1 \frac{t}{|t|}\right) |t|^{\alpha}\right], \tag{1.79}$$

where $1 \le \alpha < 2$ (we exclude the case $\alpha = 1$ from our discussion). This choice, with $c_0 > 0$, ensures that $\pi(x)$ is real, correctly normalized, and of zero mean (see Table 1.18). Furthermore, the square of $\phi(t)$, the characteristic function of the convolution of $\pi(t)$ with itself, is given by a rescaled $\phi(t)$:

$$\phi^2(t) = \exp\left[-\left(c_0 + ic_1 \frac{t}{|t|}\right) \cdot 2|t|^{\alpha}\right] = \phi(t \cdot 2^{1/\alpha}). \tag{1.80}$$

The characteristic function $\phi(t)$ is related to the probability density $\pi(x)$ by an inverse Fourier transformation:

$$\pi(x) = \frac{1}{2\pi} \int_{-\infty}^{\infty} dt\ \phi(t) e^{-ixt}.$$

This means that the empirical limiting function for the rescaled output of Alg. 1.29 (**direct-gamma**) shown in Fig. 1.46 can be described by eqn (1.79) with $\alpha = 1.25$. We need only to clarify the relationship between the Fourier parameters $\{c_0, c_1\}$ and the real-space parameters $\{A_+, A_-\}$. This point will now be investigated by means of an asymptotic expansion of a Fourier integral.

Because $\alpha > 1$, we may multiply the integrand by a small subdominant term $e^{-\epsilon|t|}$. This leaves us with a function

$$\pi_\epsilon(x) = \frac{1}{2\pi} \int_{-\infty}^{\infty} dt\ e^{-ixt} \exp\left[-\left(c_0 + ic_1 \frac{t}{|t|}\right) |t|^{\alpha}\right] e^{-\epsilon|t|}$$

$$= \underbrace{\int_{-\infty}^{0} dt\ \dots}_{\pi_\epsilon^-(x)} + \underbrace{\int_{0}^{\infty} dt\ \dots}_{\pi_\epsilon^+(x)}. \tag{1.81}$$

**Table 1.18** Basic requirements on the stable distribution $\pi(x)$ and its characteristic function $\phi(t)$

| $\pi(x)$ | $\phi(t)$ |
|---|---|
| Real | $\phi(t) = \phi(-t)^*$ |
| Normalized | $\phi(0) = 1, |\phi(t)| \le 1$ |
| Zero mean | $\phi'(0) = 1$ |
| Positive | $|c_1/c_0| < |\tan \frac{\pi\alpha}{2}|$ |
| | $c_0 \ge 0$ |

We now look in more detail at $\pi_\epsilon^+(x)$, where

$$\pi_\epsilon^+(x) = \frac{1}{2\pi} \int_0^\infty dt\; e^{-ixt} \underbrace{\exp\left[-(c_0 + ic_1)t^\alpha\right]}_{1-(c_0+ic_1)t^\alpha+\cdots} e^{-\epsilon t}. \tag{1.82}$$

After the indicated expansion of the exponential, we may evaluate the integral[9] for nonzero $\epsilon$:

$$\pi_\epsilon^+(x) \simeq \frac{1}{2\pi} \sum_{m=0}^\infty \frac{(-1)^m}{m!} \frac{\Gamma(m\alpha+1)}{(x^2+\epsilon^2)^{(m\alpha+1)/2}} (c_0 + ic_1)^m$$

$$\times \exp\left[-i\,(m\alpha+1)\arctan\frac{x}{\epsilon}\right]. \tag{1.83}$$

**Table 1.19** The function $\pi_\epsilon^+(x=10)$ for $c_0 = 1$ and $c_1 = 0$ and its approximations in eqn (1.83). All function values are multiplied by $10^3$.

| $\epsilon$ | $\pi_\epsilon^+$ | $0$ | $0, 1$ | $0, 1, 2$ |
|---|---|---|---|---|
| | | *m*-values included | | |
| 0 | 0.99 | 0 | 0.94 | 0.99 |
| 0.1 | 1.16 | 0.16 | 1.10 | 1.16 |
| 0.2 | 1.32 | 0.32 | 1.27 | 1.33 |

For concreteness, we may evaluate this asymptotic expansion term by term for fixed $x$ and $\epsilon$ and compare it with the numerical evaluation of eqn (1.82) (see Table 1.19). We can neglect the imaginary part of $\pi(x)$ (see Table 1.18) and, in the limit $\epsilon \to 0$, the $m=0$ term vanishes for $x \neq 0$, and the $m=1$ term dominates. This allows us to drop terms with $m=2$ and higher: the small-$t$ behavior of the characteristic function $\phi(t)$ governs the large-$|x|$ behavior of $\pi(x)$.

We now collect the real parts in eqn (1.83), using the decomposition of the exponential function into sines and cosines ($e^{-ixt} = \cos(xt) - i\sin(xt)$), the asymptotic behavior of $\arctan x$ ($\lim_{x\to\pm\infty}\arctan x = \pm\pi/2$), and the relations between sines and cosines ($\cos(x+\pi/2) = -\sin x$ and $\sin(x+\pi/2) = \cos x$). We then find

$$\pi^+(x) \simeq \frac{\Gamma(1+\alpha)}{2\pi x^{1+\alpha}}\left(c_0\sin\frac{\pi\alpha}{2} - c_1\cos\frac{\pi\alpha}{2}\right) \text{ for } x \to \infty.$$

The other term gives the same contribution, i.e. $\pi^-(x) = \pi^+(x)$ for large $x$. We find, analogously,

$$\pi^+(x) \simeq \frac{\Gamma(1+\alpha)}{2\pi|x|^{1+\alpha}}\left(c_0\sin\frac{\pi\alpha}{2} + c_1\cos\frac{\pi\alpha}{2}\right) \text{ for } x \to -\infty, \tag{1.84}$$

with again the same result for $\pi^-$.

The calculation from eqn (1.81) to eqn (1.84) shows that any characteristic function $\phi(t)$ whose expansion starts as in eqn (1.79) for small $t$ belongs to a probability distribution $\pi(x)$ with an asymptotic behavior

$$\pi(x) \simeq \begin{cases} A_+/x^{1+\alpha} & \text{for } x \to \infty \\ A_-/|x|^{1+\alpha} & \text{for } x \to -\infty \end{cases}, \tag{1.85}$$

where

$$A_\pm = \frac{\Gamma(1+\alpha)}{\pi}\left(c_0\sin\frac{\pi\alpha}{2} \pm c_1\cos\frac{\pi\alpha}{2}\right). \tag{1.86}$$

[9]We use

$$\int_0^\infty dt\; t^\alpha e^{-\epsilon t}\frac{\sin}{\cos}xt = \frac{\Gamma(\alpha+1)}{(\epsilon^2+x^2)^{(\alpha+1)/2}}\frac{\sin}{\cos}\left[(\alpha+1)\arctan\frac{x}{\epsilon}\right].$$

This integral is closely related to the gamma function.

Equations (1.85) and (1.86) relate the asymptotic behavior of $\pi(x)$ to the characteristic function of the corresponding stable distribution. We can test the formulas for the characteristic distribution of the $\gamma$-integral, where we find, from eqn (1.86),

$$\alpha = 1.25 \ (\gamma = -0.8): \quad \begin{bmatrix} A_+ = 1.25 \\ A_- = 0 \end{bmatrix} \Leftrightarrow \begin{bmatrix} c_0 = 1.8758 \\ c_1 = 4.5286 \end{bmatrix}. \quad (1.87)$$

This gives the following as a limit function for the rescaled sum of random variables obtained from Alg. 1.29 (`direct-gamma`), for $-1 < \gamma < -0.5$:

$$\pi(x) = \frac{1}{2\pi} \int_{-\infty}^{\infty} dt \ \exp\left[-ixt - \left(c_0 + ic_1 \frac{t}{|t|}\right)|t|^\alpha\right], \quad (1.88)$$

with parameters from eqn (1.87). Equation (1.88), an inverse Fourier transform, can be evaluated by straightforward Riemann integration, using suitable finite limits for the $t$-integration. The agreement between all of our three approaches to the Lévy distribution is excellent (see Fig. 1.48, and compare it with the rescaled averages in Fig. 1.46).

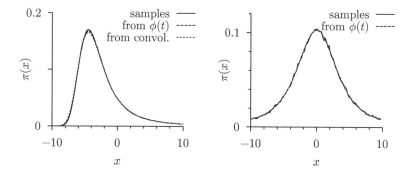

**Fig. 1.48** Stable distributions for $\alpha = 1.25$. *Left*: one-sided case ($A_- = 0$, $A_+ = 1.25$). *Right*: symmetric case ($A_- = A_+ = 1.25$).

For a second illustration of Lévy statistics, we symmetrize output of Alg. 1.29 (`direct-gamma`) and take $\Upsilon \leftarrow \mathtt{ran}\left(-\frac{1}{2}, \frac{1}{2}\right)$. Samples $x_i$ are generated as follows:

$$x_i = \begin{cases} \Upsilon^\gamma & \text{if } \Upsilon \geq 0 \\ -|\Upsilon|^\gamma & \text{if } \Upsilon < 0 \end{cases}.$$

The symmetric distribution $\pi(x_i)$ has the same asymptotic behavior for $x \to +\infty$ as the $\gamma$-integral, so that $A_+ = A_- = 1.25$. Equation (1.86) once more allows us to compute the parameters $\{c_0, c_1\}$ from $\{A_+, A_-\}$:

$$\alpha = 1.25 \ (\gamma = -0.8): \quad \begin{bmatrix} A_+ = 1.25 \\ A_- = 1.25 \end{bmatrix} \Leftrightarrow \begin{bmatrix} c_0 = 3.7516 \\ c_1 = 0 \end{bmatrix}.$$

The parameters $\{\alpha, c_0, c_1\}$ can again be entered into the inverse Fourier transform of eqn (1.88), and the result can be compared with the rescaled simulation data (see Fig. 1.48).

In conclusion, we have progressed in this subsection from a naive case study of Alg. 1.29 (`direct-gamma`) to a systematic analysis of distributions with a finite mean and infinite variance. The mathematical law and order in these very erratic random processes is expressed through the scaling in eqn (1.77), which fixes the speed of convergence towards the mean value, and through the parameters $\{\alpha, A_+, A_-\}$ or, equivalently, $\{\alpha, c_0, c_1\}$, which are related to each other by eqn (1.86).

### 1.4.5   Minimum number of samples

In this chapter, we have witnessed Monte Carlo calculations from all walks of life, and with vastly different convergence properties. With one exception, when we used Alg. 1.31 (`markov-zeta`) to sniff out singularities, all these programs attempted to estimate an observable mean $\langle \mathcal{O} \rangle$ from a sample average. The success could be outstanding, most pronouncedly in the simulation related to the crazy cobbler's needle, where a single sample gave complete information about the mean number of hits. The situation was least glamorous in the case of the $\gamma$-integral. There, typical sample averages remained different from the mean of the observable even after millions of iterations. Practical simulations are somewhere in between these extremes, even if the variance is finite: it all depends on how different the original distribution is from a Gaussian, how asymmetric it is, and how quickly it falls to zero for large absolute values of its arguments. In a quite difficult case, we saw how the situation could be saved through importance sampling.

In many problems in physics, importance sampling is incredibly efficient. This technique counterbalances the inherent slowness of Markov-chain methods, which are often the only ones available and which, in billions of iterations and months of computer time, generate only a handful of independent samples. In most cases, though, this relatively small amount of data allows one to make firm statements about the physical properties of the system studied. We can thus often get away with just a few independent samples: this is the main reason why the equilibrium Monte Carlo method is so firmly established in statistical physics.

# Exercises

## (Section 1.1)

(1.1) Implement Alg. 1.1 (`direct-pi`). Run it twenty times each for $N = 10, 100, \ldots, 1 \times 10^8$. Convince yourself that $N_{\text{hits}}/N$ converges towards $\pi/4$. Estimate the mean square deviation $\langle (N_{\text{hits}}/N - \frac{\pi}{4})^2 \rangle$ from the runs and plot it as a function of $N$. How does the mean square deviation scale with $N$?

(1.2) Implement and run Alg. 1.2 (`markov-pi`), starting from the clubhouse. Convince yourself, choosing a throwing range $\delta = 0.3$, that $N_{\text{hits}}/N$ again converges towards $\pi/4$. Next, study the precision obtained as a function of $\delta$ and check the validity of the one-half rule. Plot the mean square deviation $\langle (N_{\text{hits}}/N - \frac{\pi}{4})^2 \rangle$, for fixed large $N$, as a function of $\delta$ in the range $\delta \in [0, 3]$. Plot the rejection rate of this algorithm as a function of $\delta$. Which value of the rejection rate yields the highest precision?

(1.3) Find out whether your programming language allows you to check for the presence of a file. If so, improve handling of initial conditions in Alg. 1.2 (`markov-pi`) by including the following code fragment:

```
. . .
if (∃ initfile) then
   { input {x, y} (from initfile)
else
   { {x, y} ← {1, 1} (legal initial condition)
. . .
```

Between runs of the modified program, the output should be transferred to *initfile*, to get new initial conditions. This method for using initial conditions can be adapted to many Markov-chain programs in this book.

(1.4) Implement Alg. 1.6 (`markov-discrete-pebble`), using a subroutine for the numbering scheme and neighbor table. The algorithm should run on arbitrary rectangular pads without modification of the main program. Check that, during long runs, all sites of the pebble game are visited equally often.

(1.5) For the $3 \times 3$ pebble game, find a rejection-free local Monte Carlo algorithm (only moving up, down, left or right). If you do not succeed for the $3 \times 3$ system, consider $n \times m$ pebble games.

(1.6) Implement Alg. 1.4 (`direct-needle`), and Alg. 1.5 (`direct-needle(patch)`). Modify both programs

to allow you to simulate needles that are longer than the floorboards are wide $(a > b)$. Check that the program indeed computes the number $\pi$. Which precision can be reached with this program?

(1.7) Check by an analytic calculation that the relationship between the mean number of hits and the length of the needle (eqn (1.8)) is valid for round (cobbler's) needles of any size. Analytically calculate the function $N_{\text{hits}}(x, \phi)$ for semicircular needles, that is, for cobblers' needles broken into two equal pieces.

(1.8) Determine all eigenvalues and eigenvectors of the transfer matrix of the $3 \times 3$ pebble game in eqn (1.14) (use a standard linear algebra routine). Analogously compute the eigenvalues of $n \times n$ pebble games. How does the correlation time $\Delta_i$ depend on the pad size $n$?

## (Section 1.2)

(1.9) Sample permutations using Alg. 1.11 (`ran-perm`) and check that this algorithm generates all 120 permutations of five elements equally often. Determine the cycle representation of each permutation that is generated. For permutations of $K$ elements, determine the histogram of the probability for being in a cycle of length $l$ (see Subsection 4.2.2). Consider an alternative algorithm for generating random permutations of $K$ elements. Sort random numbers $\{x_1, \ldots, x_K\} = \{\text{ran}(0, 1), \ldots, \text{ran}(0, 1)\}$ in ascending order $\{x_{P_1} < \cdots < x_{P_K}\}$. Show that $\{P_1, \ldots, P_K\}$ is a random permutation.

(1.10) Consider the following algorithm which combines transformation and rejection techniques:

$$
\begin{aligned}
1 \quad & x \leftarrow -\log \text{ran}(0, 1) \\
& \Upsilon \leftarrow \exp\left[-\frac{(x-1)^2}{2}\right] \\
& \text{if } (\text{ran}(0, 1) \geq \Upsilon) \text{ goto } 1 \text{ (reject sample)} \\
& \text{output } x
\end{aligned}
$$

Analytically calculate the distribution function $\pi(x)$ sampled by this program, and its rejection rate. Implement the algorithm and generate a histogram to check your answers.

(1.11) Consider the following code fragment, which is part of Alg. 1.19 (`gauss(patch)`):

$$x \leftarrow \mathbf{ran}\,(-1, 1)$$
$$y \leftarrow \mathbf{ran}\,(-1, 1)$$
$$\Upsilon \leftarrow x^2 + y^2$$
**output** $\Upsilon$

Compute the distribution function $\pi(\Upsilon)$ using elementary geometric considerations. Is it true that $\Upsilon$ is uniformly distributed in the interval $\Upsilon \in [0, 1]$? Implement the algorithm and generate a histogram to check your answers.

(1.12) Implement both the naive Alg. 1.17 (**naive-gauss**) with arbitrary $K$ and the Box–Muller algorithm, Alg. 1.18 (**gauss**). For which value of $K$ can you still detect statistically significant differences between the two programs?

(1.13) Generate uniformly distributed vectors $\{x_1, \dots, x_d\}$ inside a $d$-dimensional unit sphere. Next, incorporate the following code fragment:

$$\cdots$$
$$x_{d+1} \leftarrow \mathbf{ran}\,(-1, 1)$$
$$\mathbf{if}\ \big(\textstyle\sum_{k=1}^{d+1} x_k^2 > 1\big)\ \mathbf{then}$$
$$\{\ \mathbf{output}\ \text{``reject''}$$
$$\cdots$$

Show that the acceptance rate of the modified program yields the ratio of unit-sphere volumes in $(d+1)$ and in $d$ dimensions. Determine $V_{252}(1)/V_{250}(1)$, and compare with eqn (1.39).

(1.14) Sample random vectors $\{x_1, \dots, x_d\}$ on the surface of the $d$-dimensional unit sphere, using Alg. 1.22 (**direct-surface**). Compute histograms of the variable $I_{12} = x_1^2 + x_2^2$. Discuss the special case of four dimensions ($d = 4$). Determine the distribution $\pi(I_{12})$ analytically.

(1.15) Generate three-dimensional orthonormal coordinate systems with axes $\{\hat{\mathbf{e}}_x, \hat{\mathbf{e}}_y, \hat{\mathbf{e}}_z\}$ randomly oriented in space, using Alg. 1.22 (**direct-surface**). Test your program by computing the average scalar products $\langle(\hat{\mathbf{e}}_x \cdot \hat{\mathbf{e}}_x')\rangle$, $\langle(\hat{\mathbf{e}}_y \cdot \hat{\mathbf{e}}_y')\rangle$, and $\langle(\hat{\mathbf{e}}_z \cdot \hat{\mathbf{e}}_z')\rangle$ for pairs of random coordinate systems.

(1.16) Implement Alg. 1.13 (**reject-finite**) for $K = 10\,000$ and probabilities $\pi_k = 1/k^\alpha$, where $1 < \alpha < 2$. Implement Alg. 1.14 (**tower-sample**) for the same problem. Compare the sampling efficiencies. NB: Do not recompute $\pi_{\max}$ for each sample in the rejection method; avoid recomputing $\{\Pi_0, \dots, \Pi_K\}$ for each sample in the tower-sampling algorithm.

(1.17) Use a sample transformation to derive how to generate random numbers $\phi$ distributed as $\pi(\phi) = \frac{1}{2}\sin\phi$ for $\phi \in [0, \pi]$. Likewise, determine the distribution function $\pi(x)$ for $x = \cos[\mathbf{ran}\,(0, \pi/2)]$. Test your answers with histograms.

**(Section 1.3)**

(1.18) Implement Alg. 1.25 (**binomial-convolution**). Compare the probability distribution $\pi(N_{\mathrm{hits}})$ for $N = 1000$, with histograms generated from many runs of Alg. 1.1 (**direct-pi**). Plot the probability distributions for the rescaled variables $x = N_{\mathrm{hits}}/N$ and $\tilde{x} = (x - \pi/4)/\sigma$, where $\sigma^2 = (\pi/4)(1 - \pi/4)$.

(1.19) Modify Alg. 1.26 (**ran01-convolution**) to allow you to handle more general probability distributions, which are nonzero on an arbitrary interval $x \in [a, b]$. Follow the convergence of various distributions with zero mean their convergence towards a Gaussian.

(1.20) Implement Alg. 1.28 (**data-bunch**). Test it for a single, very long, simulation of Alg. 1.2 (**markov-pi**) with throwing ranges $\delta \in \{0.03, 0.1, 0.3\}$. Test it also for output of Alg. 1.6 (**markov-discrete-pebble**) (compute the probability to be at site 1). If possible, compare with the correlation times for the $n \times n$ pebble game obtained from the second largest eigenvalue of the transfer matrix (see Exerc. 1.8).

**(Section 1.4)**

(1.21) Determine the mean value of $\mathcal{O} = x^{\gamma-\zeta}$ in a simple implementation of Alg. 1.31 (**markov-zeta**) for $\zeta > -\frac{1}{2}$. Monitor the rejection rate of the algorithm as a function of the step size $\delta$, and compute the mean square deviation of $\mathcal{O}$. Is the most precise value of $\langle\mathcal{O}\rangle$ obtained with a step size satisfying the one-half rule?

(1.22) Implement Alg. 1.29 (**direct-gamma**), subtract the mean value $1/(\gamma+1)$ for each sample, and generate histograms of the average of $N$ samples, and also of the rescaled averages, as in Fig. 1.46.

(1.23) Implement a variant of Alg. 1.29 (**direct-gamma**), in order to sample the distribution

$$\pi(x) \propto \begin{cases} (x-a)^\gamma & \text{if } x > a \\ -c|x-a|^\gamma & \text{if } x < a \end{cases}.$$

For concreteness, determine the mean of the distribution analytically as a function of $\{a, c, \gamma\}$, and subtract it for each sample. Compute the histograms of the distribution function for the rescaled sum of random variables distributed as $\pi(x)$. Compute the parameters $\{A_\pm, c_{1,2}\}$ of the Lévy distribution as a function of $\{a, c, \gamma\}$, and compare the histograms of rescaled averages to the analytic limit distribution of eqn (1.86).

# References

Aigner M., Ziegler G. M. (1992) *Proofs from THE BOOK (2nd edn)*, Springer, Berlin, Heidelberg, New York

Barbier E. (1860) Note sur le problème de l'aiguille et le jeu du joint couvert [in French], *Journal de Mathématiques Pures et Appliquées* **(2) 5**, 273–286

Gnedenko B. V., Kolmogorov A. N. (1954) *Limit Distributions for Sums of Independent Variables*, Addison-Wesley, Cambridge, MA

Krauth W., Staudacher M. (1999) Eigenvalue distributions in Yang–Mills integrals, *Physics Letters B* **453**, 253–257

Metropolis N., Rosenbluth A. W., Rosenbluth M. N., Teller A. H., Teller E. (1953) Equation of state calculations by fast computing machines, *Journal of Chemical Physics* **21**, 1087–1092

Propp J. G., Wilson D. B. (1996) Exact sampling with coupled Markov chains and applications to statistical mechanics, *Random Structures & Algorithms* **9**, 223–252

Ulam S. M. (1991) *Adventures of a Mathematician*, University of California Press, Berkeley, Los Angeles, London

# Hard disks and spheres

**2**

In the first chapter of this book, we considered simple problems in statistics: pebbles on the beach, needles falling but never rolling, and people strolling on heliports by night. We now move on to study model systems in physics—particles with positions, velocities, and interactions—that obey classical equations of motion. To understand how physical systems can be treated with the tools of statistics and simulated with Monte Carlo methods, we shall consider the hard-sphere model, which lies at the heart of statistical mechanics. Hard spheres, which are idealizations of billiard balls, in free space or in a box, behave as free particles whenever they are not in contact with other particles or with walls, and obey simple reflection rules on contact.

The hard-sphere model played a crucial role in the genesis of statistical mechanics. Since the early days of machine computing, in the 1950s, and up to the present day, the hard-sphere model has spurred the development of computer algorithms, and both the explicit numerical integration of Newton's equations and the Markov-chain Monte Carlo algorithm were first tried out on this model. We shall use such algorithms to illustrate mechanics and statistical mechanics, and to introduce the fundamental concepts of statistical mechanics: the equiprobability principle, the Boltzmann distribution, the thermodynamic temperature, and the pressure. We shall also be concerned with the practical aspects of computations and witness the problems of Markov-chain algorithms at high densities. We shall conclude the chapter with a first discussion of sophisticated cluster algorithms which are common to many fields of computational physics.

In the hard-sphere model, all configurations have the same potential energy and there is no energetic reason to prefer any configuration over any other. Only entropic effects come into play. In spite of this restriction, hard spheres and disks show a rich phenomenology and exhibit phase transitions from the liquid to the solid state. These "entropic transitions" were once quite unsuspected, and then hotly debated, before they ended up poorly understood, especially in two dimensions. The physics of entropy will appear in several places in this chapter, to be taken up again in earnest in Chapter 6.

Hard disks move about in a box much like billiard balls. The rules for wall and pair collisions are quickly programmed on a computer, allowing us to follow the time evolution of the hard-disk system (see Fig. 2.1). Given the initial positions and velocities at time $t = 0$, a simple algorithm allows us to determine the state of the system at $t = 10.37$, but the unavoidable numerical imprecision quickly explodes. This manifestation of chaos is closely related to the statistical description of hard disks and other systems, as we shall discuss in this chapter.

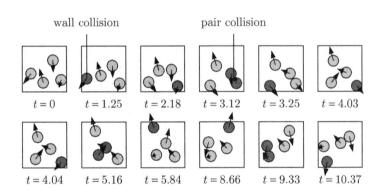

**Fig. 2.1** Event-driven molecular dynamics simulation with four hard disks in a square box.

# 2.1   Newtonian deterministic mechanics

In this section, we consider hard disks and spheres[1] colliding with each other and with walls. Instantaneous pair collisions conserve momentum, and wall collisions merely reverse one velocity component, that normal to the wall. Between collisions, disks move straight ahead, in the same manner as free particles. To numerically solve the equations of motion—that is, do a molecular dynamics simulation—we simply propagate all disks up to the next collision (the next event) in the whole system. We then compute the new velocities of the collision partners, and continue the propagation (see Fig. 2.1 and the schematic Alg. 2.1 (`event-disks`)).

**procedure event-disks**
**input** $\{\mathbf{x}_1, \ldots, \mathbf{x}_N\}, \{\mathbf{v}_1, \ldots, \mathbf{v}_N\}, t$
$\{t_{\text{pair}}, k, l\} \leftarrow$ next pair collision
$\{t_{\text{wall}}, j\} \leftarrow$ next wall collision
$t_{\text{next}} \leftarrow \min[t_{\text{wall}}, t_{\text{pair}}]$
**for** $m = 1, \ldots, N$ **do**
$\quad \{\ \mathbf{x}_m \leftarrow \mathbf{x}_m + (t_{\text{next}} - t)\mathbf{v}_m$
**if** $(t_{\text{wall}} < t_{\text{pair}})$ **then**
$\quad \{\ $ **call** `wall-collision`$(j)$
**else**
$\quad \{\ $ **call** `pair-collision`$(k, l)$
**output** $\{\mathbf{x}_1, \ldots, \mathbf{x}_N\}, \{\mathbf{v}_1, \ldots, \mathbf{v}_N\}, t_{\text{next}}$
——

**Algorithm 2.1** `event-disks`. Event-driven molecular dynamics algorithm for hard disks in a box (see Alg. 2.4 (`event-disks(patch)`)).

Our aim in the present section is to implement this event-driven molecular dynamics algorithm and to set up our own simulation of hard disks and spheres. The program is both simple and exact, because the integration of the equations of motion needs no differential calculus, and the numerical treatment contains no time discretization.

## 2.1.1   Pair collisions and wall collisions

We determine the time of the next pair collision in the box by considering all pairs of particles $\{k, l\}$ and isolating them from the rest of the system (see Fig. 2.2). This leads to the evolution equations

$$\mathbf{x}_k(t) = \mathbf{x}_k(t_0) + \mathbf{v}_k \cdot (t - t_0),$$
$$\mathbf{x}_l(t) = \mathbf{x}_l(t_0) + \mathbf{v}_l \cdot (t - t_0).$$

**Fig. 2.2** Motion of two disks from time $t_0$ to their pair collision time $t_1$.

---

[1]In this chapter, the words "disk" and "sphere" are often used synonymously. For concreteness, most programs are presented in two dimensions, for disks.

A collision occurs when the norm of the spatial distance vector

$$\underbrace{\mathbf{x}_k(t) - \mathbf{x}_l(t)}_{\Delta\mathbf{x}(t)} = \underbrace{\Delta_{\mathbf{x}}}_{\mathbf{x}_k(t_0)-\mathbf{x}_l(t_0)} + \underbrace{\Delta_{\mathbf{v}}}_{\mathbf{v}_k-\mathbf{v}_l} \cdot (t - t_0) \qquad (2.1)$$

equals twice the radius $\sigma$ of the disks (see Fig. 2.2). This can happen at two times $t_1$ and $t_2$, obtained by squaring eqn (2.1), setting $|\Delta\mathbf{x}| = 2\sigma$, and solving the quadratic equation

$$t_{1,2} = t_0 + \frac{-(\Delta_{\mathbf{x}} \cdot \Delta_{\mathbf{v}}) \pm \sqrt{(\Delta_{\mathbf{x}} \cdot \Delta_{\mathbf{v}})^2 - (\Delta_{\mathbf{v}})^2((\Delta_{\mathbf{x}})^2 - 4\sigma^2)}}{(\Delta_{\mathbf{v}})^2}. \qquad (2.2)$$

The two disks will collide in the future only if the argument of the square root is positive and if they are approaching each other ($(\Delta_{\mathbf{x}} \cdot \Delta_{\mathbf{v}}) < 0$, see Alg. 2.2 (pair-time)). The smallest of all the pair collision times obviously gives the next pair collision in the whole system (see Alg. 2.1 (event-disks)). Analogously, the parameters for the next wall collision are computed from a simple time-of-flight analysis (see Fig. 2.3, and Alg. 2.1 (event-disks) again).

**procedure pair-time**
**input** $\Delta_{\mathbf{x}}$ ($\equiv \mathbf{x}_k(t_0) - \mathbf{x}_l(t_0)$)
**input** $\Delta_{\mathbf{v}}$ ($\equiv \mathbf{v}_k - \mathbf{v}_l \neq 0$)
$\Upsilon \leftarrow (\Delta_{\mathbf{x}} \cdot \Delta_{\mathbf{v}})^2 - |\Delta_{\mathbf{v}}|^2(|\Delta_{\mathbf{x}}|^2 - 4\sigma^2)$
**if** ($\Upsilon > 0$ and $(\Delta_{\mathbf{x}} \cdot \Delta_{\mathbf{v}}) < 0$) **then**
$\quad \left\{ t_{\text{pair}} \leftarrow t_0 - \left[(\Delta_{\mathbf{x}} \cdot \Delta_{\mathbf{v}}) + \sqrt{\Upsilon}\right] / \Delta_{\mathbf{v}}^2 \right.$
**else**
$\quad \left\{ t_{\text{pair}} \leftarrow \infty \right.$
**output** $t_{\text{pair}}$
_____

**Algorithm 2.2 pair-time.** Pair collision time for two particles starting at time $t_0$ from positions $\mathbf{x}_k$ and $\mathbf{x}_l$, and with velocities $\mathbf{v}_k$ and $\mathbf{v}_l$.

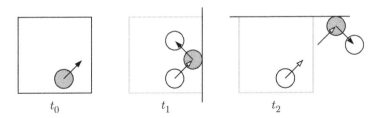

$t_0$ $\qquad\qquad$ $t_1$ $\qquad\qquad$ $t_2$

**Fig. 2.3** Computing the wall collision time $t_{\text{wall}} = \min(t_1, t_2)$.

Continuing our implementation of Alg. 2.1 (event-disks), we now compute the velocities after the collision: collisions with a wall of the box lead to a change of sign of the velocity component normal to the

**procedure** `pair-collision`
**input** $\{\mathbf{x}_k, \mathbf{x}_l\}$ (particles in contact: $|\mathbf{x}_k - \mathbf{x}_l| = 2\sigma$)
**input** $\{\mathbf{v}_k, \mathbf{v}_l\}$
$\Delta_{\mathbf{x}} \leftarrow \mathbf{x}_k - \mathbf{x}_l$
$\hat{\mathbf{e}}_\perp \leftarrow \Delta_{\mathbf{x}}/|\Delta_{\mathbf{x}}|$
$\Delta_{\mathbf{v}} \leftarrow \mathbf{v}_k - \mathbf{v}_l$
$\mathbf{v}'_k \leftarrow \mathbf{v}_k - \hat{\mathbf{e}}_\perp (\Delta_{\mathbf{v}} \cdot \hat{\mathbf{e}}_\perp)$
$\mathbf{v}'_l \leftarrow \mathbf{v}_l + \hat{\mathbf{e}}_\perp (\Delta_{\mathbf{v}} \cdot \hat{\mathbf{e}}_\perp)$
**output** $\{\mathbf{v}'_k, \mathbf{v}'_l\}$

---

**Algorithm 2.3** `pair-collision`. Computing the velocities of disks (spheres) $k$ and $l$ after an elastic collision (for equal masses).

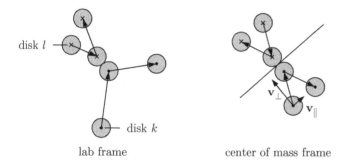

disk $l$

disk $k$

lab frame

center of mass frame

**Fig. 2.4** Elastic collision between equal disks $k$ and $l$, as seen in two different reference frames.

wall involved in the collision. Pair collisions are best analyzed in the center-of-mass frame of the two disks, where $\mathbf{v}_k + \mathbf{v}_l = 0$ (see Fig. 2.4). Let us write the velocities in terms of the perpendicular and parallel components $\mathbf{v}_\perp$ and $\mathbf{v}_\parallel$ with respect to the tangential line between the two particles when they are exactly in contact. This tangential line can be thought of as a virtual wall from which the particles rebound:

$$
\underbrace{\begin{aligned} \mathbf{v}_k &= \mathbf{v}_\parallel + \mathbf{v}_\perp \\ \mathbf{v}_l &= -\mathbf{v}_\parallel - \mathbf{v}_\perp \end{aligned}}_{\text{before collision}}, \qquad \underbrace{\begin{aligned} \mathbf{v}'_k &= \mathbf{v}_\parallel - \mathbf{v}_\perp \\ \mathbf{v}'_l &= -\mathbf{v}_\parallel + \mathbf{v}_\perp \end{aligned}}_{\text{after collision}}.
$$

The changes in the velocities of particles $k$ and $l$ are $\mp 2\mathbf{v}_\perp$. Introducing the perpendicular unit vector $\hat{\mathbf{e}}_\perp = (\mathbf{x}_k - \mathbf{x}_l)/|\mathbf{x}_k - \mathbf{x}_l|$ allows us to write $\mathbf{v}_\perp = (\mathbf{v}_k \cdot \hat{\mathbf{e}}_\perp)\hat{\mathbf{e}}_\perp$ and $2\mathbf{v}_\perp = (\Delta_{\mathbf{v}} \cdot \hat{\mathbf{e}}_\perp)\hat{\mathbf{e}}_\perp$, where $2\mathbf{v}_\perp = \mathbf{v}'_k - \mathbf{v}_k$ gives the change in the velocity of particle $k$. The formulas coded into Alg. 2.3 (`pair-collision`) follow immediately. We note that $\hat{\mathbf{e}}_\perp$ and the changes in velocities $\mathbf{v}'_k - \mathbf{v}_k$ and $\mathbf{v}'_l - \mathbf{v}_l$ are relative vectors and are thus the same in all inertial reference frames. The program can hence be used directly with the lab frame velocities.

Algorithm 2.1 (`event-disks`) is rigorous, but naive in that it computes collision times even for pairs that are far apart and also recomputes all pair collision times from scratch after each event, although they change only for pairs involving a collision partner. Improvements could easily be worked into the program, but we shall not need them. However, even our basic code can be accepted only if it runs through many billions of collisions (several hours of CPU time) without crashing. We must care about code stability. In addition, we should invest in simplifying input–output: it is a good strategy to let the program search for a file containing the initial configuration (velocities, positions, and time), possibly the final configuration of a previous calculation. If no such file is found, a legal initial configuration should be automatically generated by the program. Finally, we should plot the results in the natural units of time for the simulation, such that during a unit time interval each particle undergoes about one pair collision (and the whole system has $N/2$ pair collisions). In this way, the results become independent of the scale of the initial velocities.

Our molecular dynamics algorithm moves from one collision to the next with an essentially fixed number of arithmetic operations, requiring on the order of a microsecond for a few disks on a year 2005 laptop computer. Each collision drives the system further in physical time. Solving Newton's dynamics thus appears to be a mere question of computer budget.

### 2.1.2   Chaotic dynamics

Algorithm 2.1 (`event-disks`) solves the hard-sphere equations of motion on the assumption that the calculation of collision times, positions, velocity changes, etc. is done with infinite precision. This cannot really be achieved on a computer, but the question arises of whether it matters that numbers are truncated after 8 or 16 decimal digits. It is easiest to answer this question by pitting different versions of the same event-driven algorithm against each other, starting from identical initial conditions, but with all calculations performed at different precision levels: in one case in short format (single precision), and in the other case in long format (double precision). The standard encoding of 32-bit floating-point numbers $\pm a \times 10^b$ uses one bit for the sign, eight bits for the exponent $b$, and 23 bits for the fraction $a$, so that the precision—the ratio of neighboring numbers that can be represented on the computer— is approximately $\epsilon = 2^{-23} \simeq 1.2 \times 10^{-7}$. For a 64-bit floating point number, the precision is about $\epsilon = 1 \times 10^{-16}$. Most computer languages allow one to switch precision levels without any significant rewriting of code.

The two versions of the four-disk molecular dynamics calculation, started off from identical initial conditions (as in Fig. 2.1), get out of step after as few as 25 pair collisions (see Fig. 2.5), an extremely small number compared with the million or so collisions which we can handle per second on a year 2005 laptop computer.

This situation is quite uncomfortable: our computational results, for

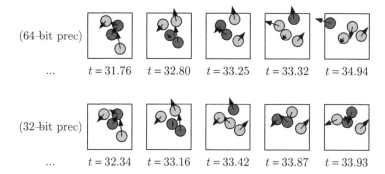

(64-bit prec)

...   $t = 31.76$   $t = 32.80$   $t = 33.25$   $t = 33.32$   $t = 34.94$

(32-bit prec)

...   $t = 32.34$   $t = 33.16$   $t = 33.42$   $t = 33.87$   $t = 33.93$

**Fig. 2.5** Calculations starting from the initial configuration of Fig. 2.1, performed with 64-bit precision (*top*) and with 32-bit precision (*bottom*).

computing times beyond a few microseconds, are clearly uncontrolled. We may drive up the precision of our calculation with special number formats that are available in many computer languages. However, this strategy cannot defeat the onset of chaos, that is, cure the extreme sensitivity to the details of the calculation. It will be impossible to control a hard-sphere molecular dynamics simulation for a few billion events.

The chaos in the hard-sphere model has its origin in the negative curvature of the spheres' surfaces, which magnifies tiny differences in the trajectory at each pair collision and causes serious rounding errors in computations and humbling experiences at the billiard table (see Fig. 2.6). On the other hand, this sensitivity to initial conditions ensures that even finite systems of disks and spheres can be described by statistical mechanics, as will be discussed later in this chapter.

(stat.)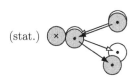

**Fig. 2.6** Magnification of a difference in trajectories through a pair collision, in the reference frame of a stationary disk.

### 2.1.3   Observables

In the previous subsections, we have integrated the equations of motion for hard spheres in a box, but have only looked at the configurations, without evaluating any observables. We shall do the latter now. For simplicity, we consider a projected density $\eta_y(a)$, essentially the fraction of time for which the $y$-coordinate of any particle is equal to $a$. More precisely, $\{\eta_y(a)\,\mathrm{d}a\}$ is the fraction of time that the $y$-coordinate of a disk center spends between $a$ and $a + \mathrm{d}a$ (see Fig. 2.7). It can be computed exactly for given particle trajectories between times $t = 0$ and $t = T$:

$$\left\{\begin{array}{c} y\text{-density} \\ \text{at } y = a \end{array}\right\} = \eta_y(a) = \frac{1}{T} \sum_{\substack{\text{intersections } i \\ \text{with gray strip} \\ \text{in Fig. 2.7}}} \frac{1}{|v_y(i)|}. \qquad (2.3)$$

In Fig. 2.7, there are five intersections (the other particles must also be considered). At each intersection, $1/|v_y|$ must be added, to take into account the fact that faster particles spend less time in the interval $[a, a + \mathrm{d}a]$, and thus contribute less to the density at $a$.

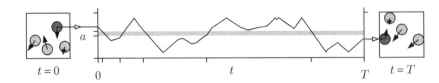

**Fig. 2.7** $y$-coordinate vs. time for one disk. The $y$-density $\eta_y(a)$ is computed from the time spent between $a$ and $a + \mathrm{d}a$ (see eqn (2.3)).

The $y$-density at $a$ corresponds to the observable $\mathcal{O}(t) = \delta\left[y(t) - a\right]$ (where we have used the Dirac $\delta$-function), and can also be obtained from the time average

$$\langle \mathcal{O}(t) \rangle_T = \frac{1}{T} \int_0^T \mathrm{d}t \; \mathcal{O}(t),$$

so that

$$\eta_y^T(a) = \frac{1}{T} \int_0^T \mathrm{d}t \; \delta\left[y(t) - a\right]. \tag{2.4}$$

Seasoned theorists can derive the formula for the $y$-density in eqn (2.3) in a split second from eqn (2.4) via the transformation rules for the $\delta$-function. Algorithm 2.1 (**event-disks**) numerically solves the equations of motion for hard disks without discretization errors or other imperfections. In addition to this best possible method of data acquisition, we may analyze the data using eqn (2.3), without losing information. In principle, we are limited only by the finite total simulation time.

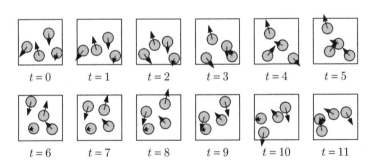

**Fig. 2.8** Stroboscopic snapshots of the simulation shown in Fig. 2.1.

It is possible to implement data analysis as in eqn (2.3), but this approach is somewhat overambitious: we can certainly obtain exact run averages, but because the number of runs is finite and the runs are not infinitely long, statistical errors would still creep into the calculation. It is thus justified to take a more leisurely approach to data analysis and simply discretize the time average (2.4):

$$\langle \mathcal{O}(t) \rangle_T \simeq \frac{1}{M} \sum_{i=1}^{M} \mathcal{O}(t_i).$$

We thus interrupt the calculation in Alg. 2.1 (`event-disks`) at fixed, regular time intervals and take stroboscopic snapshots (see Fig. 2.8 and Alg. 2.4 (`event-disks(patch)`)). The snapshots can then be incorpo-

**procedure** `event-disks(patch)`
$\cdots$
$i_{\min} \leftarrow$ `Int` $(t/\Delta_t) + 1$
$i_{\max} \leftarrow$ `Int` $(t_{\mathrm{next}}/\Delta_t)$
**for** $i = i_{\min}, \ldots, i_{\max}$ **do**
$\left\{ \begin{array}{l} t' \leftarrow i\Delta_t - t \\ \textbf{output } \{\mathbf{x}_1 + t'\mathbf{v}_1, \ldots, \mathbf{x}_N + t'\mathbf{v}_N\}, \{\mathbf{v}_1, \ldots, \mathbf{v}_N\} \end{array} \right.$
$\cdots$
_____

**Algorithm 2.4** `event-disks(patch)`. Alg. 2.1 (`event-disks`), modified to output stroboscopic snapshots (with a time interval $\Delta_t$).

rated into histograms of observables, and give results identical to an analysis of eqn (2.3) in the limit of vanishing width of the time slices or for large run times. The density at point $y$ then converges to a value independent of the initial configuration, a true time average:

$$\eta_y(a) = \lim_{T \to \infty} \frac{1}{T} \sum_k \int_0^T dt \ \delta[y_k(t) - a],$$

where $k = 1, \ldots, N$ represent the disks in the box.

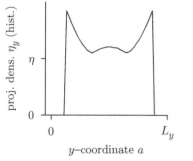 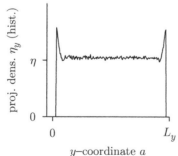

**Fig. 2.9** Projected density in a square box at density $\eta = 0.18$. *Left*: $N = 4$. *Right*: $N = 64$ (from Alg. 2.4 (`event-disks(patch)`)).

Some data obtained from Alg. 2.4 (`event-disks(patch)`), for hard-disk systems in a square box, are shown in Fig. 2.9. In the two graphs, the covering density is identical, but systems are of different size. It often surprises people that the density in these systems is far from uniform, even though the disks do not interact, other than through their hard-core. In particular, the boundaries, especially the corners, seem to attract disks. Systems with larger $\{N, L_x, L_y\}$ clearly separate into a bulk region (the inner part) and a boundary.

### 2.1.4 Periodic boundary conditions

Bulk properties (density, correlations, etc.) differ from properties near the boundaries (see Fig. 2.9). In the thermodynamic limit, almost the whole system is bulk, and the boundary reduces to nearly nothing. To compute bulk properties for an infinite system, it is often advantageous to eliminate boundaries even in simulations of small systems. To do so, we could place the disks on the surface of a sphere, which is an isotropic, homogeneous environment without boundaries. This choice is, however, problematic because at high density, we cannot place the disks in a nice hexagonal pattern, as will be discussed in detail in Chapter 7. Periodic boundary conditions are used in the overwhelming majority of cases (see Fig. 2.10): in a finite box without walls, particles that move out through the bottom are fed back at the top, and other particles come in from one side when they have just left through the opposite side. We thus identify the lower and upper boundaries with each other, and similarly the left and right boundaries, letting the simulation box have the topology of an abstract torus (see Fig. 2.11). Alternatively, we may represent a box with periodic boundary conditions as an infinite, periodically repeated box without walls, obtained by gluing together an infinite number of identical copies, as also indicated in Fig. 2.10.

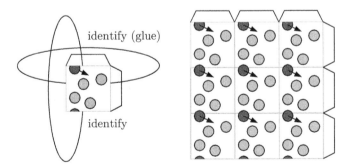

**Fig. 2.10** Sample with periodic boundary conditions, interpreted as a finite torus (*left*) or an infinite periodic system (*right*).

**Fig. 2.11** Gluing together the sides of the square in Fig. 2.10 generates an object with the topology of a torus.

For hard spheres, the concept of periodic boundary conditions is relatively straightforward. In more complicated systems, however (particles with long-range interaction potentials, quantum systems, etc.), sloppy implementation of periodic boundary conditions is frequently a source of severe inconsistencies which are difficult to detect, because they enter the program at the conception stage, not during execution. To avoid trouble, we must think of a periodic system in one of the two frameworks of Fig. 2.10: either a finite abstract torus or a replicated infinite system. We should stay away from vaguely periodic systems that have been set up with makeshift routines lacking a precise interpretation.

Algorithm 2.1 (event-disks) can be extended to the case of periodic boundary conditions with the help of just two subroutines. As one particle can be described by many different vectors (see the dark particle in the right part of Fig. 2.10), we need Alg. 2.5 (box-it) to find the representative vector inside the central simulation box, with an $x$-coordinate between 0 and $L_x$ and a $y$-coordinate between 0 and $L_y$.[2] Likewise, since many different pairs of vectors correspond to the same two particles, the difference vector between these particles takes many values. Algorithm 2.6 (diff-vec) computes the shortest difference vector between all representatives of the two particles and allows one to decide whether two hard disks overlap.

**procedure** box-it
**input x**
$x \leftarrow \mathrm{mod}\,(x, L_x)$
**if** $(x < 0)\ x \leftarrow x + L_x$
$y \leftarrow \mathrm{mod}\,(y, L_y)$
**if** $(y < 0)\ y \leftarrow y + L_y$
**output x**
————

**Algorithm 2.5** box-it. Reducing a vector $\mathbf{x} = \{x, y\}$ into a periodic box of size $L_x \times L_y$.

**procedure** diff-vec
**input** $\{\mathbf{x}, \mathbf{x}'\}$
$\Delta_\mathbf{x} \leftarrow \mathbf{x}' - \mathbf{x}\ (\Delta_\mathbf{x} \equiv \{x_\Delta, y_\Delta\})$
**call** box-it $(\Delta_\mathbf{x})$
**if** $(x_\Delta > L_x/2)\ x_\Delta \leftarrow x_\Delta - L_x$
**if** $(y_\Delta > L_y/2)\ y_\Delta \leftarrow y_\Delta - L_y$
**output** $\Delta_\mathbf{x}$
————

**Algorithm 2.6** diff-vec. The difference $\Delta_\mathbf{x} = \{x_\Delta, y_\Delta\}$ between vectors $\mathbf{x}$ and $\mathbf{x}'$ in a box of size $L_x \times L_y$ with periodic boundary conditions.

Periodic systems have no confining walls, and thus no wall collisions.

[2]Some computer languages allow the output of the mod () function to be negative (e.g. mod $(-1, 3) = -1$). Others implement mod () as a nonnegative function (so that, for example, mod $(-1, 3) = 2$). The **if ( )** statements in Alg. 2.5 (box-it) are then unnecessary.

Unlike the case for the planar box considered at the beginning of this section, the pair collision time $t_\text{pair}$ is generally finite: even two particles moving apart from each other eventually collide, after winding several times around the simulation box (see Fig. 2.12), although this is relevant only for low-density systems. We have to decide on a good strategy to cope with this: either carefully program the situation in Fig. 2.12, or stay with our initial attitude (in Alg. 2.2 (`pair-time`)) that particles moving away from each other never collide—a dangerous fudge that must be remembered when we are running the program for a few small disks in a large box.

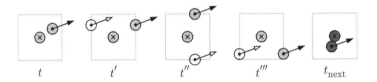

**Fig. 2.12** Pair collision in a box with periodic boundary conditions.

The complications of Fig. 2.12 are also absent in Sinai's system of two relatively large disks in a small periodic box, such that disks cannot pass by each other (Sinai, 1970). Figure 2.13 shows event frames generated by Alg. 2.1 (`event-disks`), using the (noninertial) stationary-disk reference frame that was introduced in Fig. 2.6. Sinai's hard disks can be simulated

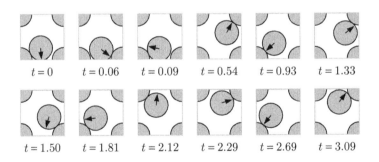

**Fig. 2.13** Time evolution of two disks in a square box with periodic boundary conditions, in the stationary-disk reference frame.

directly as a single point in the stationary-disk reference frame.

## 2.2   Boltzmann's statistical mechanics

We could set up the event-driven algorithm of Section 2.1 in a few hours and follow the ballet of disks (spheres) approaching and flying away from each other along intricate, even unpredictable trajectories. In doing so, however, we engage in a computational project which in many respects is far too complicated. In the limit $t \to \infty$, detailed timing information, for example the ordering of the snapshots in Fig. 2.8, does not enter into the density profiles, spatial correlation functions, thermodynamic properties, etc. We need to know only how often a configuration $a$ appears during an infinitely long molecular dynamics calculation. For hard spheres, it is the quintessence of Boltzmann's statistical mechanics that any two legal configurations $a$ and $b$ have the same probability to appear: $\pi(a) = \pi(b)$ (see Fig. 2.14).

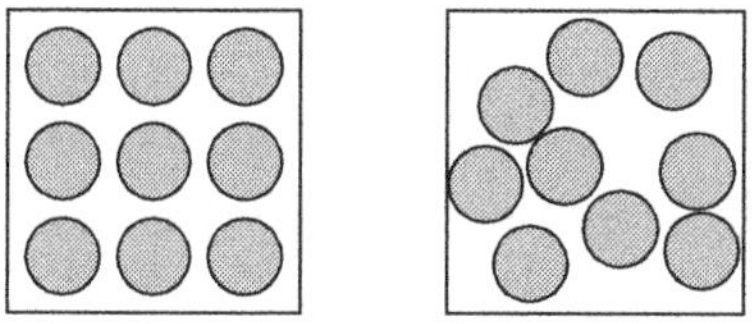

$a$            $b$

**Fig. 2.14** Equiprobability principle for hard disks: $\pi(a) = \pi(b)$.

More precisely (for an infinitely long simulation), this means the following:

$$\left\{ \begin{array}{l} \text{probability of configuration with} \\ [\mathbf{x}_1, \mathbf{x}_1 + d\mathbf{x}_1], \ldots, [\mathbf{x}_N, \mathbf{x}_N + d\mathbf{x}_N] \end{array} \right\} \propto \pi(\mathbf{x}_1, \ldots, \mathbf{x}_N)\, d\mathbf{x}_1, \ldots, d\mathbf{x}_N,$$

where

$$\pi(\mathbf{x}_1, \ldots, \mathbf{x}_N) = \begin{cases} 1 & \text{if configuration legal} \\ 0 & \text{otherwise} \end{cases}. \tag{2.5}$$

In the presence of an energy $E$ which contains kinetic and potential terms, eqn (2.5) takes the form of the equiprobability principle $\pi(a) = \pi(E(a))$, where $a$ refers to positions and velocities (see Subsection 2.2.4). In the hard-sphere model, the potential energy vanishes for all legal configurations, and we get back to eqn (2.5). This hypothesis can be illustrated by molecular dynamics simulations (see Fig. 2.15), but the equal probability of all configurations $a$ and $b$ is an axiom in statistical mechanics, and does not follow from simple principles, such as micro-reversibility or detailed balance. Its verification from outside of statistical mechanics, by solving Newton's equations of motion, has presented a formidable mathematical challenge. Modern research in mathematics has gone far in actually proving ergodicity, the equivalence between Newton's deterministic mechanics and Boltzmann's statistical mechanics, for the special case of hard spheres. The first milestone result of Sinai (1970), a 50-page proof, shows rigorously that the two-disk problem of Fig. 2.13 is ergodic. Several decades later it has become possible to prove that general hard-disk and hard-sphere systems are indeed ergodic, under very mild assumptions (Simanyi 2003, 2004).

**Fig. 2.15** Trajectory of the system shown in Fig. 2.13 after 18, 50, and 500 collisions.

Besides the mathematically rigorous approach, many excellent arguments plead in favor of the equiprobability principle, for hard spheres in particular, and for statistical physical systems in general. One of the most discussed is Jaynes' information-theoretical principle, which essentially states (for hard spheres) that the equal-probability choice is an unbiased guess. In Fig. 2.15, we show the trajectory of Sinai's two-disk system in the stationary-disk reference frame. It has been mathematically proven that for almost all initial conditions, the central area is swept out evenly. This is the simplest pattern, the one containing the least information about the system, and the one corresponding to Jaynes' principle. The latter principle is closely related to Bayesian statistics (see Subsection 1.3.4), with a most interesting difference. In Bayesian statistics, one is hampered by the arbitrariness of defining an unbiased (flat) a priori probability: what is flat with a given choice of variables acquires structure under a coordinate transformation. In physics, the problem can be avoided because there exists a special coordinate system—Cartesian positions and velocities. Boltzmann's equal-probability choice is to be understood with respect to Cartesian coordinates, as indicated in eqn (2.5).

**Fig. 2.16** Hard disks (*left*) and planets orbiting the sun (*right*): classical dynamic systems with greatly different behavior.

The foundations of statistical mechanics would be simpler if all physical systems (finite or infinite, with an arbitrary energy) fell under the reign of equiprobability (eqn (2.5)) and its generalizations. However, this is not the case. A notorious counterexample to equal probability of states with equal energy is the weakly interacting system of a few planets of mass $m$ orbiting the sun, of mass $M$ (see Fig. 2.16). If we neglect the planet–planet interactions altogether, in the limit $m/M \to 0$, the planets orbit the sun on classical Kepler ellipses, the solutions of the two-body problem of classical mechanics. Small planet–planet interactions modify the trajectories only slightly, even in the limit of infinite times, as was shown by the seminal Kolmogorov–Arnold–Moser theorem (see Thirring (1978) for a rigorous textbook discussion). As a consequence, for small planet masses, the trajectories remain close to the unperturbed trajectories. This is totally different from what happens for small disks in a box, which usually fly off on chaotic trajectories after a few violent collisions. Statistical mechanics applies, however, for a large number of bodies (e.g. for describing a galaxy).

## 2.2.1 Direct disk sampling

Boltzmann's statistical mechanics calls for all legal configurations to be generated with the same statistical weight. This can be put into practice by generating all configurations—legal and illegal—with the same probability, and then throwing away (rejecting) the illegal ones. What remains are hard-disk configurations, and they are clearly generated with equal probability (see Alg. 2.7 (direct-disks) and Fig. 2.17). It is wasteful to bring up illegal configurations only to throw them away later, but a better solution has not yet been found. The direct-sampling algorithm can

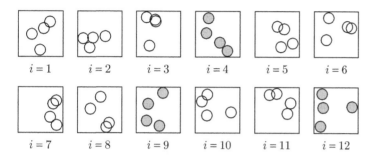

$i = 1$  $i = 2$  $i = 3$  $i = 4$  $i = 5$  $i = 6$

$i = 7$  $i = 8$  $i = 9$  $i = 10$  $i = 11$  $i = 12$

**Fig. 2.17** Direct sampling for four hard disks in a box. The frames $i = 4, 9, 12$ contain legal configurations (from Alg. 2.7 (direct-disks)).

**procedure** direct-disks
1   **for** $k = 1, \ldots, N$ **do**
$\left\{ \begin{array}{l} x_k \leftarrow \mathbf{ran}\,(x_{\min}, x_{\max}) \\ y_k \leftarrow \mathbf{ran}\,(y_{\min}, y_{\max}) \\ \textbf{for } l = 1, \ldots, k - 1 \textbf{ do} \\ \quad \{ \textbf{ if } (\mathbf{dist}\,(\mathbf{x}_k, \mathbf{x}_l) < 2\sigma) \textbf{ goto } 1 \text{ (reject sample—tabula rasa)} \end{array} \right.$
**output** $\{\mathbf{x}_1, \ldots, \mathbf{x}_N\}$

——

**Algorithm 2.7** direct-disks. Direct sampling for $N$ disks of radius $\sigma$ in a fixed box. The values of $\{x_{\min}, x_{\max}\}$, etc., depend on the system.

be written much faster than the molecular dynamics routine (Alg. 2.1 (event-disks)), as we need not worry about scalar products, collision subroutines etc. The output configurations $\{\mathbf{x}_1, \ldots, \mathbf{x}_N\}$ are produced with the same probability as the snapshots of Alg. 2.1 (event-disks) and lead to the same histograms as in Fig. 2.9 (because the physical system is ergodic). We get a flavor of the conceptual and calculational simplifications brought about by statistical mechanics.

The tabula-rasa rejection in Alg. 2.7 (direct-disks) often leads to confusion: instead of sweeping away the whole configuration after the generation of an overlap, one may be tempted to lift up the offending disk only, and try again (see Fig. 2.18). In this procedure of random sequential deposition, any correctly placed disk stays put, whereas incorrectly positioned disks are moved away.

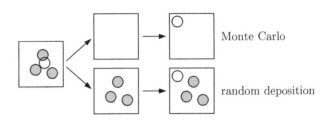

**Fig. 2.18** Difference between the direct-sampling Monte Carlo method and random sequential deposition.

Random sequential deposition is an important model for adhesion and catalysis (see the discussion in Chapter 7) but not for equilibrium: all configurations are not equally probable. A simplified one-dimensional discrete hard-rod model allows us to compute deposition probabilities explicitly and show that they are not the same (see Fig. 2.19): we suppose that rods may be deposited onto five sites, with a minimum distance of three sites between them. After placing the first rod (with the same probability on all sites), we try again and again until a two-rod configuration without overlap is found.

The configurations $a$ and $b$ are thus generated with half the probability of their parent, whereas the configuration $c$, a unique descendant, inherits all the probability of its parent. We obtain

$$\pi(a) = \pi(b) = \pi(e) = \pi(f) = \frac{1}{4} \times \frac{1}{2} = \frac{1}{8},$$

whereas the configurations $c$ and $d$ have a probability

$$\pi(c) = \pi(d) = \frac{1}{4}.$$

In Fig. 2.19, we distinguish rods according to when they were placed ($a$ is different from $d$), but the deposition probabilities are nonequivalent even if we identify $a$ with $d$, $b$ with $e$, and $c$ with $f$. The counterexample of Fig. 2.19 proves that random deposition is incorrect for hard disks and spheres in any dimensionality, if the aim is to generate configurations with equal probabilities.

The direct-sampling Monte Carlo algorithm, as discussed before, generates all 25 legal and illegal configurations with probability $1/25$. Only the six configurations $\{a, \ldots, f\}$ escape rejection, and their probabilities are $\pi(a) = \cdots = \pi(f) = 1/6$.

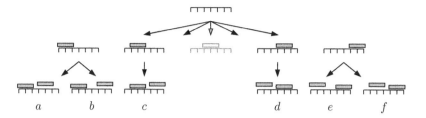

**Fig. 2.19** Random deposition of discrete hard rods.

## 2.2.2 Partition function for hard disks

Direct sampling would solve the simulation problem for disks and spheres if its high rejection rate did not make it impractical for all but the smallest and least dense systems. To convey how serious a problem the rejections become, we show in Fig. 2.20 the configurations, of which there are only six, returned by one million trials of Alg. 2.7 (`direct-disks`) with $N = 16$ particles and a density $\eta = \pi\sigma^2 N/V = 0.3$. The acceptance rate is next to zero. It deteriorates further with increasing particle number or density.

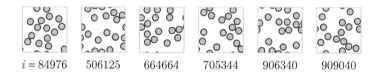

$i = 84976$     506125     664664     705344     906340     909040

**Fig. 2.20** The six survivors from $10^6$ trials of Alg. 2.7 (`direct-disks`) ($N = 16$, $\eta = 0.3$, periodic boundary conditions).

Although Alg. 2.7 (`direct-disks`) does not appear to be a very useful program, we shall treat it as a VIP[3] and continue analyzing it. We shall come across a profound link between computation and physics: the acceptance rate of the algorithm is proportional to the partition function, the number of configurations of disks with a finite radius.

We could compute the acceptance rate of the direct-sampling algorithm from long runs at many different values of the radius $\sigma$, but it is better to realize that each sample of random positions $\{\mathbf{x}_1, \ldots, \mathbf{x}_N\}$ gives a hard-disk configuration for all disk radii from zero up to half the minimum distance between the vectors or for all densities smaller than a limiting $\eta_{\max}$ (we consider periodic boundary conditions; see Alg. 2.8 (`direct-disks-any`)). Running this algorithm a few million times gives the probability distribution $\pi(\eta_{\max})$ and the acceptance rate of Alg. 2.7

---

[3]VIP: Very Important Program.

(`direct-disks`) for all densities (see Fig. 2.21):

$$\underbrace{p_{\mathrm{accept}}(\eta)}_{\substack{\text{acceptance rate of} \\ \text{Alg. 2.7 (direct-disks)}}} = 1 - \underbrace{\int_0^\eta d\eta_{\max}\,\pi(\eta_{\max})}_{\substack{\text{integrated histogram} \\ \text{of Alg. 2.8 (direct-disks-any)}}}. \tag{2.6}$$

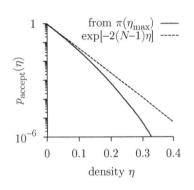

from $\pi(\eta_{\max})$ ——
$\exp[-2(N-1)\eta]$ - - - -

$p_{\mathrm{accept}}(\eta)$

density $\eta$

**Fig. 2.21** Acceptance rate of Alg. 2.7 (`direct-disks`) for 16 disks in a square box (from Alg. 2.8 (`direct-disks-any`), using eqn (2.6)).

**procedure direct-disks-any**
**input** $\{N, L_x, L_y\}$
**for** $k = 1, \ldots, N$ **do**
$\left\{ \begin{array}{l} x_k \leftarrow \mathrm{ran}\,(0, L_x) \\ y_k \leftarrow \mathrm{ran}\,(0, L_y) \end{array} \right.$
$\sigma \leftarrow \frac{1}{2}\min_{k \ne l}\left[\mathrm{dist}\,(\mathbf{x}_k, \mathbf{x}_l)\right]$
$\eta_{\max} \leftarrow \pi\sigma^2 N/(L_x L_y)$ (limiting density)
**output** $\eta_{\max}$
——

**Algorithm 2.8 direct-disks-any.** The limiting hard-disk density for $N$ random vectors in an $L_x \times L_y$ box with periodic boundary conditions.

The acceptance rate is connected to the number of configurations, that is, the partition function $Z(\eta)$ for disks of covering density $\eta$. For zero radius, that is, for an ideal gas, we have

$$\left\{ \begin{array}{c} \text{number of} \\ \text{configurations} \\ \text{for density 0} \end{array} \right\} : Z(\eta = 0) = \int d\mathbf{x}_1 \ldots \int d\mathbf{x}_N = V^N.$$

The partition function $Z(\eta)$ for disks with a finite radius and a density $\eta$ is related to $Z(0)$ via

$$\left\{ \begin{array}{c} \text{number of} \\ \text{configurations} \\ \text{for density } \eta \end{array} \right\} : Z(\eta) = \int \ldots \int d\mathbf{x}_1 \ldots d\mathbf{x}_N \underbrace{\pi(\mathbf{x}_1, \ldots, \mathbf{x}_N)}_{\text{for disks of finite radius}}$$

$$= Z(0)p_{\mathrm{accept}}(\eta).$$

This expression resembles eqn (1.38), where the volume of the unit sphere in $d$ dimensions was related to the volume of a hypercube via the acceptance rate of a direct-sampling algorithm.

We shall now determine $p_{\mathrm{accept}}(\eta)$ for the hard-disk system, and its partition function, for small densities $\eta$ in a box with periodic boundary conditions, using a basic concept in statistical physics, the virial expansion. Clearly,

$$Z(\eta) = \int \ldots \int d\mathbf{x}_1 \ldots d\mathbf{x}_N$$
$$\underbrace{[1 - \Upsilon(\mathbf{x}_1, \mathbf{x}_2)]}_{\substack{\text{no overlap} \\ \text{between 1 and 2}}} [1 - \Upsilon(\mathbf{x}_1, \mathbf{x}_3)] \cdots [1 - \Upsilon(\mathbf{x}_{N-1}, \mathbf{x}_N)], \tag{2.7}$$

where

$$\Upsilon(\mathbf{x}_k, \mathbf{x}_l) = \begin{cases} 1 & \text{if } \mathrm{dist}(\mathbf{x}_k, \mathbf{x}_l) < 2\sigma \\ 0 & \text{otherwise} \end{cases}.$$

The product in eqn (2.7) can be multiplied out. The dominant term collects a "1" in each of the $N(N-1)/2$ parentheses, the next largest term (for small $\sigma$) picks up a single term $\Upsilon(\mathbf{x}_k, \mathbf{x}_l)$, etc. Because

$$\int \int \mathrm{d}\mathbf{x}_k \, \mathrm{d}\mathbf{x}_l \, \Upsilon(\mathbf{x}_k, \mathbf{x}_l) = V \underbrace{\int \mathrm{d}\mathbf{x}_l \, \Upsilon(\mathbf{x}_k, \mathbf{x}_l)}_{\substack{\text{volume of} \\ \text{excluded region for } \mathbf{x}_l}} = V^2 \cdot 4\pi \frac{\sigma^2}{V},$$

(the area of a disk of radius $2\sigma$ appears; see Fig. 2.22), we obtain

$$Z(\eta) = V^N \left( 1 - 4\pi\sigma^2 \frac{N(N-1)}{2V} \right) \simeq V^N \underbrace{\exp\left[-2(N-1)\eta\right]}_{p_{\text{accept}}(\eta)}. \qquad (2.8)$$

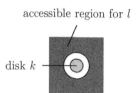

accessible region for $l$

disk $k$

**Fig. 2.22** Excluded and accessible regions for two disks of radius $\sigma$.

This implies that the probability for randomly chosen disks $k$ and $l$ not to overlap,

$$\left\{ \begin{array}{c} \text{probability that} \\ \text{disks } k \text{ and } l \\ \text{do not overlap} \end{array} \right\} = 1 - \frac{4\pi\sigma^2}{V} \simeq \exp\left(-\frac{4\pi\sigma^2}{V}\right), \qquad (2.9)$$

is uncorrelated at low density. For 16 disks, the function in eqn (2.8) yields $p_{\text{accept}} = Z(\eta)/V^N \simeq e^{-30\eta}$. This fits very well the empirical acceptance rate obtained from Alg. 2.8 (`direct-disks-any`). At low density, it is exact (see Fig. 2.21).

We have computed in eqn (2.8) the second virial coefficient $B$ of the hard-disk gas in two dimensions, the first correction term in $1/V$ beyond the ideal-gas expression for the equation of state:

$$\frac{PV}{NRT} = \frac{V}{N} \frac{\partial \log Z}{\partial V} = 1 + B\frac{1}{V} + C\frac{1}{V^2} + \cdots,$$

which, from eqn (2.8), where $\eta = \pi\sigma^2 N/V$, is equal to

$$1 + \underbrace{2(N-1)\pi\sigma^2}_{B} \frac{1}{V}.$$

This is hardly a state-of-the-art calculation in the twenty-first century, given that in 1874, Boltzmann had already computed the fourth virial coefficient, the coefficient of $V^{-3}$ in the above expansion, for three-dimensional spheres. The virial expansion was once believed to give systematic access to the thermodynamics of gases and liquids at all densities up to close packing, in the same way that, say, the expansion of the exponential function $e^x = 1 + x + x^2/2! + x^3/3! + \cdots$ converges for all real and complex $x$, but it becomes unwieldy at higher orders. More fundamentally, this perturbation approach cannot describe phase transitions: there is important physics beyond virial expansions around $\eta = 0$, and beyond the safe harbor of direct-sampling algorithms.

The relation between our algorithms and the partition functions of statistical mechanics holds even for the Markov-chain algorithm in Subsection 2.2.3, which concentrates on a physical system in a small window corresponding to the throwing range. This algorithm thus overcomes the direct-sampling algorithm's limitation to low densities or small particle numbers, but has difficulties coping with large-scale structures, which no longer allow cutting up into small systems.

### 2.2.3  Markov-chain hard-sphere algorithm

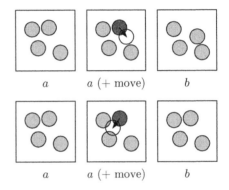

Fig. 2.23 Accepted (*top*) and rejected (*bottom*) Monte Carlo moves for a hard-disk system.

Direct sampling for hard disks works only at low densities and small particle numbers, and we thus switch to a more general Markov-chain Monte Carlo algorithm (see Alg. 2.9 (`markov-disks`)). Disks are moved

**procedure** `markov-disks`
**input** $\{\mathbf{x}_1, \ldots, \mathbf{x}_N\}$ (configuration $a$)
$k \leftarrow \mathbf{nran}\,(1, N)$
$\delta \mathbf{x}_k \leftarrow \{\mathbf{ran}\,(-\delta, \delta), \mathbf{ran}\,(-\delta, \delta)\}$
**if** (disk $k$ can move to $\mathbf{x}_k + \delta \mathbf{x}_k$) $\mathbf{x}_k \leftarrow \mathbf{x}_k + \delta \mathbf{x}_k$
**output** $\{\mathbf{x}_1, \ldots, \mathbf{x}_N\}$ (configuration $b$)

———

**Algorithm 2.9** `markov-disks`. Generating a hard-disk configuration $b$ from configuration $a$ using a Markov-chain algorithm (see Fig. 2.23).

analogously to the way adults wander between pebble positions on the Monaco heliport, and attempts to reach illegal configurations with overlaps are rejected (see Fig. 2.23). Detailed balance between configurations holds for the same reason as in Alg. 1.2 (`markov-pi`). The Markov-chain hard-disk algorithm resembles the adults' game on the heliport (see

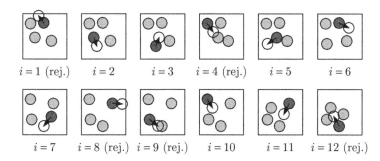

$i = 1$ (rej.)  $i = 2$  $i = 3$  $i = 4$ (rej.)  $i = 5$  $i = 6$

$i = 7$  $i = 8$ (rej.)  $i = 9$ (rej.)  $i = 10$  $i = 11$  $i = 12$ (rej.)

**Fig. 2.24** Markov-chain Monte Carlo algorithm for hard disks in a box without periodic boundary conditions (see Alg. 2.9 (`markov-disks`)).

Fig. 2.24), but although we again drive disks (pebbles) across a two-dimensional box (the landing pad), the $2N$-dimensional configuration space is not at all easy to characterize. We must, for example, understand whether the configuration space is simply connected, so that any configuration can be reached from any other by moving disks, one at a time, with an infinitesimal step size. Simple connectivity of the configuration space implies that the Monte Carlo algorithm is ergodic, a crucial requirement. Ergodicity, in the sense just discussed, can indeed be broken for small $N$ or for very high densities, close to jammed configurations (see the discussion of jamming in Chapter 7). For our present purposes, the question of ergodicity is best resolved within the constant-pressure ensemble, where the box volume may fluctuate (see Subsection 2.3.4), and the hard-sphere Markov-chain algorithm is trivially ergodic.

The Markov-chain algorithm allows us to generate independent snapshots of configurations for impressively large system sizes, and at high density. These typical configurations are liquid-like at low and moderate densities, but resemble a solid beyond a phase transition at $\eta \simeq 0.70$. This transition was discovered by Alder and Wainwright (1957) using the molecular dynamics approach of Section 2.1. This was very surprising because, in two dimensions, a liquid–solid transition was not expected to exist. A rigorous theorem (Mermin and Wagner 1966) even forbids positional long-range order for two-dimensional systems with short-range interactions, a class to which hard disks belong. An infinitely large system thus cannot have endless patterns of disks neatly aligned as in the right frame of Fig. 2.25. Nevertheless, in two dimensions, long-range order is possible for the orientation of links between neighbors, that is, the angles, which are approximately $0, 60, 120, 180$, and $240$ degrees in the right frame of Fig. 2.25, can have long-range correlations across an infinite system. A detailed discussion of crystalline order in two dimensions would go beyond the scope of this book, but the transition itself will be studied again in Subsection 2.4, in the constant-pressure ensemble.

It is interesting to interpret the changes in the configuration space when the system passes through the phase transition. Naively, we would

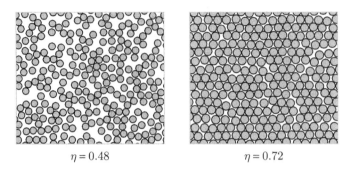

$\eta = 0.48$                    $\eta = 0.72$

**Fig. 2.25** Snapshots of 256 hard disks in a box of size $1 \times \sqrt{3}/2$ with periodic boundary conditions (from Alg. 2.9 (`markov-disks`)).

suppose that below the critical density only liquid-like configurations exist, and above the transition only solid ones. This first guess is wrong at low density because a crystalline configuration at high density obviously also exists at low density; it suffices to reduce the disk radii. Disordered configurations (configurations without long-range positional or orientational order) also exist right through the transition and up to the highest densities; they can be constructed from large, randomly arranged, patches of ordered disks. Liquid-like, disordered configurations and solid configurations of disks thus do not disappear as we pass through the liquid–solid phase transition density one way or the other; it is only the balance of statistical weights which is tipped in favor of crystalline configurations at high densities, and in favor of liquid configurations at low densities.

The Markov-chain hard-disk algorithm is indeed very powerful, because it allows us to sample configurations at densities and particle numbers that are far out of reach for direct-sampling methods. However, it slows down considerably upon entering the solid phase. To see this in a concrete example, we set up a particular tilted initial condition for a long simulation with Alg. 2.9 (`markov-disks`) (see Fig. 2.26). Even 25 billion moves later, that is, one hundred million sweeps (attempted moves per disk), the initial configuration still shows through in the state of the system. A configuration independent of the initial configuration has not yet been sampled.

We can explain—but should not excuse—the slow convergence of the hard-disk Monte Carlo algorithm at high density by the slow motion of single particles (in the long simulation of Fig. 2.26, the disk $k$ has only moved across one-quarter of the box). However, an equilibrium Monte Carlo algorithm is not meant to simulate time evolution, but to generate, as quickly as possible, configurations $a$ with probability $\pi(a)$ for all $a$ making up the configuration space. Clearly, at a density $\eta = 0.72$, Alg. 2.9 (`markov-disks`) fails at this task, and Markov-chain sampling slows down dangerously.

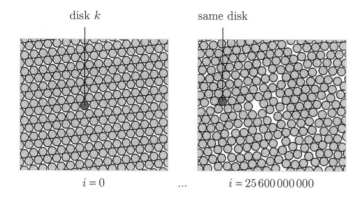

disk $k$         same disk

$i = 0$     ...     $i = 25\,600\,000\,000$

**Fig. 2.26** Initial and final configurations of a Monte Carlo simulation for 256 disks (size $1 \times \sqrt{3}/2$, periodic boundary conditions, $\eta = 0.72$).

## 2.2.4   Velocities: the Maxwell distribution

Molecular dynamics concerns positions and velocities, whereas Alg. 2.7 (`direct-disks`) and Alg. 2.9 (`markov-disks`), only worry about positions. Why the velocities disappear from the Monte Carlo programs deserves a most thorough answer (and is a No. 1 exam question).

To understand velocities in statistical mechanics, we again apply the equiprobability principle, not to particle positions within a box, but to the velocities themselves. This principle calls for all legal sets of hard-sphere velocities to appear with the same probability:

$$\pi(\mathbf{v}_1, \ldots, \mathbf{v}_N) = \begin{cases} 1 & \text{if velocities legal} \\ 0 & \text{if forbidden} \end{cases}.$$

For concreteness, we consider hard disks in a box. A set $\{\mathbf{v}_1, \ldots, \mathbf{v}_N\}$ of velocities is legal if it corresponds to the correct value of the kinetic energy

$$E_{\text{kin}} = \frac{1}{2}m \cdot \left(\mathbf{v}_1^2 + \cdots + \mathbf{v}_N^2\right) \quad \text{(fixed)}.$$

Each squared velocity in this equation has two components, that is, $\mathbf{v}_k^2 = \mathbf{v}_{x,k}^2 + \mathbf{v}_{y,k}^2$, and any legal set of velocities corresponds to a point on a $2N$-dimensional sphere with $r^2 = 2E_{\text{kin}}/m$. The equiprobability principle thus calls for velocities to be random vectors on the surface of this $2N$-dimensional sphere (see Fig. 2.27).

We recall from Subsection 1.2.6 that random points on the surface of a hypersphere can be sampled with the help of $2N$ independent Gaussian random numbers. The algorithm involves a rescaling, which becomes unnecessary in high dimensions if the Gaussians' variance is chosen correctly (see the discussion of Alg. 1.22 (`direct-surface`)). In our case, the correct scaling is

$$\pi(v_x) = \frac{1}{\sqrt{2\pi}\sigma} \exp\left(-\frac{v_x^2}{2\sigma^2}\right), \quad \text{etc.,}$$

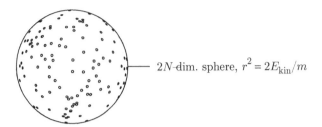

**Fig. 2.27** Legal sets of velocities for $N$ hard disks in a box.

where

$$\sigma = \sqrt{\frac{2}{m}\frac{E_{\mathrm{kin}}}{dN}}.$$

This is the Maxwell distribution of velocities in $d$ dimensions; $E_{\mathrm{kin}}/(dN)$ is the mean kinetic energy per degree of freedom and is equal to $\frac{1}{2}k_{\mathrm{B}}T$, where $T$ is the temperature (in kelvin), and $k_{\mathrm{B}}$ is the Boltzmann constant. We find that the variance of the Gaussian describing the velocity distribution function is $\sigma^2 = k_{\mathrm{B}}T/m$, and we finally arrive at the following expressions for the probability distribution of a single component of the velocity:

$$\pi(v_x)\,\mathrm{d}v_x = \sqrt{\frac{m}{2\pi k_{\mathrm{B}}T}}\exp\left(-\frac{1}{2}\frac{mv_x^2}{k_{\mathrm{B}}T}\right)\mathrm{d}v_x.$$

In two dimensions, we use the product of distributions, one for $v_x$, and another for $v_y$. We also take into account the fact that the volume element can be written as $\mathrm{d}v_x\,\mathrm{d}v_y = \mathrm{d}\phi\,v\,\mathrm{d}v = 2\pi v\,\mathrm{d}v$:

$$\pi(v)\,\mathrm{d}v = \frac{m}{k_{\mathrm{B}}T}v\exp\left(-\frac{1}{2}\frac{mv^2}{k_{\mathrm{B}}T}\right)\mathrm{d}v.$$

In three dimensions, we do the same with $\{v_x, v_y, v_z\}$ and find

$$\pi(v)\,\mathrm{d}v = \sqrt{\frac{2}{\pi}}\left(\frac{m}{k_{\mathrm{B}}T}\right)^{3/2}v^2\exp\left(-\frac{1}{2}\frac{mv^2}{k_{\mathrm{B}}T}\right)\mathrm{d}v.$$

Here $v$ is equal to $\sqrt{v_x^2 + v_y^2}$ in two dimensions and to $\sqrt{v_x^2 + v_y^2 + v_z^2}$ in three dimensions.

We can compare the Maxwell distribution with the molecular dynamics simulation results for four disks in a box, and check that the distribution function for each velocity component is Gaussian (see Fig. 2.28). Even for these small systems, the difference between Gaussians and random points on the surface of the hypersphere is negligible.

In conclusion—and in response to the exam question at the beginning of this subsection—we see that particle velocities drop out of the Monte

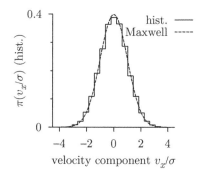 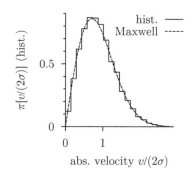

**Fig. 2.28** Histograms of a velocity component $v_x$ (*left*) and of $v = \sqrt{v_x^2 + v_y^2}$ (*right*) for four disks in a box (from Alg. 2.1 (`event-disks`)).

Carlo approach of Alg. 2.7 (`direct-disks`) and its generalizations because they form an independent sampling problem of random points on the surface of a hypersphere, solved by Alg. 1.22 (`direct-surface`), which is naturally connected to the Maxwell distribution of particle velocities.

## 2.2.5   Hydrodynamics: long-time tails

The direct-sampling algorithms for the positions and the velocities of hard spheres (Algs 2.7 (`direct-disks`) and 1.22 (`direct-surface`)) implement a procedure analogous to the molecular-dynamics approach, which also determines positions and velocities. In this subsection, we again scrutinize the relationship of this Monte Carlo approach to hard spheres with simulations using molecular dynamics. The theorems of Sinai and Simanyi assure us that molecular dynamics, the solution of Newton's equations, converges towards equilibrium, meaning that during an infinitely long simulation, all legal sets of positions and velocities come up with the same probability.

In a related context, concerning random processes, simple convergence towards the stationary probability distribution has proved insufficient (see Subsection 1.1.4). We needed exponential convergence, with a timescale, the correlation time, for practical equivalence of the Markov-chain approach to direct sampling. This timescale was provided by the second largest eigenvalue of the transfer matrix and it allowed us to distinguish between short and long simulations: a Markov chain that had run several times longer than the correlation time could be said to be practically in equilibrium.

As we shall see, the molecular dynamics of hard disks and spheres lacks an analogous timescale, that is, a typical time after which Alg. 2.1 (`event-disks`) would move from one set of positions and velocities to another independent set. Convergence is guaranteed by theorems in mathematics, but it is not exponential. The inescapable consequence of this absence of a scale is that statistical mechanics cannot capture all there is to molecular dynamics. Another discipline of physics, hydrodynamics, also has its word to say here.

We shall follow this discussion using a special quantity, the velocity autocorrelation function, whose choice we shall first describe the motivation for. We imagine a configuration of many hard disks, during a molecular dynamics simulation, in equilibrium from time $t' = 0$ to time $t$. Each particle moves from position $\mathbf{x}(0)$ to $\mathbf{x}(t)$, where

$$\mathbf{x}(t) - \mathbf{x}(0) = \int_0^t dt' \, \mathbf{v}(t')$$

(we omit particle indices in this and the following equations). We may average this equation over all possible initial conditions, but it is better to first square it to obtain the mean squared particle displacement

$$\langle (\mathbf{x}(t) - \mathbf{x}(0))^2 \rangle = \int_0^t dt' \int_0^t dt'' \underbrace{\langle (\mathbf{v}(t') \cdot \mathbf{v}(t'')) \rangle}_{C_{\mathbf{v}}(t'' - t')}.$$

For a rapidly decaying velocity autocorrelation function, the autocorrelation function $C_{\mathbf{v}}$ will be concentrated in a strip around $\tau = t'' - t' = 0$ (see Fig. 2.29). In the limit $t \to \infty$, we can then extend the integration, as shown, let the strip extension go to $\infty$, and obtain the following, where $\tau = t'' - t'$:

$$\frac{1}{t} \langle (\mathbf{x}(t) - \mathbf{x}(0))^2 \rangle \simeq \frac{1}{t} \underset{\substack{\text{strip in} \\ \text{Fig. 2.29}}}{\iint} dt' \, d\tau \, C_{\mathbf{v}}(\tau)$$

$$\xrightarrow[\substack{\text{decay of } C(\tau) \\ \text{faster than } 1/\tau}]{t \to \infty} 2 \int_0^\infty d\tau \, C_{\mathbf{v}}(\tau) = 2D, \quad (2.10)$$

We see that the mean square displacement during the time interval from 0 to $t$ is proportional to $t$ (not to $t^2$, as for straight-line motion). This is a hallmark of diffusion, and $D$ in the above equation is the diffusion constant. Equation (2.10) relates the diffusion constant to the integrated velocity autocorrelation function. Exponential decay of the autocorrelation function causes diffusive motion, at least for single particles, and would show that molecular dynamics is practically identical to statistical mechanics on timescales much larger than the correlation time.

The simple version of Alg. 2.1 (**event-disks**) (with none of the refinements sketched in Subsection 2.1.1) allows us to compute the velocity autocorrelation function in the interesting time regime for up to $\simeq 1000$ disks. It is best to measure time in units of the collision time: between time 0 and time $t'$, each particle should have undergone on average $t'$ collisions—the total number of pair collisions in the system should be $\simeq \frac{1}{2} N t'$. Even with a basic program that tests far more collisions than necessary, we can redo a calculation similar that of Alder and Wainwright

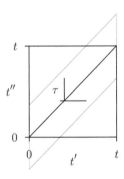

**Fig. 2.29** Integration domain for the velocity autocorrelation function (*square*), and strip chosen (*gray*), equivalent for $t \to \infty$.

(1970), and produce the clear-cut result for the correlation between the velocity $\mathbf{v}_k(t)$ of particle $k$ at time $t$ and the velocity $\mathbf{v}_k(t+\tau)$ at some later time (see Fig. 2.30). There is no indication of an exponential decay of the autocorrelation function (which would give a timescale); instead, the velocity autocorrelation function of the two-dimensional hard-sphere gas decays as a power law $1/\tau$. In $d$ dimensions, the result is

$$C_{\mathbf{v}}(\tau) = \langle(\mathbf{v}(0) \cdot \mathbf{v}(\tau))\rangle \propto \frac{1}{\tau^{d/2}}. \qquad (2.11)$$

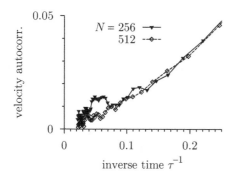

**Fig. 2.30** Velocity autocorrelation for disks vs. inverse time (in collisions per disk; $\eta = 0.302$, square box, from Alg. 2.1 (`event-disks`)).

The disks in the simulation are in equilibrium: snapshots of velocities give only the Maxwell distribution. It makes sense, however, to consider the direction of motion of a particle during a certain time interval. In order to move in that direction, a particle has to get other particles (in front of it) out of the way, and these, in turn, form a kind of eddy that closes on itself behind the original particle, pushing it to continue moving in its original direction. Theoretically, long-time tails are very well understood, and detailed calculations confirm the picture of eddies and the results of straightforward molecular dynamics (for an introductory discussion of long-time tails, see Pomeau and Résibois (1975)).

Long-time tails are most pronounced in two dimensions, basically because particles that are pushed away in front of a target disk have only two directions to go. In this case, of two dimensions, this constriction has dangerous effects on the diffusion constants: the mean square displacement, for large time intervals $t$, is not proportional to $t$ as in diffusive motion, but to $t \log t$. (This follows from entering eqn (2.11) into eqn (2.10).) All this, however, does not imply that, in two dimensions, diffusive motion does not exist and that, for example, a colloidal Brownian particle on a water surface in a big shallow trough moves faster and faster as time goes on. For very slow motion, thermal coupling to the outside world restores a finite diffusion constant (for an example of thermal coupling, see Subsection 2.3.1). It is certainly appropriate to treat a Brownian particle on a water surface with statistical mechanics. Some other systems, such as the earth's atmosphere, are in relatively fast motion, with limited thermal exchange. Such systems are described partly by statistical physics, but mostly by hydrodynamics, as is finally quite natural. We seen in this subsection that hydrodynamics remains relevant to mechanical systems even in the long-time (equilibrium) limit, in the absence of thermal coupling to the outside world.

## 2.3   Pressure and the Boltzmann distribution

Equilibrium statistical mechanics contains two key concepts. The first and foremost is equiprobability, the principle that configurations with the same energy are equally probable. This is all we need for hard disks and spheres in a box of fixed volume. In this section, we address the second key concept, the Boltzmann distribution $\pi(a) \propto e^{-\beta E(a)}$, which relates the probabilities $\pi(a)$ and $\pi(b)$ of configurations $a$ and $b$ with different energies. It comes up even for hard spheres if we allow variations in the box volume (see Fig. 2.31) and allow exchange of energy with an external bath. We thus consider a box at constant temperature and pressure rather than at constant temperature and volume.

For hard spheres, the constant-pressure ensemble allows density fluctuations on length scales larger than the fixed confines of the simulation box. The absence of such fluctuations is one of the major differences between a large and a periodic small system (see Fig. 2.32; the small system has exactly four disks per box, the large one between two and six). Our Monte Carlo algorithm for hard spheres at constant pressure is quite different from Alg. 2.7 (`direct-disks`), because it allows us to carry over some elements of a direct sampling algorithm for ideal particles. We shall discuss this non-interacting case first.

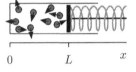

**Fig. 2.31** A box with disks, closed off by a piston exerting a constant force.

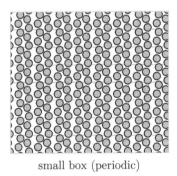
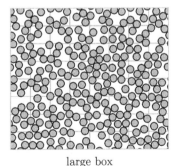

small box (periodic)　　　　　　　large box

**Fig. 2.32** 256 disks at density $\eta = 0.48$. *Left*: periodically replicated system of four particles. *Right*: large periodic box.

## 2.3.1 Bath-and-plate system

To familiarize ourselves with the concept of pressure, we consider a box
filled with hard disks and closed off by a piston, at position $x = L$. A
spring pushes the piston to the left with constant force, independent
of $x$ (see Fig. 2.31). The particles and the piston have kinetic energy.
The piston has also potential energy, which is stored in the spring. The
sum of the two energies is constant. If the piston is far to the right, the
particles have little kinetic energy, because potential energy is stored in
the spring. In contrast, at small $L$, the particles are compressed and they
have a higher kinetic energy. As the average kinetic energy is identified
with the temperature (see Subsection 2.2.4), the disks are not only at
variable volume but also at nonconstant temperature.

It is preferable to keep the piston–box system at constant temperature.
We thus couple it to a large bath of disks through a loose elastic plate,
which can move along the $x$-direction over a very small distance $\Delta$ (see
Fig. 2.34). By zigzagging in this interval, the plate responds to hits from
both the bath and the system. For concreteness, we suppose that the
particles in the system and in the bath, and also the plate, all have a
mass $m = 1$ ( the spring itself is massless). All components are perfectly
elastic. Head-on collisions between elastic particles of the same mass
exchange the velocities (see Fig. 2.33), and the plate, once hit by a bath
particle with an $x$-component of its velocity $v_x$ will start vibrating with
a velocity $\pm v_x$ inside its small interval (over the small distance $\Delta$) until
it eventually transfers this velocity to another particle, either in the box
or in the bath.

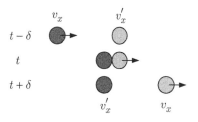

Fig. 2.33 Elastic head-on collision between equal-mass objects (case $v'_x = 0$
shown).

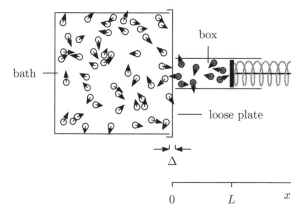

Fig. 2.34 The box containing particles shown in Fig. 2.31, coupled to
an infinite bath through a loose plate.

The plate's velocity distribution—the fraction of time it spends at
velocity $v_x$—is not the same as the Maxwell distribution of one velocity component for the particles. This is most easily seen for a bath of
Maxwell-distributed noninteracting point particles (hard disks with zero

radius): fast particles zigzag more often between the plate and the left boundary of the bath than slow particles, biasing the distribution by a factor $|v_x|$:

$$\pi(v_x)\,\mathrm{d}v_x \propto |v_x|\exp\left(-\beta v_x^2/2\right)\mathrm{d}v_x. \qquad (2.12)$$

We note that the Maxwell distribution for one velocity component lacks the $|v_x|$ term of eqn (2.12), and it is finite at $v_x = 0$. The biased distribution, however, must vanish at $v_x = 0$: to acquire zero velocity, the plate must be hit by a bath particle which itself has velocity zero (see Fig. 2.33). However, these particles do not move, and cannot get to the plate. This argument for a biased Maxwell distribution can be applied to a small layer of finite-size hard disks close to the plate, and eqn (2.12) remains valid.

The relatively infrequent collisions of the plate with box particles play no role in establishing the probability distribution of the plate velocity, and we may replace the bath and the plate exactly by a generator of biased Gaussian random velocities (with $v_x > 0$; see Fig. 2.33). The distribution in eqn (2.12) is formally equivalent to the Maxwell distribution for the absolute velocity in two dimensions, and so we can sample it with two independent Gaussians as follows:

$$\{\Upsilon_1, \Upsilon_2\} \leftarrow \{\mathbf{gauss}\,(1/\sqrt{\beta}), \mathbf{gauss}\,(1/\sqrt{\beta})\},$$
$$v_x \leftarrow \sqrt{\Upsilon_1^2 + \Upsilon_2^2}. \qquad (2.13)$$

Alternatively, the sample transformation of Subsection 1.2.4 can also be applied to this problem:

$$\int_0^1 \mathrm{d}\Upsilon = c \int_0^\infty \mathrm{d}u\ \exp\left(-u\right) = c'\int_0^\infty \mathrm{d}v_x\ v_x\exp\left(-\beta v_x^2/2\right).$$

The leftmost integral is sampled by $\Upsilon = \mathbf{ran}\,(0,1)$. The substitutions $\exp\left(-u\right) = \Upsilon$ and $\beta v_x^2/2 = u$ yield

$$v_x \leftarrow \sqrt{\frac{-2\log\left[\mathbf{ran}\,(0,1)\right]}{\beta}}.$$

This routine is implemented in Alg. 2.10 (`maxwell-boundary`). It exactly replaces—integrates out—the infinite bath.

Maxwell boundary

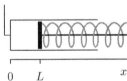

**Fig. 2.35** A piston with Maxwell boundary conditions at $x = 0$.

**procedure** `maxwell-boundary`
**input** $\{v_x, v_y\}$ (disk in contact with plate)
$\Upsilon \leftarrow \mathbf{ran}\,(0,1)$
$v_x \leftarrow \sqrt{-2\mathbf{log}\,(\Upsilon)\,/\beta}$
**output** $\{v_x, v_y\}$
——

**Algorithm 2.10** `maxwell-boundary`. Implementing Maxwell boundary conditions.

In conclusion, to study the box–bath–piston system, we need not set up a gigantic molecular dynamics simulation with particles on either side of the loose plate. The bath can be integrated out exactly, to leave us with a pure box–piston model with Maxwell boundary conditions. These boundary conditions are of widespread use in real-life simulations, notably when the temperature varies through the system.

## 2.3.2 Piston-and-plate system

We continue our analysis of piston-and-plate systems, without a computer, for a piston in a box without disks, coupled to an infinite bath represented by Maxwell boundary conditions (see Fig. 2.35). The piston hits the plate $L = 0$ at times $\{\ldots, t_i, t_{i+1}, \ldots\}$. Between these times, it obeys Newton's equations with the *constant force* generated by the spring. The piston height satisfies $L(t - t_i) = v_0 \cdot (t - t_i) - \frac{1}{2}(t - t_i)^2$ (see Fig. 2.36). We take the piston mass and restoring force to be equal to one, and find

$$\underbrace{t_{i+1} - t_i}_{\substack{\text{time of flight} \\ \text{in Fig. 2.36}}} = 2v_0.$$

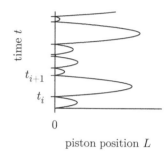

We see that the time the piston spends on a trajectory with initial velocity $v_0$ is proportional to $v_0$. We thus find the following:

$$\begin{Bmatrix} \text{fraction of time spent} \\ \text{at initial velocities} \\ [v_0, v_0 + \mathrm{d}v_0] \end{Bmatrix} \propto \overbrace{v_0}^{\substack{\text{time of} \\ \text{flight}}} \overbrace{v_0 \exp\left(-\beta v_0^2/2\right) \mathrm{d}v_0}^{\text{Maxwell boundary cond.}}.$$

**Fig. 2.36** Trajectory $\{L, t\}$ of a piston coupled to a plate with Maxwell boundary conditions at $L = 0$.

During each flight, the energy is constant, and we can translate what we have found into a probability distribution of the energy $E$. Because $\mathrm{d}E = v_0\,\mathrm{d}v_0$, the time the piston spends in the interval of energies $[E, E+\mathrm{d}E]$—the probability $\pi(E)\,\mathrm{d}E$—is

$$\pi(E)\,\mathrm{d}E \propto \sqrt{E}\mathrm{e}^{-\beta E}\,\mathrm{d}E. \tag{2.14}$$

The factor $\sqrt{E}$ in eqn (2.14) is also obtained by considering the phase space of the moving piston, spanned by the variables $L$ and $v$ (see Fig. 2.37). States with an energy smaller than $E$ are below the curve

$$L(E, v) = E - \frac{v^2}{2}.$$

The volume $\mathcal{V}(E)$ of phase space for energies $\leq E$ is given by

$$\mathcal{V}(E) = \int_0^{\sqrt{2E}} \mathrm{d}v \left[E - \frac{v^2}{2}\right] = \left[Ev - \frac{v^3}{6}\right]_0^{\sqrt{2E}} = \sqrt{2}\frac{2}{3}E^{3/2}.$$

It follows that the density of states for an energy $E$, that is, the number of phase space elements with energies between $E$ and $E + \mathrm{d}E$, is given by

$$\mathcal{N}(E) = \frac{\partial}{\partial E}\mathcal{V}(E) = \sqrt{2E}.$$

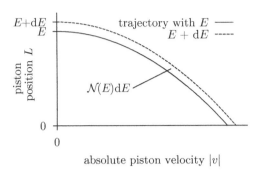

**Fig. 2.37** Piston–plate system. The phase space for an energy in the interval $[E, E + dE]$ has a volume $\mathcal{N}(E)\,dE$.

So, we expect to have a probability $\pi(E)$ as follows:

$$\pi(E)\,dE = \mathcal{N}(E)\,e^{-\beta E}\,dE.$$

It also follows that the system satisfies the equiprobability principle, that is, it spends equal amounts of time in the interval $dx\,dv$, whatever $v$ is. This follows simply from the fact that $dx$ is proportional to $dE$ and $dv$ is proportional to $dt$:

$$\pi(x, v)\,dx\,dv = \exp\left[-\beta E(x, v)\right]\,dx\,dv.$$

This is the Boltzmann distribution, and we have derived it, as promised, from the equiprobability principle. We can also obtain the Boltzmann distribution for a piston–plate model with modified springs, for example with a potential energy $E(L) = L^\alpha$ with arbitrary positive $\alpha$ different from the case $\alpha = 1$ treated in this subsection (see Exerc. 2.15).

Our solution of the piston–plate model of Fig. 2.35 can be generalized to the case of a box containing particles in addition to the piston, the spring, and the vibrating plate. With more than one disk, we can no longer solve the equations of motion of the coupled system analytically, and have to suppose that for a fixed piston position and velocity of the piston all disk positions and velocities are equally probable, and also that for fixed disks, the parameters of the piston obey the Boltzmann distribution. The argument becomes quite involved. In the remainder of this book, we rather take for granted the two pillars of statistical physics, namely the equiprobability principle ($\pi(a) = \pi(E(a))$) and the Boltzmann distribution, and study their consequences, moving away from the foundations of statistical mechanics to what has been built on top of them.

### 2.3.3  Ideal gas at constant pressure

In this subsection, we work out some sampling methods for a one-dimensional gas of point particles interacting with a piston. What we learn here can be put to work for hard spheres and many other systems.

In this gas, particles at positions $\{x_1, \ldots, x_N\}$ on the positive axis may move about and pass through each other, but must satisfy $x_k < L$, where $L$ is again the piston position, the one-dimensional volume of the box. The energy of the piston at position $L$ is $PL$, where $P$ is the pressure (see Fig. 2.38).

The system composed of the $N$ particles and the piston may be treated by Boltzmann statistical mechanics with a partition function

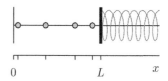

Fig. **2.38** Particles in a line, with a piston enforcing constant pressure, a restoring force independent of $L$.

$$Z = \int_0^\infty \mathrm{d}L \; \mathrm{e}^{-\beta PL} \int_0^L \mathrm{d}x_1 \; \ldots \int_0^L \mathrm{d}x_N, \qquad (2.15)$$

which we can evaluate analytically:

$$Z = \int_0^\infty \mathrm{d}L \; \mathrm{e}^{-\beta PL} L^N = \frac{N!}{(\beta P)^{N+1}}.$$

We can use this to compute the mean volume $\langle L \rangle$ of our system,

$$\langle \text{Volume} \rangle = \langle L \rangle = \frac{\int_0^\infty \mathrm{d}L \; L^{N+1} \mathrm{e}^{-\beta PL}}{\int_0^\infty \mathrm{d}L \; L^N \mathrm{e}^{-\beta PL}} = \frac{N+1}{\beta P},$$

which gives essentially the ideal-gas law $PV = Nk_{\mathrm{B}}T$.

We may sample the integral in eqn (2.15) by a two-step approach. First, we fix $L$ and directly sample particle positions to the left of the piston. Then, we fix the newly obtained particle positions $\{x_1, \ldots, x_N\}$ and sample a new piston position $L$, to the right of all particles, using the Metropolis algorithm (see Alg. 2.11 (**naive-piston-particles**) and Fig. 2.39). To make sure that it is correct to proceed in this way, we may write the partition function given in eqn (2.15) without $L$-dependent boundaries for the $x$-integration:

$$Z = \int_0^\infty \mathrm{d}L \int_0^\infty \mathrm{d}x_1 \; \ldots \int_0^\infty \mathrm{d}x_N \; \mathrm{e}^{-\beta PL} \; \{L > \{x_1, \ldots, x_N\}\}. \quad (2.16)$$

The integrals over the positions $\{x_1, \ldots, x_N\}$ no longer have an $L$-dependent upper limit, and Alg. 2.11 (**naive-piston-particles**) is thus correct. The naive piston–particle algorithm can be improved: for fixed particle positions, the distribution of $L$, from eqn (2.16), is $\pi(L) \propto \mathrm{e}^{-\beta PL}$ for $L > x_{\max}$, so that $\Delta_L = L - x_{\max}$ can be sampled directly (see Alg. 2.12 (**naive-piston-particles(patch)**)). This Markov-chain algorithm consists of two interlocking direct-sampling algorithms which exchange the current values of $x_{\max}$ and $L$: one algorithm generates particle positions for a given $L$, and the other generates piston positions for given $\{x_1, \ldots, x_N\}$.

```
procedure naive-piston-particles
input L
{x₁,...,xₙ} ← {ran (0, L),...,ran (0, L)}  (all indep.)
xmax ← max (x₁,...,xₙ)
ΔL ← ran (−δ, δ)
ϒ ← exp (−βPΔL)
if (ran (0, 1) < ϒ and L + ΔL > xmax) then
   { L ← L + ΔL
output L, {x₁,...,xₙ}
```

**Algorithm 2.11 naive-piston-particles.** Markov-chain algorithm for one-dimensional point particles at pressure $P$ (see patch in Alg. 2.12).

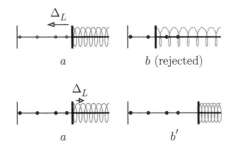

**Fig. 2.39** Piston moves in Alg. 2.11 (**naive-piston-particles**). The move $a \rightarrow b'$ is accepted with probability $\exp(-\beta P \Delta_L)$.

We can construct a direct-sampling algorithm by a simple change of variables in the partition function $Z$:

$$Z = \int_0^\infty \mathrm{d}L \; L^N \int_0^L \frac{\mathrm{d}x_1}{L} \cdots \int_0^L \frac{\mathrm{d}x_N}{L} \; \mathrm{e}^{-\beta PL} \qquad (2.17)$$

$$= \underbrace{\int_0^\infty \mathrm{d}L \; L^N \mathrm{e}^{-\beta PL}}_{\substack{\text{sample from} \\ \text{gamma distribution}}} \underbrace{\int_0^1 \mathrm{d}\alpha_1 \cdots \int_0^1 \mathrm{d}\alpha_N}_{\substack{\text{sample as } \alpha_k = \text{ran}(0,1) \\ \text{for } k = 1, \ldots, N}} . \qquad (2.18)$$

The integration limits for the variables $\{\alpha_1, \ldots, \alpha_N\}$ no longer depend on $L$, and the piston and particles are decoupled. The first integral in eqn (2.18) is a rescaled gamma distribution $\pi(x) \propto x^N \mathrm{e}^{-x}$ with $x = \beta PL$ (see Fig. 2.40), and gamma-distributed random numbers can be directly sampled as a sum of $N + 1$ exponential random numbers. For $N = 0$, $\pi(x)$ is a single exponential random variable. For $N = 1$, it is sampled by the sum of two independent exponential random numbers, whose

**procedure** `naive-piston-particles(patch)`
**input** $L$
$\{x_1, \ldots, x_N\} \leftarrow \{\mathtt{ran}\,(0, L), \ldots, \mathtt{ran}\,(0, L)\}$
$x_{\max} \leftarrow \max(x_1, \ldots, x_N)$
$\Delta_L \leftarrow -\log(\mathtt{ran}\,(0, 1))/(\beta P)$
$L \leftarrow x_{\max} + \Delta_L$
**output** $L, \{x_1, \ldots, x_N\}$

———

**Algorithm 2.12** `naive-piston-particles(patch)`. Implementing direct sampling of $L$ into the Markov-chain algorithm for $\{x_1, \ldots, x_N\}$.

distribution, the convolution of the original distributions, is given by

$$\pi(x) = \int_0^x \mathrm{d}y \; \mathrm{e}^{-y}\mathrm{e}^{-(x-y)}$$
$$= \mathrm{e}^{-x}\int_0^x \mathrm{d}y \; = x\mathrm{e}^{-x}$$

(see Subsection 1.3.1). More generally, a gamma-distributed random variable taking values $x$ with probability $\Gamma_N(x)$ can be sampled by the sum of logarithms of $N+1$ random numbers, or, better, by the logarithm of the product of the random numbers, to be computed alongside the $\alpha_k$. It remains to rescale the gamma-distributed sample $x$ into the size of the box, and the random numbers $\{\alpha_1, \ldots, \alpha_N\}$ into particle positions (see Alg. 2.13 (`direct-piston-particles`)).

**procedure** `direct-piston-particles`
$\Upsilon \leftarrow \mathtt{ran}\,(0, 1)$
**for** $k = 1, \ldots, N$ **do**
$\left\{ \begin{array}{l} \alpha_k \leftarrow \mathtt{ran}\,(0, 1) \\ \Upsilon \leftarrow \Upsilon\mathtt{ran}\,(0, 1) \end{array} \right.$
$L \leftarrow -\log(\Upsilon)/(\beta P)$
**output** $L, \{\alpha_1 L, \ldots, \alpha_N L\}$

———

**Algorithm 2.13** `direct-piston-particles`. Direct sampling of one-dimensional point particles and a piston at pressure $P$.

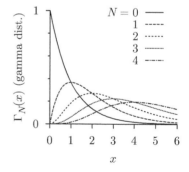

**Fig. 2.40** Gamma distribution $\Gamma_N(x) = x^N\mathrm{e}^{-x}/N!$, the distribution of the sum of $N + 1$ exponentially distributed random numbers.

### 2.3.4   Constant-pressure simulation of hard spheres

It takes only a few moments to adapt the direct-sampling algorithm for one-dimensional particles to hard spheres in a $d$-dimensional box of variable volume $V$ (and fixed aspect ratio) with $\pi(V) \propto \exp(-\beta PV)$. We simply replace the piston by a rescaling of the box volume and take into account the fact that the sides of the box scale with the $d$th root of the volume. We then check whether the output is a legal hard-sphere configuration (see Alg. 2.14 (`direct-p-disks`)). This direct-sampling algorithm mirrors Alg. 2.7 (`direct-disks`) (see Fig. 2.41).

```
procedure direct-p-disks
1   Υ ← ran (0, 1)
    for k = 1, ..., N do
      { αₖ ← {ran (0, 1), ran (0, 1)}
      { Υ ← Υran (0, 1)
    L ← √(−log (Υ)/(βP))
    for k = 1, ..., N do
      { xₖ ← Lαₖ
    if ({{x₁, ..., x_N}, L} not a legal configuration) goto 1
    output L, {x₁, ..., x_N}
```

**Algorithm 2.14** `direct-p-disks`. Direct sampling for $N$ disks in a square box with periodic boundary conditions at pressure $P$.

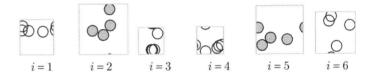

$i = 1$ $\qquad$ $i = 2$ $\qquad$ $i = 3$ $\qquad$ $i = 4$ $\qquad$ $i = 5$ $\qquad$ $i = 6$

**Fig. 2.41** Direct sampling for four hard disks at constant pressure (from Alg. 2.14 (`direct-p-disks`)).

We again have to migrate to a Markov-chain Monte Carlo algorithm, allowing for changes in the volume and for changes in the particle positions, the variables $\{\alpha_1, \dots, \alpha_N\}$. Although we cannot hope for a rejection-free direct-sampling algorithm for hard spheres, we shall see that the particle rescaling and the direct sampling of the volume carry over to this interacting system.

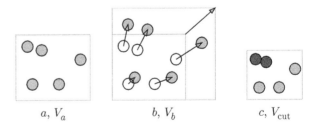

$a, V_a$ $\qquad\qquad$ $b, V_b$ $\qquad\qquad$ $c, V_{\text{cut}}$

**Fig. 2.42** Disk configurations with fixed $\{\alpha_1, \dots, \alpha_N\}$. Configuration $c$ is at the lower cutoff volume.

Let us consider a fixed configuration $\alpha = \{\alpha_1, \dots, \alpha_N\}$. It can exist at any box dimension which is sufficiently big to make it into a legal hard-sphere configuration. What happens for different volumes at fixed $\alpha$ is shown in Fig. 2.42: the particle positions are blown up together with the box, but the radii of the disks remain the same. There is an $\alpha$-dependent lower cutoff (minimum) volume, $V_{\text{cut}}$, below which configurations are rejected.

Above $V_{\text{cut}}$, the rescaled volume $x = \beta PV$ (with $x_{\text{cut}} = \beta PV_{\text{cut}}$) is distributed with what is called the gamma-cut distribution:

$$\pi(x) = \Gamma_N^{\text{cut}}(x, x_{\text{cut}}) \propto \begin{cases} x^N e^{-x} & \text{for } x > x_{\text{cut}} \\ 0 & \text{otherwise} \end{cases}.$$

As always in one dimension, the gamma-cut distribution can be directly sampled. We can compare it above $x_{\text{cut}} > N$ with an exponential:

$$\underline{\Gamma_N^{\text{cut}}(x, x_{\text{cut}})} \propto \pi_\Gamma(x, x_{\text{cut}}) = \left(\frac{x}{x_{\text{cut}}}\right)^N \exp\left[-(x - x_{\text{cut}})\right]$$

$$= \exp\left[-(x - x_{\text{cut}}) + N \log \frac{x}{x_{\text{cut}}}\right]$$

$$= \exp\left[-(x - x_{\text{cut}}) + N \underbrace{\log\left(1 + \frac{x - x_{\text{cut}}}{x_{\text{cut}}}\right)}_{<(x-x_{\text{cut}})/x_{\text{cut}}}\right]$$

$$< \overbrace{\exp\left[-(1 - N/x_{\text{cut}})(x - x_{\text{cut}})\right]}^{\pi_{\exp}(x, x_{\text{cut}})}. \quad (2.19)$$

To sample the gamma-cut distribution, we adapt the rejection method of Subsection 1.2.4, and rather sample the exponential distribution, which is everywhere larger. We thus throw uniformly distributed pebbles into the region delimited by $x \in [x_{\text{cut}}, \infty]$ and $y \in [0, \pi_{\exp}(x)]$ (see Fig. 2.43). $x$ can be sampled from $\pi_{\exp}(x, x_{\text{cut}})$, and $y$ can be sampled as a random number between 0 and $\pi_{\exp}(x)$. We must reject the pebble if $y = \text{ran}(0, \pi_{\exp}(x))$ is above the gamma-cut distribution, in other words if $y = \pi_{\exp}(x)\text{ran}(0, 1) > \pi_\Gamma(x)$ (see Alg. 2.15 (gamma-cut)).

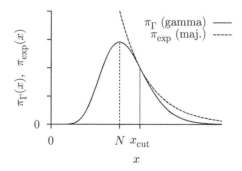

**Fig. 2.43** Gamma distribution $\Gamma_N(x) \propto \pi_\Gamma = x^N e^{-x}$, and its exponential majoration $\pi_{\exp}$, which allows us to sample the gamma-cut distribution.

**procedure** gamma-cut
**input** $x_{\text{cut}}$
$x^* \leftarrow 1 - N/x_{\text{cut}}$
**if** $(x^* < 0)$ **exit**
1   $\Delta_x \leftarrow -\log\left(\text{ran}\,(0,1)\right)/x^*$
    $x \leftarrow x_{\text{cut}} + \Delta_x$
    $\Upsilon' \leftarrow (x/x_{\text{cut}})^N \exp\left[-(1-x^*)\Delta_x\right]$
    **if** $(\text{ran}\,(0,1) > \Upsilon')$ **goto** 1 (reject sample)
    **output** $x_{\text{cut}} + \Delta_x$
———

**Algorithm 2.15** gamma-cut. Sampling the Gamma distribution for $x > x_{\text{cut}} > N$.

Alg. 2.15 (gamma-cut) rapidly samples the gamma-cut distribution for any $N$ and, after rescaling, a legal box volume for a fixed configuration $\alpha$ (see Alg. 2.16 (rescale-volume)). The algorithm is due to Wood (1968). It must be sandwiched in between runs of constant-volume Monte Carlo calculations, and provides a powerful hard-sphere algorithm in the $NPT$ ensemble, with the particle number, the pressure, and the temperature all kept constant.

**procedure** rescale-volume
**input** $\{L_x, L_y\}, \{\mathbf{x}_1, \ldots, \mathbf{x}_N\}$
$V \leftarrow L_x L_y$
$\sigma_{\text{cut}} \leftarrow \min_{k,l}\left[\text{dist}\,(\mathbf{x}_k, \mathbf{x}_l)\right]$
$x_{\text{cut}} \leftarrow \beta P V \cdot (\sigma/\sigma_{\text{cut}})^2$
$V_{\text{new}} \leftarrow \left[\text{gamma-cut}(N, x_{\text{cut}})\right]/(\beta P)$
$\Upsilon \leftarrow \sqrt{V_{\text{new}}/V}$
**output** $\{\Upsilon L_x, \Upsilon L_y\}, \{\Upsilon \mathbf{x}_1, \ldots, \Upsilon \mathbf{x}_N\}$
———

**Algorithm 2.16** rescale-volume. Sampling and rescaling the box dimensions and particle coordinates for hard disks at constant $P$.

Finally, we note that, for hard spheres, the pressure $P$ and inverse temperature $\beta = 1/(k_{\text{B}}T)$ always appear as a product $\beta P$ in the Boltzmann factor $e^{-\beta P V}$. For hard spheres at constant volume, the pressure is thus proportional to the temperature, as was clearly spelled out by Daniel Bernoulli in the first scientific work on hard spheres, in 1733, long before the advent of the kinetic theory of gases and the statistical interpretation of the temperature. Bernoulli noticed, so to speak, that if a molecular dynamics simulation is run at twice the original speed, the particles will hit the walls twice as hard and transfer a double amount of the original momentum to the wall. But this transfer takes place in half the original time, so that the pressure must be four times larger. This implies that the pressure is proportional to $v^2 \propto T$.

# 2.4 Large hard-sphere systems

Daily life accustoms us to phase transitions between different forms of matter, for example in water, between ice (solid), liquid, and gas. We usually think of these phase transitions as resulting from the antagonistic interplay between the interactions and the temperature. At low temperature, the interactions win, and the atoms or the ions settle into crystalline order. Materials turn liquid when the atoms' kinetic energy increases with temperature, or when solvents screen the interionic forces. Descriptions of phase transitions which focus on the energy alone are over-simplified for regular materials. They certainly do not explain phase transitions in hard spheres because there simply are no forces; all configurations have the same energy. However, the transition between the disordered phase and the ordered phase still takes place.

Our understanding of these entropic effects will improve in Chapter 6, but we start here by describing the phase transitions of hard disks more quantitatively than by just contemplating snapshots of configurations. To do so, we shall compute the equation of state, the relationship between volume and pressure.

When studying phase transitions, we are naturally led to simulating large systems. Throughout this book, and in particular during the present section, we keep to basic versions of programs. However, we should be aware of engineering tricks which can considerably speed up the execution of programs without changing in any way the output created. We shall discuss these methods in Subsection 2.4.1.

## 2.4.1 Grid/cell schemes

In this subsection, we discuss grid/cell techniques which allow one to decide in a constant number of operations whether, in a system of $N$ particles, a disk $k$ overlaps any other disk. This task comes up when we must decide whether a configuration is illegal, or whether a move is to be rejected. This can be achieved faster than by our naive checks of the $N-1$ distances from all other disks in the system (see for example Alg. 2.9 (`markov-disks`)). The idea is to assign all disks to an appropriate grid with cells large enough that a disk in one cell can only overlap with particles in the same cell or in the adjacent ones. This reduces the overlap checks to a neighborhood (see Fig. 2.44). Of course, particles may move across cell boundaries, and cell occupancies must be kept consistent (see Fig. 2.45).

There are number of approaches to setting up grid/cell schemes and to handling the bookkeeping involved in them. One may simply keep all disk numbers of cell $k$ in a table. A move between cells $k$ and $l$ has us locate the disk index in the table for cell $k$, swap it with the last element

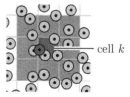

**Fig. 2.44** Grid/cell scheme with large cells: a disk in cell $k$ can only overlap with disks in the same cell or in adjacent cells.

of that table, and then reduce the number of elements:

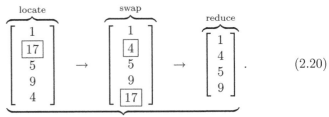

$$(2.20)$$

disk 17 leaving a cell containing disks $\{1, 17, 5, 9, 4\}$.

In the case of the target cell $l$, we simply append the disk's index to the table, and increment its number of elements.

There are other solutions for grid/cell schemes. They may involve linked lists rather than tables, that is, a data structure where element 1 in eqn (2.20) points to (is linked to) element 17, which itself points to disk 5, etc. Disk 4 would point to an "end" mark, with a "begin" mark pointing to disk 1. In that case, disk 17 is eliminated by redirecting the pointer of 1 from 17 directly to 5. Besides using linked lists, it is also possible to work with very small cells containing no more than one particle, at the cost of having to check more than just the adjacent cells for overlaps. Any of these approaches can be programmed in several subroutines and requires only a few instructions per bookkeeping operation. The extra memory requirements are no issue with modern computers. Grid/cell schemes reduce running times by a factor of $\alpha N$, where $\alpha < 1$ because of the bookkeeping overhead. We must also consider the human time that it takes to write and debug the modified code. The computations in this book have been run without improvement schemes on a year 2005 laptop computer, but some of them (in Subsections 2.2.5 and 2.4.2) approach a limit where the extra few hours—more realistically a few days—needed for implementing them were well spent.

Improved codes are easily tested against naive implementations, which we always write first and always keep handy. The output of molecular dynamics runs or Monte Carlo codes should be strictly equivalent between basic and improved versions of a program, even after going through billions of configurations. This frame-to-frame equivalence between two programs is easier to check than statistical equivalence, say, between direct sampling and Markov-chain sampling, where we can only compare average quantities, and only up to statistical uncertainties.

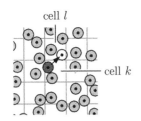

cell $l$

cell $k$

**Fig. 2.45** Moving a particle between boxes involves bookkeeping.

## 2.4.2   Liquid–solid transitions

In this subsection, we simulate hard disks at constant pressure, in order to obtain the equation of state of a system of hard disks, that is, the relationship between the pressure and the volume of the system. For concreteness, we restrict ourselves to a system of 100 disks in a box with aspect ratio $\sqrt{3}/2$. For this system, we do not need to implement the grid/cell schemes of Subsection 2.4.1. Straight simulation gives the following curve shown in Fig. 2.46 for the mean volume per particle. At

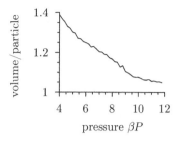 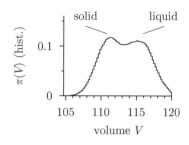

**Fig. 2.46** Equation of state for 100 disks (*left*), and histogram of $V$ at $\beta P = 8.4177$ (*right*) ($\sigma = \frac{1}{2}$, $L_x/L_y = \sqrt{3}/2$).

small pressure, the volume is naturally large, and the configurations are liquid-like. At high pressure, the configurations are crystalline, as was already discussed in Subsection 2.2.3. It is of fundamental importance that the volume as a function of pressure behaves differently above and below the transition separating the two regimes, but this cannot be seen very clearly in the equation of state of small systems. It is better to trace the histogram of volumes visited (see Fig. 2.46 again). Both at low pressure and at high pressure, this histogram has a single peak. In the transition region $\beta P \simeq 8.4177$, the histogram has two peaks. The two types of configurations appear (see Fig. 2.47), configurations that are solid-like (at small volume) and configurations that are liquid-like (at large volume). In this same region, the Monte Carlo simulation using the local algorithm slows down enormously. While it mixes up liquid configurations without any problem, it has a very hard time moving from a solid configuration (as the left configuration in Fig. 2.47) to a liquid-like configuration (as the right one in that same figure, see Lee and Strandburg 1992).

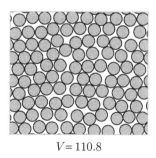 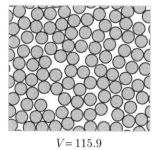

$V = 110.8$ $\qquad\qquad$ $V = 115.9$

**Fig. 2.47** Typical configuration for 100 disks of radius $\frac{1}{2}$ at pressure $\beta P = 8.4177$. *Left*: solid configuration (at small volume); *right*: liquid configuration (at large volume).

At the time of writing of this book, the nature of the transition in two-dimensional hard disks (in the thermodynamic limit) has not been cleared up. It might be a first-order phase transition, or a continuous Kosterlitz–Thouless transition. It is now well understood that in two dimensions the nature of transition depends on the details of the microscopic model. The phase transition in hard disks could be of first order, but a slight softening-up of the interparticle potential would make the transition continuous. The question about the phase transition in hard disks—although it is highly model-specific—would have been cleared up a long time ago if only we disposed of Monte Carlo algorithms that, while respecting detailed balance, allowed us to move in a split second between configurations as different as the two configurations in Fig. 2.47. However, this is not the case. On a year 2005 laptop computer, we have to wait several minutes before moving from a crystalline configuration to a liquid one, for 100 particles, even if we used the advanced methods described in Subsection 2.4.1. These times get out of reach of any simulation for the larger systems that we need to consider in to understand the finite-size effects at the transition. We conclude that simulations of hard disks do not converge in the transition region (for systems somewhat larger than those considered in Fig. 2.47). The failure of computational approaches keeps us from answering an important question about the phase transition in one of the fundamental models of statistical physics. (For the transition in three-dimensional hard spheres, see Hoover and Ree (1968).)

## 2.5   Cluster algorithms

Local simulation algorithms using the Metropolis algorithm and molecular dynamics methods allow one to sample independent configurations for large systems at relatively high densities. This gives often very important information on the system from the inside, so to speak, because the samples represent the system that one wants to study. In contrast, analytical methods are often forced to extrapolate from the noninteracting system (see the discussion of virial expansions, in Subsection 2.2.2). Even the Monte Carlo algorithm, however, runs into trouble at high density, when any single particle can no longer move, so that the Markov chain of configurations effectively gets stuck during long times (although it remains, strictly speaking, ergodic).

In the present section, we start to explore more sophisticated Monte Carlo algorithms that are not inspired by the physical process of single-particle motion. Instead of moving one particle after the other, these methods construct coordinated moves of several particles at a time. This allows one to go from one configuration, $a$, to a very different configuration, $b$, even though single particles cannot really move by themselves. These algorithm methods can no longer be proven correct by common sense alone, but by the proper use of a priori probabilities. The algorithms generalize the triangle algorithm of Subsection 1.1.6, which first went beyond naive pebble-throwing on the Monte Carlo heliport. In many fields of statistical mechanics, coordinated cluster moves—the displacement of many disks at once, simultaneous flips of spins in a region of space, collective exchange of bosons, etc.—have overcome the limitations of local Monte Carlo algorithms. The pivot cluster algorithm of Subsection 2.5.2 (Dress and Krauth 1995) is the simplest representative of this class of methods.

## 2.5.1   Avalanches and independent sets

By definition, the local hard-sphere Monte Carlo algorithm rejects all moves that produce overlaps (see Fig. 2.23). We now study an algorithm, which accepts the move of an independent disk even if it generates overlaps. It then simply moves the overlapped particles out of place, and starts an avalanche, where many disks are constrained to move and, in turn, tip off other disks. Disks that must move but which entail no other moves are called "terminal". For simplicity, we suppose that the displacement vector is the same for all disks (see Fig. 2.48 and Alg. 2.17 (`naive-avalanche`)).

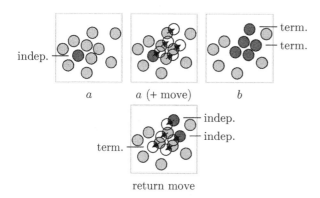

**Fig. 2.48** Avalanche move $a \to b$ and its return move, with independent and terminal disks.

Detailed balance stipulates that, for hard spheres, the return move be proposed with the same probability as the forward move. From Fig. 2.48, it follows that the forward and the return move swap the labels of the independent and the terminal disks. In that example, the return move has two independent disks, and hence is never proposed by Alg. 2.17 (`naive-avalanche`), so that $a \to b$ must be rejected also. Only avalanche moves with a single terminal disk can be accepted. This happens very rarely: avalanches usually gain breadth when they build up, and do not taper into a single disk.

Algorithm 2.17 (`naive-avalanche`) has a tiny acceptance rate for all but the smallest displacement vectors, and we thus need to generalize it, by allowing more than one independent disk to kick off the avalanche. For

```
procedure naive-avalanche
input {x₁,...,x_N}
k ← nran (1, N)
δ ← {ran (−δ, δ) , ran (−δ, δ)}
construct move (involving disks {k₁,...,k_M})
if (move has single terminal disk) then
    ⎰ for l = 1,...,M do
    ⎱   { x_{k_l} ← x_{k_l} + δ
output {x₁,...,x_N}
```
————

**Algorithm 2.17** naive-avalanche. Avalanche cluster algorithm for hard disks, with a low acceptance rate unless $|\delta|$ is small.

concreteness, we suppose that avalanches must be connected and that they must have a certain disk $l$ in common. Under this condition, there

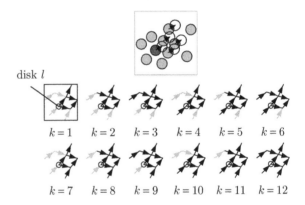

disk $l$

$k=1$  $k=2$  $k=3$  $k=4$  $k=5$  $k=6$

$k=7$  $k=8$  $k=9$  $k=10$  $k=11$  $k=12$

**Fig. 2.49** The move $a \to b$, and all avalanches containing the disk $l$ for a displacement $\delta$. The avalanche $k = 1$ realizes the move.

are now 12 connected avalanches containing disk $l$ (see Fig. 2.49). This means that the a priori probability of selecting the frame $k = 1$, rather than another one, is $1/12$. As in the study of the triangle algorithm, the use of a priori probabilities obliges us to analyze the return move. In the configuration $b$ of Fig. 2.48, with a return displacement $-\delta$, 10 connected avalanches contain disk $l$, of which one (the frame $k = 8$) realizes the return move (see Fig. 2.50). The a priori probability of selecting this return move is $1/10$. We thus arrive at

$$\mathcal{A}(a \to b) = \frac{1}{12} \quad \left\{ \begin{array}{c} \text{one of the 12 avalanches} \\ \text{in Fig. 2.49} \end{array} \right\},$$

$$\mathcal{A}(b \to a) = \frac{1}{10} \quad \left\{ \begin{array}{c} \text{one of the 10 avalanches} \\ \text{in Fig. 2.50} \end{array} \right\}.$$

These a priori probabilities must be entered into the detailed-balance condition

$$\underbrace{\mathcal{A}(a \to b)}_{\text{propose}} \underbrace{\mathcal{P}(a \to b)}_{\text{accept}} = \underbrace{\mathcal{A}(b \to a)}_{\text{propose}} \underbrace{\mathcal{P}(b \to a)}_{\text{accept}}.$$

Detailed balance is realized by use of the generalized Metropolis algorithm

$$\mathcal{P}(a \to b) = \min\left(1, \frac{12}{10}\right) = 1.$$

It follows that the move $a \to b$ in Fig. 2.48 must be accepted with probability 1.

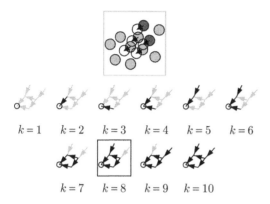

$k = 1$   $k = 2$   $k = 3$   $k = 4$   $k = 5$   $k = 6$

$k = 7$   $k = 8$   $k = 9$   $k = 10$

**Fig. 2.50** Return move $b \to a$ and all avalanches containing disk $l$ for a displacement $-\delta$. The avalanche $k = 8$ realizes the return move.

The algorithm we have discussed in this subsection needs to count avalanches, and to sample them. For small systems, this can be done by enumeration methods. It is unclear, however, whether a general efficient implementation exists for counting and sampling avalanches, because this problem is related to the task of enumerating the number of independent sets of a graph, a notoriously difficult problem. (The application to the special graphs that arise in our problem has not been studied.) We shall find a simpler solution in Subsection 2.5.2.

## 2.5.2 Hard-sphere cluster algorithm

The avalanche cluster algorithm of Subsection 2.5.1 suffers from an imbalance between forward and return moves because labels of independent and terminal disks are swapped between the two displacements, and because their numbers are not normally the same. In the present subsection, we show how to avoid imbalances, rejections, and the complicated calculations of Subsection 2.5.1 by use of a pivot: rather than displacing each particle by a constant vector $\delta$, we choose a symmetry operation that, when applied twice to the same disk, brings it back to the original position. For concreteness, we consider reflections about a vertical line, but other pivots (symmetry axes or reflection points) could also be used.

With periodic boundary conditions, we can scroll the whole system such that the pivot comes to lie in the center of the box (see Fig. 2.51).

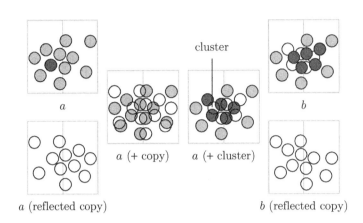

**Fig. 2.51** Hard-sphere cluster algorithm. Some particles are swapped between the original configuration and the copy, i.e., they exchange color.

Let us consider a copy of the original system with all disks reflected about the symmetry axis, together with the original configuration $a$. The original and the copy may be superimposed, and disks in the combined systems form connected clusters. We may pick one cluster and exchange in it the disks of the copy with those in the original configuration (see configuration $b$ in Fig. 2.51). For the return move, we may superimpose the configuration $b$ and a copy of $b$ obtained through the same reflection that was already used for $a$. We may pick a cluster which, by symmetry, makes us come back to the configuration $a$. The move $a \to b$ and the return move $b \to a$ satisfy the detailed-balance condition. All these transformations map the periodic simulation box onto itself, avoiding problems at the boundaries. Furthermore, ergodicity holds under the same conditions as for the local algorithm: any local move can be seen as a point reflection about the midpoint between the old and the new disk configuration. The basic limitation of the pivot cluster algorithm is that, for high covering density, almost all particles end up in the same cluster, and will be moved together. Flipping this cluster essentially reflects the whole system. Applications of this rejection-free method will be presented in later chapters.

To implement the pivot cluster algorithm, we need not work with an original configuration and its copy, as might be suggested by Fig. 2.51. After sampling the symmetry transformation (horizontal or vertical reflection line, pivot, etc.), we can work directly on particles (see Fig. 2.52). The transformation is applied to a first particle. From then on, we keep track (by keeping them in a "pocket") of all particles which still have to be moved (one at a time) in order to arrive at a legal configuration (see Alg. 2.18 (**hard-sphere-cluster**)). A single iteration of the pivot

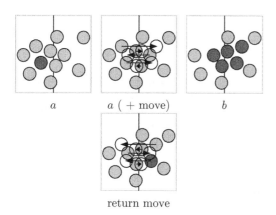

$a$      $a$ ( + move)      $b$

return move

**Fig. 2.52** Hard-sphere cluster algorithm without copies. There is one independent particle. Constrained particles can be moved one at a time.

cluster algorithm consists in all the operations until the pocket is empty. The inherent symmetry guarantees that the process terminates.

In conclusion, we have considered in this subsection a rejection-free cluster algorithm for hard spheres and related systems, which achieves perfect symmetry between the forward and the return move because the clusters in the two moves are the same. This algorithm does not work very well for equal disks or spheres, because the densities of interest (around the liquid–solid transition) are rather high, so that the clusters usually comprise almost the entire system. This algorithm has, however, been used in numerous other contexts, and generalized in many ways, as will be discussed in later chapters.

**procedure** hard-sphere-cluster
**input** $\{\mathbf{x}_1, \ldots, \mathbf{x}_N\}$
$k \leftarrow \mathbf{nran}\,(1, N)$
$\mathcal{P} \leftarrow \{k\}$ (the "pocket")
$\mathcal{A} \leftarrow \{1, \ldots, N\} \setminus \{k\}$ (other particles)
**while** $(\mathcal{P} \neq \{\})$ **do**
$\left\{\begin{array}{l} i \leftarrow \text{any element of } \mathcal{P} \\ \mathbf{x}_i \leftarrow T(\mathbf{x}_i) \\ \textbf{for } \forall \, j \, \in \, \mathcal{A} \textbf{ do} \\ \quad \left\{\begin{array}{l} \textbf{if } (i \text{ overlaps } j) \textbf{ then} \\ \quad \left\{\begin{array}{l} \mathcal{A} \leftarrow \mathcal{A} \setminus \{j\} \\ \mathcal{P} \leftarrow \mathcal{P} \cup \{j\} \end{array}\right. \end{array}\right. \\ \mathcal{P} \leftarrow \mathcal{P} \setminus \{i\} \end{array}\right.$
**output** $\{\mathbf{x}_1, \ldots, \mathbf{x}_N\}$

**Algorithm 2.18** hard-sphere-cluster. Cluster algorithm for $N$ hard spheres. $T$ is a random symmetry transformation.

# Exercises

### (Section 2.1)

(2.1) Implement Alg. 2.2 (`pair-time`) and incorporate it into a test program generating two random positions $\{\mathbf{x}_1, \mathbf{x}_2\}$ with $|\Delta_{\mathbf{x}}| > 2\sigma$, and two random velocities with $(\Delta_{\mathbf{x}} \cdot \Delta_{\mathbf{v}}) < 0$. Propagate both disks up to the time $t_{\text{pair}}$, if finite, and check that they indeed touch. Otherwise, if $t_{\text{pair}} = \infty$, propagate the disks up to their time of closest encounter and check that, then, $(\Delta_{\mathbf{x}} \cdot \Delta_{\mathbf{v}}) = 0$. Run this test program for at least $1 \times 10^7$ iterations.

(2.2) Show that Alg. 2.3 (`pair-collision`) is correct in the center-of-mass frame, but that it can also be used in the lab frame. Implement it and incorporate it into a test program generating two random positions $\{\mathbf{x}_1, \mathbf{x}_2\}$ with $|\Delta_{\mathbf{x}}| = 2\sigma$, and two random velocities $\{\mathbf{v}_1, \mathbf{v}_2\}$ with $(\Delta_{\mathbf{x}} \cdot \Delta_{\mathbf{v}}) < 0$. Check that the total momentum and energy are conserved (so that initial velocities $\mathbf{v}_{1,2}$ and final velocities $\mathbf{v}'_{1,2}$ satisfy $\mathbf{v}_1 + \mathbf{v}_2 = \mathbf{v}'_1 + \mathbf{v}'_2$ and $\mathbf{v}_1^2 + \mathbf{v}_2^2 = \mathbf{v}_1'^2 + \mathbf{v}_2'^2$). Run this test program for at least $1 \times 10^7$ iterations.

(2.3) (Uses Exerc. 2.1 and 2.2.) Implement Alg. 2.1 (`event-disks`) for disks in a square box without periodic boundary conditions. Start from a legal initial condition. If possible, handle the initial conditions as discussed in Exerc. 1.3. Stop the program at regular time-intervals (as in Alg. 2.4 (`event-disks(patch)`)). Generate a histogram of the projected density in one of the coordinates. In addition, generate histograms of velocities $v_x$ and of the absolute velocity $v = \sqrt{v_x^2 + v_y^2}$.

(2.4) Consider Sinai's system of two large disks in a square box with periodic boundary conditions ($L_x/4 < \sigma < L_x/2$). Show that in the reference frame of a stationary disk (remaining at position $\{0,0\}$), the center of the moving disk reflects off the stationary one as in geometric optics, with the incoming angle equal to the outgoing angle. Implement the time evolution of this system, with stops at regular time intervals. Compute the two-dimensional histogram of positions, $\pi(x, y)$, and determine from it the histogram of projected densities.

(2.5) Find out whether your programming language allows you to treat real variables and constants using different precision levels (such as single precision and double precision). Then, for real variables

$x = 1$ and $y = 2^{-k}$, compute $x + y$ in both cases, where $k \in \{\ldots, -2, -1, 0, 1, 2, \ldots\}$. Interpret the results of this basic numerical operation in the light of the discussion of numerical precision, in Subsection 2.1.2.

### (Section 2.2)

(2.6) Directly sample, using Alg. 2.7 (`direct-disks`), the positions of four disks in a square box without periodic boundary conditions, for different covering densities. Run it until you have data for a high-quality histogram of $x$-values (this determines the projected densities in $x$, compare with Fig. 2.9). If possible, confront this histogram to data from your own molecular dynamics simulation (see Exerc. 2.3), thus providing a simple "experimental" test of the equiprobability principle.

(2.7) Write a version of Alg. 2.7 (`direct-disks`) with periodic boundary conditions. First implement Alg. 2.5 (`box-it`) and Alg. 2.6 (`diff-vec`), and check them thoroughly. Verify correctness of the program by running it for Sinai's two-disk system: compute histograms $\pi(x_k, y_k)$ for the position $\mathbf{x}_k = \{x_k, y_k\}$ of disk $k$, and for the distance $\pi(\Delta_x, \Delta_y)$ between the two particles.

(2.8) Implement Alg. 2.9 (`markov-disks`) for four disks in a square box without periodic boundary conditions (use the same covering densities as in Exerc. 2.6). Start from a legal initial condition. If possible, implement initial conditions as in Exerc. 1.3. Generate a high-quality histogram of $x$-values. If possible, compare this histogram to the one obtained by molecular dynamics (see Exerc. 2.3), or by direct sampling (see Exerc. 2.6).

(2.9) Implement Alg. 2.8 (`direct-disks-any`), in order to determine the acceptance rate of Alg. 2.7 (`direct-disks`). Modify the algorithm, with the aim of avoiding use of histograms (which lose information). Sort the $N$ output samples for $\eta_{\max}$, such that $\eta_{\max,1} \leq \cdots \leq \eta_{\max,N}$. Determine the rejection rate of Alg. 2.7 (`direct-disks`) directly from the ordered vector $\{\eta_{\max,1}, \ldots, \eta_{\max,N}\}$.

(2.10) Implement Alg. 2.9 (`markov-disks`), with periodic boundary conditions, for four disks. If possible, use the subroutines tested in Exerc. 2.7. Start your simulation from a legal initial condition. Demonstrate explicitly that histograms of projected densities generated by Alg. 2.7 (direct-disks), Alg. 2.9, and by Alg. 2.1 agree for very long simulation times.

(2.11) Implement Alg. 2.9 (markov-disks) with periodic boundary conditions, as in Exerc. 2.10, but for a larger number of disks, in a box with aspect ratio $L_x/L_y = \sqrt{3}/2$. Set up a subroutine for generating initial conditions from a hexagonal arrangement. If possible, handle initial conditions as in Exerc. 1.3 (subsequent runs of the program start from the final output of a previous run). Run this program for a very large number of iterations, at various densities. How can you best convince yourself that the hard-disk system undergoes a liquid–solid phase transition?
NB: The grid/cell scheme of Subsection 2.4.1 need not be implemented.

**(Section 2.3)**

(2.12) Sample the gamma distribution $\Gamma_N(x)$ using the naive algorithm contained in Alg. 2.13 (direct-piston-particles). Likewise, implement Alg. 2.15 (gamma-cut). Generate histograms from long runs of both programs to check that the distribution sampled are indeed $\Gamma_N(x)$ and $\Gamma_N^{\mathrm{cut}}(x, x_{\mathrm{cut}})$. Histograms have a free parameter, the number of bins. For the same set of data, generate one histogram with very few bins and another one with very many bins, and discuss merits and disadvantages of each choice. Next, analyze data by sorting $\{x_1, \dots, x_N\}$ in ascending order $\{\tilde{x}_1, \dots, \tilde{x}_k, \dots, \tilde{x}_N\}$ (compare with Exerc. 2.9). Show that the plot of $k/N$ against $\tilde{x}_k$ can be compared directly with the integral of the considered distribution function, without any binning. Look up information about the Kolmogorov–Smirnov test, the standard statistical test for integrated probability distributions.

(2.13) Implement Alg. 2.11 (naive-piston-particles) and Alg. 2.12 (naive-piston-particles(patch)), and compare these Markov-chain programs to Alg. 2.13 (direct-piston-particles). Discuss whether outcomes should be identical, or whether small differences can be expected. Back up your conclusion with high-precision calculations of the mean box volume $\langle L \rangle$, and with histograms of $\pi(L)$ from all three programs. Generalize the direct-sampling algorithm to the case of two-dimensional hard disks with periodic boundary conditions (see Alg. 2.14 (direct-p-disks)). Plot the equation of

state (mean volume vs. pressure) to sufficient precision to see deviations from the ideal gas law.

(2.14) (Uses Exerc. 2.3.) Verify that Maxwell boundary conditions can be implemented with the sum of Gaussian random numbers, as in eqn (2.13), or alternatively by rescaling an exponentially distributed variable, as in Alg. 2.10 (maxwell-boundary). Incorporate Maxwell boundary conditions into a molecular dynamics simulation with Alg. 2.1 (event-disks) in a rectangular box of sides $\{L_x, L_y\}$, a tube with $L_x \gg L_y$ (see Exerc. 2.3). Set up regular reflection conditions on the horizontal walls (the sides of the tube), and Maxwell boundary conditions on the vertical walls (the ends of the tube). Make sure that positive $x$-velocities are generated at the left wall, and negative $x$-velocities at the other one. Let the two Maxwell conditions correspond to different temperatures, $T_{\mathrm{left}}$, and $T_{\mathrm{right}}$. Can you measure the temperature distribution along $x$ in the tube?

(2.15) In the piston-and-plate system of Subsection 2.3.2, the validity of the Boltzmann distribution was proved for a piston subject to a constant force (constant pressure). Prove the validity of the Boltzmann distribution for a more general piston energy

$$E(L) = L^\alpha$$

(earlier we used $\alpha = 1$). Specifically show that the piston is at position $L$, and at velocity $v$ with the Boltzmann probability

$$\pi(L, v)\, \mathrm{d}L\, \mathrm{d}v \propto \exp\left[-\beta E(L, v)\right] \mathrm{d}L\, \mathrm{d}v.$$

First determine the time of flight, and compute the density of state $\mathcal{N}(E)$ in the potential $h^\alpha$. Then show that at constant energy, each phase space element $\mathrm{d}L\, \mathrm{d}v$ appears with equal probability.
NB: A general formula for the time of flight follows from the conservation of energy $E = \frac{1}{2}v^2 + L^\alpha$:

$$\frac{\mathrm{d}L}{\mathrm{d}t} = \sqrt{2\left(E - L^\alpha\right)},$$

which can be integrated by separation of variables.

**(Section 2.4)**

(2.16) (Uses Exerc. 2.11.) Include Alg. 2.15 (gamma-cut) (see Exerc. 2.12) into a simulation of hard disks at a constant pressure. Use this program to compute the equation of state. Concentrate most of the computational effort at high pressure.

# References

Alder B., Wainwright T. E. (1957) Phase transition for a hard sphere system, *Journal of Chemical Physics* **27**, 1208–1209

Alder B. J., Wainwright T. E. (1962) Phase transition in elastic disks, *Physical Review* **127**, 359–361

Alder B. J., Wainwright T. E. (1970) Decay of the velocity autocorrelation function, *Physical Review A* **1**, 18–21

Dress C., Krauth W. (1995) Cluster algorithm for hard spheres and related systems, *Journal of Physics A* **28**, L597–L601

Hoover W. G., Ree F. H. (1968) Melting transition and communal entropy for hard spheres, *Journal of Chemical Physics* **49**, 3609–3617

Lee J. Y., Strandburg K. J. (1992) 1st-order melting transition of the hard-disk system, *Physical Review B* **46**, 11190–11193

Mermin N. D., Wagner H. (1966) Absence of ferromagnetism or antiferromagnetism in one- or two-dimensional isotropic Heisenberg models, *Physical Review Letters* **17**, 1133–1136

Metropolis N., Rosenbluth A. W., Rosenbluth M. N., Teller A. H., Teller E. (1953) Equation of state calculations by fast computing machines, *Journal of Chemical Physics* **21**, 1087–1092

Pomeau Y., Résibois P. (1975) Time dependent correlation functions and mode–mode coupling theory, *Physics Reports* **19**, 63–139

Simanyi N. (2003) Proof of the Boltzmann–Sinai ergodic hypothesis for typical hard disk systems, *Inventiones Mathematicae* **154**, 123–178

Simanyi N. (2004) Proof of the ergodic hypothesis for typical hard ball systems, *Annales de l'Institut Henri Poincaré* **5**, 203–233

Sinai Y. G. (1970) Dynamical systems with elastic reflections, *Russian Mathematical Surveys* **25**, 137–189

Thirring W. (1978) *A Course in Mathematical Physics. 1. Classical Dynamical Systems*, Springer, New York

Wood W. W. (1968) Monte Carlo calculations for hard disks in the isothermal–isobaric ensemble, *Journal of Chemical Physics* **48**, 415–434

# Density matrices and path integrals

**3**

In this chapter, we continue our parallel exploration of statistical and computational physics, but now move to the field of quantum mechanics, where the density matrix generalizes the classical Boltzmann distribution. The density matrix constructs the quantum statistical probabilities from their two origins: the quantum mechanical wave functions and the Boltzmann probabilities of the energy levels. The density matrix can thus be defined in terms of the complete solution of the quantum problem (wave functions, energies), but all this information is available only in simple examples, such as the free particle or the harmonic oscillator.

A simple general expression exists for the density matrix only at high temperature. However, a systematic convolution procedure allows us to compute the density matrix at any given temperature from a product of two density matrices at higher temperature. By iteration, we reach the density matrix at any temperature from the high-temperature limit. We shall use this procedure, matrix squaring, to compute the quantum-statistical probabilities for particles in an external potential.

The convolution procedure, and the connection it establishes between classical and quantum systems, is the basis of the Feynman path integral, which really opens up the field of finite-temperature quantum statistics to computation. We shall learn about path integrals in more and more complicated settings. As an example of interacting particles, we shall come back to the case of hard spheres, which model quantum gases and quantum liquids equally well. The path integral allows us to conceive of simulations in interacting quantum systems with great ease. This will be studied in the present chapter for the case of two hard spheres, before taking up the many-body problem in Chapter 4.

Classical particles are related to points, the pebbles of our first chapter. Analogously, quantum particles are related to one-dimensional objects, the aforementioned paths, which have important geometric properties. Path integrals, like other concepts of statistical mechanics, have spread to areas outside their field of origin, and even beyond physics. They are for example used to describe stock indices in financial mathematics. The final section of this chapter introduces the geometry of paths and related objects. This external look at our subject will foster our understanding of quantum physics and of the path integral. We shall reach analytical and algorithmic insights complementing those of earlier sections of the chapter.

A quantum particle in a harmonic potential is described by energies and wave functions that we know exactly (see Fig. 3.1). At zero temperature, the particle is in the ground state; it can be found with high probability only where the ground-state wave function differs markedly from zero. At finite temperatures, the particle is spread over more states, and over a wider range of $x$-values. In this chapter, we discuss exactly how this works for a particle in a harmonic potential and for more difficult systems. We also learn how to do quantum statistics if we ignore everything about energies and wave functions.

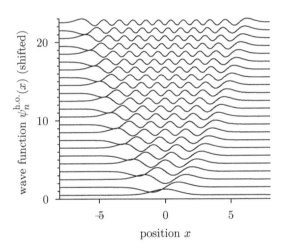

**Fig. 3.1** Harmonic-oscillator wave functions $\psi_n^{\text{h.o.}}(x)$ shifted by $E_n$ (from Alg. 3.1 (`harmonic-wavefunction`)).

# 3.1 Density matrices

Quantum systems are described by wave functions and eigenvalues, solutions of the Schrödinger equation. We shall show for the one-dimensional harmonic oscillator how exactly the probabilities of quantum physics combine with the statistical weights in the Boltzmann distribution, before moving on to more general problems.

## 3.1.1 The quantum harmonic oscillator

The one-dimensional quantum mechanical harmonic oscillator, which consists of a particle of mass $m$ in a potential

$$V(x) = \frac{1}{2}m\omega^2 x^2,$$

is governed by the Schrödinger equation

$$H\psi_n^{\text{h.o.}} = \left(-\frac{\hbar^2}{2m}\frac{\partial^2}{\partial x^2} + \frac{1}{2}m\omega^2 x^2\right)\psi_n^{\text{h.o.}} = E_n\psi_n^{\text{h.o.}}. \qquad (3.1)$$

---

**procedure** `harmonic-wavefunction`
**input** $x$
$\psi_{-1}^{\text{h.o.}}(x) \leftarrow 0$ (unphysical, starts recursion)
$\psi_0^{\text{h.o.}}(x) \leftarrow \pi^{-1/4}\exp\left(-x^2/2\right)$ (ground state)
**for** $n = 1, 2, \ldots$ **do**
$\left\{\; \psi_n^{\text{h.o.}}(x) \leftarrow \sqrt{\frac{2}{n}}x\psi_{n-1}^{\text{h.o.}}(x) - \sqrt{\frac{n-1}{n}}\psi_{n-2}^{\text{h.o.}}(x)\right.$
**output** $\{\psi_0^{\text{h.o.}}(x), \psi_1^{\text{h.o.}}(x), \ldots\}$

---

**Algorithm 3.1** `harmonic-wavefunction`. Eigenfunctions of the one-dimensional harmonic oscillator (with $\hbar = m = \omega = 1$).

In general, the wave functions $\{\psi_0, \psi_1, \ldots\}$ satisfy a completeness condition

$$\sum_{n=0}^{\infty} \psi_n^*(x)\psi_n(y) = \delta(x-y),$$

where $\delta(x-y)$ is the Dirac $\delta$-function, and form an orthonormal set:

$$\int_{-\infty}^{\infty} \mathrm{d}x \; \psi_n^*(x)\psi_m(x) = \delta_{nm} = \begin{cases} 1 & \text{if } n = m \\ 0 & \text{otherwise} \end{cases}, \qquad (3.2)$$

where $\delta_{nm}$ is the discrete Kronecker $\delta$-function. The wave functions of the harmonic oscillator can be computed recursively[1] (see Alg. 3.1 (`harmonic-wavefunction`)). We can easily write down the first few of them, and verify that they are indeed normalized and mutually orthogonal, and that $\psi_n^{\text{h.o.}}$ satisfies the above Schrödinger equation (for $m = \hbar = \omega = 1$) with $E_n = n + \frac{1}{2}$.

---

[1]In most formulas in this chapter, we use units such that $\hbar = m = \omega = 1$.

In thermal equilibrium, a quantum particle occupies an energy state $n$ with a Boltzmann probability proportional to $\mathrm{e}^{-\beta E_n}$, and the partition function is therefore

$$Z^{\mathrm{h.o.}}(\beta) = \sum_{n=0}^{\infty} \mathrm{e}^{-\beta E_n} = \mathrm{e}^{-\beta/2} + \mathrm{e}^{-3\beta/2} + \mathrm{e}^{-5\beta/2} + \cdots$$

$$= \mathrm{e}^{-\beta/2} \left( \frac{1}{1 - \mathrm{e}^{-\beta}} \right) = \frac{1}{\mathrm{e}^{\beta/2} - \mathrm{e}^{-\beta/2}} = \frac{1}{2 \sinh{(\beta/2)}}. \quad (3.3)$$

The complete thermodynamics of the harmonic oscillator follows from eqn (3.3). The normalized probability of being in energy level $n$ is

$$\left\{ \begin{array}{c} \text{probability of being} \\ \text{in energy level } n \end{array} \right\} = \frac{1}{Z}\, \mathrm{e}^{-\beta E_n}.$$

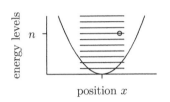

**Fig. 3.2** A quantum particle at position $x$, in energy level $n$. The density matrix only retains information about positions.

When it is in energy level $n$ (see Fig. 3.2), a quantum system is at a position $x$ with probability $\psi_n^*(x)\psi_n(x)$. (The asterisk stands for the complex conjugate; for the real-valued wave functions used in most of this chapter, $\psi^* = \psi$.) The probability of being in level $n$ at position $x$ is

$$\left\{ \begin{array}{c} \text{probability of being} \\ \text{in energy level } n \\ \text{at position } x \end{array} \right\} = \frac{1}{Z}\, \mathrm{e}^{-\beta E_n} \psi_n(x)\psi_n^*(x). \quad (3.4)$$

This expression generalizes the Boltzmann distribution to quantum physics. However, the energy levels and wave functions are generally unknown for complicated quantum systems, and eqn (3.4) is not useful for practical computations. To make progress, we discard the information about the energy levels and consider the (diagonal) density matrix

$$\pi(x) = \left\{ \begin{array}{c} \text{probability of being} \\ \text{at position } x \end{array} \right\} \propto \rho(x, x, \beta) = \sum_n \mathrm{e}^{-\beta E_n} \psi_n(x)\psi_n^*(x),$$

as well as a more general object, the nondiagonal density matrix (in the position representation)

$$\left\{ \begin{array}{c} \text{density} \\ \text{matrix} \end{array} \right\} : \rho(x, x', \beta) = \sum_n \psi_n(x) \mathrm{e}^{-\beta E_n} \psi_n^*(x'), \quad (3.5)$$

which is the central object of quantum statistics. For example, the partition function $Z(\beta)$ is the trace of the density matrix, i.e. the sum or the integral of its diagonal terms:

$$Z(\beta) = \mathrm{Tr}\, \rho = \int \mathrm{d}x\, \rho(x, x, \beta). \quad (3.6)$$

We shall compute the density matrix in different settings, and often without knowing the eigenfunctions and eigenvalues. For the case of the harmonic oscillator, however, we have all that it takes to compute the density matrix from our solution of the Schrödinger equation (see Alg. 3.2 (**harmonic-density**) and Fig. 3.3). The output of this program will allow us to check less basic approaches.

**procedure** `harmonic-density`
**input** $\{\psi_0^{\text{h.o.}}(x),\dots,\psi_N^{\text{h.o.}}(x)\}$ (from Alg. 3.1 (`harmonic-wavefunction`))
**input** $\{\psi_0^{\text{h.o.}}(x'),\dots,\psi_N^{\text{h.o.}}(x')\}$
**input** $\{E_n=n+\frac{1}{2}\}$
$\rho^{\text{h.o.}}(x,x',\beta)\leftarrow 0$
**for** $n=0,\dots,N$ **do**
$\quad\{\ \rho^{\text{h.o.}}(x,x',\beta)\leftarrow\rho^{\text{h.o.}}(x,x',\beta)+\psi_n^{\text{h.o.}}(x)\psi_n^{\text{h.o.}}(x')e^{-\beta E_n}$
**output** $\{\rho^{\text{h.o.}}(x,x',\beta)\}$
———

**Algorithm 3.2** `harmonic-density`. Density matrix for the harmonic oscillator obtained from the lowest-energy wave functions (see eqn (3.5)).

## 3.1.2 Free density matrix

We move on to our first analytic calculation, a prerequisite for further developments: the density matrix of a free particle, with the Hamiltonian

$$H^{\text{free}}\psi = -\frac{1}{2}\frac{\partial^2}{\partial x^2}\psi = E\psi.$$

We put the particle in a box of length $L$ with periodic boundary conditions, that is, a torus. The solutions of the Schrödinger equation in a periodic box are plane waves that are periodic in $L$:

$$\psi_n^{\text{a}}(x) = \sqrt{\frac{2}{L}}\sin\left(2n\pi\frac{x}{L}\right)\ (n=1,2,\dots), \tag{3.7}$$

$$\psi_n^{\text{s}}(x) = \sqrt{\frac{2}{L}}\cos\left(2n\pi\frac{x}{L}\right)\ (n=0,1,\dots) \tag{3.8}$$

(see Fig. 3.4), where the superscripts denote wave functions that are antisymmetric and symmetric with respect to the center of the interval $[0,L]$. Equivalently, we can use complex wave functions

$$\psi_n^{\text{per},L}(x) = \sqrt{\frac{1}{L}}\exp\left(\mathrm{i}2n\pi\frac{x}{L}\right)\ (n=-\infty,\dots,\infty), \tag{3.9}$$

$$E_n = \frac{2n^2\pi^2}{L^2}, \tag{3.10}$$

which give

$$\rho^{\text{per},L}(x,x',\beta) = \sum_n \psi_n^{\text{per},L}(x)e^{-\beta E_n}\left[\psi_n^{\text{per},L}(x')\right]^*$$

$$= \frac{1}{L}\sum_{n=-\infty}^{\infty}\exp\left[\mathrm{i}2\pi n\frac{x-x'}{L}\right]\exp\left(-\frac{\beta 2n^2\pi^2}{L^2}\right). \tag{3.11}$$

We now let $L$ tend to infinity (the exact expression for finite $L$ is discussed in Subsection 3.1.3). In this limit, we can transform the sum in eqn (3.11) into an integral. It is best to introduce a dummy parameter

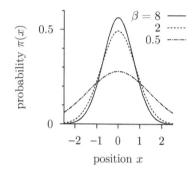

**Fig. 3.3** Probability to be at position $x$, $\pi(x) = \rho^{\text{h.o.}}(x,x,\beta)/Z$ (from Alg. 3.2 (`harmonic-density`); see also eqn (3.39)).

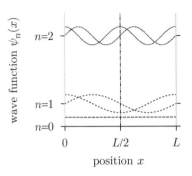

**Fig. 3.4** Wave functions of a one-dimensional free particle in a torus of length $L$ (shifted; see eqn (3.8)).

$\Delta_n = 1$, the difference between two successive $n$-values:

$$\rho^{\text{per},L}(x, x', \beta) = \frac{1}{L} \sum_{n=-\infty}^{\infty} \overbrace{\Delta_n}^{=1} \exp\left[\text{i}2n\pi \frac{x - x'}{L}\right] \exp\left(-\frac{\beta 2 n^2 \pi^2}{L^2}\right).$$

Changing variables from $n$ to $\lambda$, where $\lambda = 2n\pi/L$, and thus $\Delta_\lambda = 2\pi\Delta_n/L$, gives the term-by-term equivalent sum

$$\rho^{\text{per},L}(x, x', \beta) = \frac{1}{2\pi} \sum_{\lambda=\ldots,-\frac{2\pi}{L},0,\frac{2\pi}{L},\ldots} \Delta_\lambda \exp\left[\text{i}\lambda(x - x')\right] \exp\left(-\frac{\beta}{2}\lambda^2\right)$$

$$\xrightarrow[L\to\infty]{} \frac{1}{2\pi} \int_{-\infty}^{\infty} \text{d}\lambda \, \exp\left[\text{i}\lambda(x - x')\right] \exp\left(-\frac{\beta}{2}\lambda^2\right).$$

We use

$$\int_{-\infty}^{\infty} \text{d}\lambda \, \exp\left(-\frac{1}{2}\frac{\lambda^2}{\sigma^2} \pm \lambda\tilde{c}\right)$$

$$= \sqrt{2\pi}\sigma \exp\left(\frac{1}{2}\tilde{c}^2\sigma^2\right), \quad (3.12)$$

which follows, when we take $\tilde{c} = c/\sigma^2$, from the Gaussian integral

$$\int_{-\infty}^{\infty} \frac{\text{d}\lambda}{\sqrt{2\pi}\sigma} \exp\left[-\frac{1}{2}\frac{(\lambda \pm c)^2}{\sigma^2}\right] = 1.$$

This Gaussian integral can be evaluated (see eqn (3.12)). We arrive at the free density matrix, the periodic density matrix in the limit of an infinite torus:

$$\rho^{\text{free}}(x, x', \beta) = \sqrt{\frac{m}{2\pi\hbar^2\beta}} \exp\left[-\frac{m(x - x')^2}{2\hbar^2\beta}\right], \quad (3.13)$$

where we have reinserted the Planck constant and the particle mass. In the limit of high temperature, $\rho^{\text{free}}(x, x', \beta)$ is infinitely peaked at $x = x'$, and it continues to satisfy $\int_{-\infty}^{\infty} \text{d}x \, \rho^{\text{free}}(x, x', \beta) = 1$. This means that it realizes the Dirac $\delta$-function:

$$\lim_{\beta\to 0} \rho^{\text{free}}(x, x', \beta) \to \delta(x - x'). \quad (3.14)$$

For the perturbation theory of Subsection 3.2.1, we need the fact that the density matrix is generally given by the operator

$$\rho = \text{e}^{-\beta H} = 1 - \beta H + \frac{1}{2}\beta^2 H^2 - \cdots. \quad (3.15)$$

The matrix elements of $H$ in an arbitrary basis are $H_{kl} = \langle k|H|l \rangle$, and the matrix elements of $\rho$ are

$$\langle k|\rho|l \rangle = \langle k|e^{-\beta H}|l \rangle = \delta_{kl} - \beta H_{kl} + \frac{1}{2}\beta^2 \sum_n H_{kn} H_{nl} - \cdots .$$

Knowing the density matrix in any basis allows us to compute $\rho(x, x', \beta)$, i.e. the density matrix in the position representation:

$$\rho(x, x', \beta) = \sum_{kl} \underbrace{\langle x|k \rangle}_{\psi_k(x)} \langle k|\rho|l \rangle \underbrace{\langle l|x' \rangle}_{\psi_l^*(x')} .$$

This equation reduces to eqn (3.5) for the diagonal basis where $H_{nm} = E_n \delta_{nm}$ and $H_{nm}^2 = E_n^2 \delta_{nm}$, etc.

### 3.1.3   Density matrices for a box

The density matrix $\rho(x, x', \beta)$ is as fundamental as the Boltzmann weight $e^{-\beta E}$ of classical physics, but up to now we know it analytically only for the free particle (see eqn (3.13)). In the present subsection, we determine the density matrix of a free particle in a finite box, both with walls and with periodic boundary conditions. The two results will be needed in later calculations. For efficient simulations of a finite quantum system, say with periodic boundary conditions, it is better to start from the analytic solution of the noninteracting particle in a periodic box rather than from the free density matrix in the thermodynamic limit. (For the analytic expression of the density matrix for the harmonic oscillator, see Subsection 3.2.2.)

We first finish the calculation of the density matrix for a noninteracting particle with periodic boundary conditions, which we have evaluated only in the limit $L \to \infty$. We arrived in eqn (3.11) at the exact formula

$$\rho^{\mathrm{per},L}(x, x', \beta) = \sum_{n=-\infty}^{\infty} \psi_n^{\mathrm{per},L}(x) e^{-\beta E_n} \left[\psi_n^{\mathrm{per},L}(x')\right]^*$$

$$= \frac{1}{L} \sum_{n=-\infty}^{\infty} \underbrace{\exp\left[\mathrm{i}2n\pi\frac{x-x'}{L}\right] \exp\left(-\frac{\beta n^2 \pi^2}{L^2}\right)}_{g(n)}. \quad (3.16)$$

This infinite sum can be transformed using the Poisson sum formula

$$\sum_{n=-\infty}^{\infty} g(n) = \sum_{w=-\infty}^{\infty} \int_{-\infty}^{\infty} \mathrm{d}\phi\, g(\phi) e^{\mathrm{i}2\pi w\phi} \quad (3.17)$$

and the Gaussian integral rule (eqn (3.12)), to obtain

$$
\rho^{\mathrm{per},L}\left(x, x', \beta\right)
$$
$$
= \sum_{w=-\infty}^{\infty} \int_{-\infty}^{\infty} \mathrm{d}\phi \ \exp\left(\mathrm{i}2\pi\phi\frac{x - x' + Lw}{L}\right) \exp\left(-\frac{\beta\pi^2\phi^2}{L^2}\right)
$$
$$
= \sqrt{\frac{1}{2\pi\beta}} \sum_{w=-\infty}^{\infty} \exp\left[-\frac{(x - x' - wL)^2}{2\beta}\right]
$$
$$
= \sum_{w=-\infty}^{\infty} \rho^{\mathrm{free}}\left(x, x' + wL, \beta\right). \quad (3.18)
$$

The index $w$ stands for the winding number in the context of periodic boundary conditions (see Subsection 2.1.4). The Poisson sum formula (eqn (3.17)) is itself derived as follows. For a function $g(\phi)$ that decays rapidly to zero at infinity, the periodic sum function $G(\phi) = \sum_{k=-\infty}^{\infty} g(\phi + k)$ can be expanded into a Fourier series:

$$
G(\phi) = \underbrace{\sum_{k=-\infty}^{\infty} g(\phi + k)}_{\text{periodic in } [0,1]} = \sum_{w=-\infty}^{\infty} c_w \mathrm{e}^{\mathrm{i}2\pi w\phi}.
$$

The Fourier coefficients $c_w = \int_0^1 \mathrm{d}\phi \ G(\phi)\mathrm{e}^{-\mathrm{i}2\pi w\phi}$ can be written as

$$
c_w = \int_0^1 \mathrm{d}\phi \sum_{k=-\infty}^{\infty} g(\phi + k)\mathrm{e}^{-\mathrm{i}2\pi w\phi} = \int_{-\infty}^{\infty} \mathrm{d}\phi \ g(\phi)\mathrm{e}^{-\mathrm{i}2\pi w\phi}.
$$

For $\phi = 0$, we obtain $G(0) = \sum_w c_w$, which gives eqn (3.17).

From eqn (3.18), we obtain the diagonal density matrix for a box with periodic boundary conditions as a sum over diagonal and nondiagonal free-particle density matrices:

$$
\rho^{\mathrm{per},L}\left(x, x, \beta\right) = \sqrt{\frac{1}{2\pi\beta}} \left\{1 + \exp\left(-\frac{L^2}{2\beta}\right) + \exp\left[-\frac{(2L)^2}{2\beta}\right] + \cdots\right\}.
$$

Using eqn (3.6), that is, integrating from 0 to $L$, we find a new expression for the partition function (see eqn (3.6)), differing in inspiration from the sum over energy eigenvalues but equivalent to it in outcome:

$$
Z^{\mathrm{per},L} = \underbrace{\sum_{n=-\infty}^{\infty} \exp\left(-\beta\frac{2n^2\pi^2}{L^2}\right)}_{\text{sum over energy eigenvalues}} = \underbrace{\frac{L}{\sqrt{2\pi\beta}} \sum_{w=-\infty}^{\infty} \exp\left(-\frac{w^2 L^2}{2\beta}\right)}_{\text{sum over winding numbers}}. \quad (3.19)
$$

We next determine the density matrix of a free particle in a box of length $L$ with hard walls rather than with periodic boundary conditions. Again, the free Hamiltonian is solved by plane waves, but they must vanish at $x = 0$ and at $x = L$:

$$
\psi_n^{\mathrm{box},[0,L]}(x) = \sqrt{\frac{2}{L}} \sin\left(n\pi\frac{x}{L}\right) \quad (n = 1,\ldots,\infty) \quad (3.20)
$$

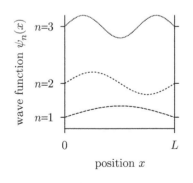

**Fig. 3.5** Eigenfunctions of a one-dimensional free particle in a box (shifted); (see eqn (3.20), and compare with Fig. 3.4).

(see Fig. 3.5). The normalized and mutually orthogonal sine functions in eqn (3.20) have a periodicity $2L$, not $L$, unlike the wave functions in a box with periodic boundary conditions.

Using the plane waves in eqn (3.20) and their energies $E_n = \frac{1}{2}(n\pi/L)^2$, we find the following for the density matrix in the box:

$$\rho^{\text{box},[0,\,L]}(x, x', \beta)$$

$$= \frac{2}{L} \sum_{n=1}^{\infty} \sin\left(\frac{\pi}{L}nx\right) \exp\left(-\beta\frac{\pi^2 n^2}{2L^2}\right) \sin\left(\frac{\pi}{L}nx'\right)$$

$$= \frac{1}{L} \sum_{n=-\infty}^{\infty} \sin\left(n\pi\frac{x}{L}\right) \exp\left(-\beta\frac{\pi^2 n^2}{2L^2}\right) \sin\left(n\pi\frac{x'}{L}\right). \quad (3.21)$$

We can write the product of the two sine functions as

$$\sin\left(n\pi\frac{x}{L}\right) \sin\left(n\pi\frac{x'}{L}\right)$$

$$= \frac{1}{4}\left[\exp\left(in\pi\frac{x - x'}{L}\right) + \exp\left(-in\pi\frac{x - x'}{L}\right)\right.$$

$$\left. - \exp\left(in\pi\frac{x + x'}{L}\right) - \exp\left(-in\pi\frac{x + x'}{L}\right)\right]. \quad (3.22)$$

Using eqn (3.22) in eqn (3.21), and comparing the result with the formula for the periodic density matrix in eqn (3.11), which is itself expressed in terms of the free density matrix, we obtain

$$\rho^{\text{box},[0,\,L]}(x, x', \beta) = \rho^{\text{per},2L}(x, x', \beta) - \rho^{\text{per},2L}(x, -x', \beta)$$

$$= \sum_{w=-\infty}^{\infty} \left[\rho^{\text{free}}(x, x' + 2wL, \beta) - \rho^{\text{free}}(x, -x' + 2wL, \beta)\right]. \quad (3.23)$$

This intriguing sum over winding numbers is put together from terms different from those of eqn (3.21). It is, however, equivalent, as we can easily check in an example (see Fig. 3.6).

In conclusion, we have derived in this subsection the density matrix of a free particle in a box either with periodic boundary conditions or with walls. The calculations were done the hard way—explicit mathematics—using the Poisson sum formula and the representation of products of sine functions in terms of exponentials. The final formulas in eqns (3.18) and (3.23) can also be derived more intuitively, using path integrals, and they are more generally valid (see Subsection 3.3.3).

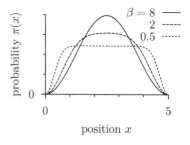

**Fig. 3.6** Probability of being at position $x$ for a free particle in a box with walls (from eqn (3.21) or eqn (3.23), with $L = 5$).

## 3.1.4 Density matrix in a rotating box

In Subsection 3.1.3, we determined the density matrix of a free quantum particle in a box with periodic boundary conditions. We now discuss the physical content of boundary conditions and the related counter-intuitive behavior of a quantum system under slow rotations. Our discussion is a prerequisite for the treatment of superfluidity in quantum liquids (see Subsection 4.2.6), but also for superconductivity in electronic systems, a subject beyond the scope of this book.

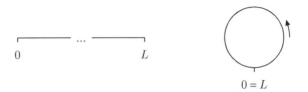

**Fig. 3.7** A one-dimensional quantum system on a line of length $L$ (*left*), and rolled up into a slowly rotating ring (*right*).

Let us imagine for a moment a complicated quantum system, for example a thin wire rolled up into a closed circle of circumference $L$, or a circular tube filled with a quantum liquid, ideally $^4\text{He}$ (see Fig. 3.7). Both systems are described by wave functions with periodic boundary conditions $\psi_n(0) = \psi_n(L)$. More precisely, the wave functions in the lab system for $N$ conduction electrons or $N$ helium atoms satisfy

$$\psi_n^{\text{lab}}(x_1, \ldots, x_k + L, \ldots, x_N) = \psi_n^{\text{lab}}(x_1, \ldots, x_k, \ldots, x_N). \quad (3.24)$$

These boundary conditions for the wave functions in the laboratory frame continue to hold if the ring rotates with velocity $v$, because a quantum system—even if parts of it are moving—must be described everywhere by a single-valued wave function. The rotating system can be described in the laboratory reference frame using wave functions $\psi_n^{\text{lab}}(x, t)$ and the time-dependent Hamiltonian $H^{\text{lab}}(t)$, which represents the rotating crystal lattice or the rough surface of the tube enclosing the liquid.

The rotating system is more conveniently described in the corotating reference frame, using coordinates $x^{\text{rot}}$ rather than the laboratory coordinates $x^{\text{lab}}$ (see Table 3.1). In this reference frame, the crystal lattice or the container walls are stationary, so that the Hamiltonian $H^{\text{rot}}$ is time independent. For very slow rotations, we can furthermore neglect centripetal forces and also magnetic fields generated by slowly moving charges. This implies that the Hamiltonian $H^{\text{rot}}$ (with coordinates $x^{\text{rot}}$) is the same as the laboratory Hamiltonian $H^{\text{lab}}$ at $v = 0$. However, we shall see that the boundary conditions on the corotating wave functions $\psi^{\text{rot}}$ are nontrivial.

To discuss this point, we now switch back from the complicated quantum systems to a free particle on a rotating ring, described by a Hamiltonian

$$H^{\text{rot}} = -\frac{1}{2}\frac{\partial^2}{(\partial x^{\text{rot}})^2}.$$

We shall go through the analysis of $H^{\text{rot}}$, but keep in mind that the very distinction between rotating and laboratory frames is problematic for the

**Table 3.1** Galilei transformation for a quantum system. $E^{\text{rot}}$, $p_n^{\text{rot}}$, and $x^{\text{rot}}$ are defined in the moving reference frame.

| Reference frame | |
|---|---|
| Lab. | Rot. |
| $x^{\text{lab}} = x^{\text{rot}} + vt$ | $x^{\text{rot}}$ |
| $p_n^{\text{lab}} = p_n^{\text{rot}} + mv$ | $p_n^{\text{rot}}$ |
| $E_n^{\text{lab}} = \frac{1}{2m}(p_n^{\text{lab}})^2$   $E_n^{\text{rot}} = \frac{1}{2m}(p_n^{\text{rot}})^2$ | |

noninteracting system, because of the missing contact with the crystal lattice or the container boundaries. Strictly speaking, the distinction is meaningful for a noninteracting system only because the wave functions of the interacting system can be expanded in a plane-wave basis. In the rotating system, plane waves can be written as

$$\psi_n^{\mathrm{rot}}(x^{\mathrm{rot}}) = \exp\left(ip^{\mathrm{rot}}x^{\mathrm{rot}}\right).$$

This same plane wave can also be written, at all times, in the laboratory frame, where it must be periodic (see eqn (3.24)). The momentum of the plane wave in the laboratory system is related to that of the rotating system by the Galilei transform of Table 3.1.

$$p_n^{\mathrm{lab}} = \frac{2\pi}{L}n \implies p_n^{\mathrm{rot}} = \frac{2\pi}{L}n - mv \quad (n = -\infty, \dots, \infty). \quad (3.25)$$

It follows from the Galilei transform of momenta and energies that the partition function $Z^{\mathrm{rot}}(\beta)$ in the rotating reference frame is

$$Z^{\mathrm{rot}}(\beta) = \sum_{n=-\infty}^{\infty} \exp\left[-\beta E_n^{\mathrm{rot}}\right] = \sum_{-\infty}^{\infty} \exp\left[-\frac{\beta}{2}\left(-v + 2n\pi/L\right)^2\right].$$

In the rotating reference frame, each plane wave with momentum $p_n^{\mathrm{rot}}$ contributes velocity $p_n^{\mathrm{rot}}/m$. The mean velocity measured in the rotating reference frame is

$$\left\langle v^{\mathrm{rot}}(\beta) \right\rangle = \frac{1}{mZ^{\mathrm{rot}}(\beta)} \sum_{n=-\infty}^{\infty} p_n^{\mathrm{rot}} \exp\left(-\beta E_n^{\mathrm{rot}}\right), \quad (3.26)$$

with energies and momenta given by eqn (3.25) that satisfy $E_n^{\mathrm{rot}} = (p_n^{\mathrm{rot}})^2/(2m)$. In the limit of zero temperature, only the lowest-lying state contributes to this rotating reference-frame particle velocity. The momentum of this state is generally nonzero, because the lowest energy state, the $n = 0$ state at very small rotation, $E_0^{\mathrm{rot}}$, has nonzero momentum (it corresponds to particle velocity $-v$). The particle velocity in the laboratory frame is

$$\left\langle v^{\mathrm{lab}} \right\rangle = \left\langle v^{\mathrm{rot}} \right\rangle + v. \quad (3.27)$$

At low temperature, this particle velocity differs from the velocity of the ring (see Fig. 3.8).

The particle velocity in a ring rotating with velocity $v$ will now be obtained within the framework of density matrices, similarly to the way we obtained the density matrix for a stationary ring, in Subsection 3.1.3. Writing $x$ for $x^{\mathrm{rot}}$ (they are the same at time $t = 0$, and we do not have to compare wave functions at different times),

$$\psi_n^{\mathrm{rot}}(x) = \sqrt{\frac{1}{L}} \exp\left(ip_n^{\mathrm{rot}}x\right) \quad (n = -\infty, \dots, \infty),$$

$$E_n^{\mathrm{rot}} = \frac{(p_n^{\mathrm{rot}})^2}{2m}.$$

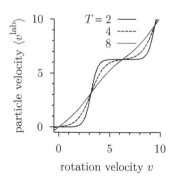

**Fig. 3.8** Mean lab-frame velocity of a particle in a one-dimensional ring (from eqns (3.26) and (3.27) with $L = m = 1$).

The density matrix at positions $x$ and $x'$ in the rotating system is:

$$\rho^{\text{rot}}(x, x', \beta) = \sum_{n=-\infty}^{\infty} e^{-\beta E_n^{\text{rot}}} \psi_n^{\text{rot}}(x) \left[\psi_n^{\text{rot}}(x')\right]^*$$

$$= \frac{1}{L} \sum_{n=-\infty}^{\infty} \underbrace{\exp\left[-\frac{\beta}{2}\left(p_n^{\text{rot}}\right)^2\right] \exp\left[ip_n^{\text{rot}}(x - x')\right]}_{g(n), \text{ compare with eqn (3.16)}}$$

$$= \sum_{w=-\infty}^{\infty} \int \frac{\mathrm{d}\phi}{L} \exp\left[-\frac{\beta}{2}\left(\frac{2\pi\phi}{L} - v\right)^2 + i\left(\frac{2\pi\phi}{L} - v\right)(x - x') + i2\pi w\phi\right],$$

where we have again used the Poisson sum formula (eqn (3.17)). Setting $\phi' = 2\pi\phi/L - v$ and changing $w$ into $-w$, we reach

$$\rho^{\text{rot}}(x, x', \beta) = \sum_{w=-\infty}^{\infty} e^{-iLwv} \int_{-\infty}^{\infty} \frac{\mathrm{d}\phi}{2\pi} \exp\left[-\frac{\beta}{2}\phi'^2 + i\phi'(x - x' - Lw)\right]$$

$$= \sum_{w=-\infty}^{\infty} e^{-iLwv} \rho^{\text{free}}(x, x' + Lw, \beta).$$

Integrating the diagonal density matrix over all the positions $x$, from 0 to $L$, yields the partition function in the rotating system:

$$Z^{\text{rot}}(\beta) = \text{Tr}\,\rho^{\text{rot}}(x, x, \beta)$$

$$= \sum_{w=-\infty}^{\infty} e^{-iLvw} \int_0^L \mathrm{d}x\,\rho^{\text{free}}(x, x + wL, \beta) \quad (3.28)$$

(see eqn (3.6)). For small velocities, we expand the exponential in the above equation. The zeroth-order term gives the partition function in the stationary system, and the first-order term vanishes because of a symmetry for the winding number $w \to -w$. The term proportional to $w^2$ is nontrivial. We multiply and divide by the laboratory partition function at rest, $Z_{v=0}^{\text{lab}}$, and obtain

$$Z^{\text{rot}}(\beta) = \underbrace{\sum_{w=-\infty}^{\infty} \int_0^L \mathrm{d}x\,\rho^{\text{free}}(x, x + wL, \beta)}_{Z_{v=0}^{\text{lab}}, \text{ see eqn (3.18)}}$$

$$-\frac{1}{2}v^2 L^2 \underbrace{\sum_{w=-\infty}^{\infty} \frac{w^2 \rho^{\text{free}}(x, x + wL, \beta)}{Z_{v=0}^{\text{lab}}} Z_{v=0}^{\text{lab}}}_{\langle w^2 \rangle_{v=0}} + \cdots$$

$$= \underline{Z_{v=0}^{\text{lab}}\left(1 - \frac{1}{2}v^2 L^2 \langle w^2 \rangle_{v=0}\right)}.$$

Noting that the free energy is $F = -\log(Z)/\beta$, we obtain the following fundamental formula:

$$F^{\text{rot}} = F_{v=0}^{\text{lab}} + \frac{v^2 L^2 \langle w^2 \rangle_{v=0}}{2\beta} + \cdots. \quad (3.29)$$

This is counter-intuitive because we would naively expect the slowly rotating system to rotate along with the reference system and have the same free energy as the laboratory system at rest, in the same way as a bucket of water, under very slow rotation, simply rotates with the walls and satisfies $F^{\mathrm{rot}} = F^{\mathrm{lab}}_{v=0}$. However, we saw the same effect in the previous calculation, using wave functions and eigenvalues. Part of the system simply does not rotate with the bucket. It remains stationary in the laboratory system.

The relation between the mean squared winding number and the change of free energies upon rotation is due to Pollock and Ceperley (1986). Equation (3.29) contains no more quantities specific to noninteracting particles; it therefore holds also in an interacting system, at least for small velocities. We shall understand this more clearly by representing the density matrix as a sum of paths, which do not change with interactions, and only get reweighted. Equation (3.29) is very convenient for evaluating the superfluid fraction of a liquid, that part of a very slowly rotating system which remains stationary in the laboratory frame.

In conclusion, we have computed in this subsection the density matrix for a free particle in a rotating box. As mentioned several times, this calculation has far-reaching consequences for quantum liquids and superconductors. Hess and Fairbank (1967) demonstrated experimentally that a quantum liquid in a slowly rotating container rotates more slowly than the container, or even stops to rest. The reader is urged to study the experiment, and Leggett's classic discussion (Leggett 1973).

## 3.2 Matrix squaring

A classical equilibrium system is at coordinates $\mathbf{x}$ with a probability $\pi(\mathbf{x})$ given by the Boltzmann distribution. In contrast, a quantum statistical system is governed by the diagonal density matrix, defined through wave functions and energy eigenvalues. The problem is that we usually do not know the solutions of the Schrödinger equation, so that we need other methods to compute the density matrix. In this section, we discuss matrix squaring, an immensely practical approach to computing the density matrix at any temperature from a high-temperature limit with the help of a convolution principle, which yields the density matrix at low temperature once we know it at high temperature.

Convolutions of probability distributions have already been discussed in Chapter 1. Their relation to convolutions of density matrices will become evident in the context of the Feynman path integral (see Section 3.3).

### 3.2.1 High-temperature limit, convolution

In the limit of high temperature, the density matrix of a quantum system described by a Hamiltonian $H = H^{\mathrm{free}} + V$ is given by a general expression known as the Trotter formula:

$$\rho(x, x', \beta) \xrightarrow[\beta \to 0]{} \mathrm{e}^{-\frac{1}{2}\beta V(x)} \rho^{\mathrm{free}}(x, x', \beta) \; \mathrm{e}^{-\frac{1}{2}\beta V(x')}. \tag{3.30}$$

To verify eqn (3.30), we expand the exponentials of operators, but observe that terms may not commute, that is, $H_{\mathrm{free}} V \neq V H_{\mathrm{free}}$. We then compare the result with the expansion in eqn (3.15) of the density matrix $\rho = \mathrm{e}^{-\beta H}$. The above Trotter formula gives

$$\left(\underbrace{1 - \frac{\beta}{2}V}_{a} + \underbrace{\frac{\beta^2}{8}V^2}_{b} \cdots\right)\left[\underbrace{1 - \beta H^{\mathrm{free}}}_{c} + \underbrace{\frac{\beta^2}{2}(H^{\mathrm{free}})^2}_{d} \cdots\right]\left(\underbrace{1 - \frac{\beta}{2}V}_{e} + \underbrace{\frac{\beta^2}{8}V^2}_{f} \cdots\right),$$

which yields

$$1 - \underbrace{\beta(V + H^{\mathrm{free}})}_{a+e+c} + \frac{\beta^2}{2}\left[\underbrace{V^2}_{ae+b+f} + \underbrace{VH^{\mathrm{free}}}_{ac} + \underbrace{H^{\mathrm{free}}V}_{ce} + \underbrace{(H^{\mathrm{free}})^2}_{d}\right] - \cdots.$$

This agrees up to order $\beta^2$ with the expansion of

$$\mathrm{e}^{-\beta(V + H^{\mathrm{free}})} = 1 - \beta(V + H^{\mathrm{free}}) + \frac{\beta^2}{2}\underbrace{(V + H^{\mathrm{free}})(V + H^{\mathrm{free}})}_{V^2 + VH^{\mathrm{free}} + H^{\mathrm{free}}V + (H^{\mathrm{free}})^2} + \cdots.$$

The above manipulations are generally well defined for operators and wave functions in a finite box.

Any density matrix $\rho(x, x', \beta)$ possesses a fundamental convolution property:

$$\int \mathrm{d}x' \, \rho(x, x', \beta_1) \, \rho(x', x'', \beta_2) \tag{3.31}$$

$$= \int \mathrm{d}x' \sum_{n,m} \psi_n(x)\mathrm{e}^{-\beta_1 E_n}\psi_n^*(x')\psi_m(x')\mathrm{e}^{-\beta_2 E_m}\psi_m^*(x'')$$

$$= \sum_{n,m} \psi_n(x)\mathrm{e}^{-\beta_1 E_n} \underbrace{\int \mathrm{d}x' \, \psi_n^*(x')\psi_m(x')}_{\delta_{nm}, \text{ see eqn (3.2)}} \mathrm{e}^{-\beta_2 E_m}\psi_m^*(x'')$$

$$= \sum_{n} \psi_n(x)\mathrm{e}^{-(\beta_1 + \beta_2)E_n}\psi_n^*(x'') = \rho(x, x'', \beta_1 + \beta_2). \tag{3.32}$$

We can thus express the density matrix in eqn (3.32) at the inverse temperature $\beta_1 + \beta_2$ (low temperature) as an integral (eqn (3.31)) over density matrices at higher temperatures corresponding to $\beta_1$ and $\beta_2$. Let us suppose that the two temperatures are the same ($\beta_1 = \beta_2 = \beta$) and that the positions $x$ are discretized. The integral in eqn (3.31) then turns into a sum $\sum_l$, and $\rho(x, x', \beta)$ becomes a discrete matrix $\rho_{kl}$. The convolution turns into a product of a matrix with itself, a matrix squared:

$$\int \mathrm{d}x' \, \rho(x, x', \beta) \, \rho(x', x'', \beta) = \rho(x, x'', 2\beta).$$

$$\begin{array}{ccccc} \updownarrow & \updownarrow & \updownarrow & & \updownarrow \\ \sum_l & \rho_{kl} & \rho_{lm} & = & (\rho^2)_{km} \end{array}$$

Matrix squaring was first applied by Storer (1968) to the convolution of density matrices. It can be iterated: after computing the density matrix at $2\beta$, we go to $4\beta$, then to $8\beta$, etc., that is, to lower and lower temperatures. Together with the Trotter formula, which gives a high-temperature approximation, we thus have a systematic procedure for computing the low-temperature density matrix. The procedure works for any Hamiltonian provided we can evaluate the integral in eqn (3.31) (see Alg. 3.3 (`matrix-square`)). We need not solve for eigenfunctions and eigenvalues of the Schrödinger equation. To test the program, we may iterate Alg. 3.3 (`matrix-square`) several times for the harmonic oscillator, starting from the Trotter formula at high temperature. With some trial and error to determine a good discretization of $x$-values and a suitable initial temperature, we can easily recover the plots of Fig. 3.3.

---

**procedure** `matrix-square`
**input** $\{x_0, \ldots, x_K\}, \{\rho(x_k, x_l, \beta)\}$ (grid with step size $\Delta_x$)
**for** $x = x_0, \ldots, x_K$ **do**
$\left\{ \begin{array}{l} \text{\textbf{for} } x' = x_0, \ldots, x_K \text{ \textbf{do}} \\ \quad \{ \ \rho(x, x', 2\beta) \leftarrow \sum_k \Delta_x \rho(x, x_k, \beta) \, \rho(x_k, x', \beta) \end{array} \right.$
**output** $\{\rho(x_k, x_l, 2\beta)\}$

---

**Algorithm 3.3** `matrix-square`. Density matrix at temperature $1/(2\beta)$ obtained from that at $1/\beta$ by discretizing the integral in eqn (3.32).

## 3.2.2  Harmonic oscillator (exact solution)

Quantum-statistics problems can be solved by plugging the high-temperature approximation for the density matrix into a matrix-squaring routine and iterating down to low temperature (see Subsection 3.2.1). This strategy works for anything from the simplest test cases to complicated quantum systems in high spatial dimensions, interacting particles, bosons, fermions, etc. How we actually do the integration inside the matrix-squaring routine depends on the specific problem, and can involve saddle point integration or other approximations, Riemann sums, Monte Carlo sampling, etc. For the harmonic oscillator, all the integrations can be done analytically. This yields an explicit formula for the density matrix for a harmonic oscillator at arbitrary temperature, which we shall use in later sections and chapters.

The density matrix at high temperature,

$$\rho^{\text{h.o.}}(x, x', \beta) \xrightarrow[\beta \to 0]{\substack{\text{from} \\ \text{eqn (3.30)}}} \sqrt{\frac{1}{2\pi\beta}} \exp\left[-\frac{\beta}{4}x^2 - \frac{(x-x')^2}{2\beta} - \frac{\beta}{4}x'^2\right],$$

can be written as

$$\rho^{\text{h.o.}}(x, x', \beta) = c(\beta) \exp\left[-g(\beta)\frac{(x-x')^2}{2} - f(\beta)\frac{(x+x')^2}{2}\right], \quad (3.33)$$

where

$$f(\beta) \xrightarrow[\beta \to 0]{} \frac{\beta}{4},$$

$$g(\beta) \xrightarrow[\beta \to 0]{} \frac{1}{\beta} + \frac{\beta}{4}, \qquad (3.34)$$

$$c(\beta) \xrightarrow[\beta \to 0]{} \sqrt{\frac{1}{2\pi\beta}}.$$

The convolution of two Gaussians is again a Gaussian, so that the harmonic-oscillator density matrix at inverse temperature $2\beta$,

$$\rho^{\text{h.o.}}(x, x'', 2\beta) = \int_{-\infty}^{\infty} \mathrm{d}x' \; \rho^{\text{h.o.}}(x, x', \beta) \, \rho^{\text{h.o.}}(x', x'', \beta),$$

must also have the functional form of eqn (3.33). We recast the exponential in the above integrand,

$$-\frac{f}{2}\left[(x + x')^2 + (x' + x'')^2\right] - \frac{g}{2}\left[(x - x')^2 + (x' - x'')^2\right]$$

$$= \underbrace{-\frac{f + g}{2}\left(x^2 + x''^2\right)}_{\text{independent of }x'} \underbrace{-2(f + g)\frac{x'^2}{2} - (f - g)(x + x'')x'}_{\text{Gaussian in }x', \text{ variance } \sigma^2 = (2f + 2g)^{-1}},$$

and obtain, using eqn (3.12),

$$\rho^{\text{h.o.}}(x, x'', 2\beta)$$
$$= c(2\beta) \exp\left[-\frac{f + g}{2}\left(x^2 + x''^2\right) + \frac{1}{2}\frac{(f - g)^2}{f + g}\frac{(x + x'')^2}{2}\right]. \qquad (3.35)$$

The argument of the exponential function in eqn (3.35) is

$$-\underbrace{\left[\frac{f + g}{2} - \frac{1}{2}\frac{(f - g)^2}{f + g}\right]}_{f(2\beta)}\frac{(x + x'')^2}{2} - \underbrace{\left(\frac{f + g}{2}\right)}_{g(2\beta)}\frac{(x - x'')^2}{2}.$$

We thus find

$$f(2\beta) = \frac{f(\beta) + g(\beta)}{2} - \frac{1}{2}\frac{[f(\beta) - g(\beta)]^2}{f(\beta) + g(\beta)} = \frac{2f(\beta)g(\beta)}{f(\beta) + g(\beta)},$$

$$g(2\beta) = \frac{f(\beta) + g(\beta)}{2},$$

$$c(2\beta) = c^2(\beta)\sqrt{\frac{2\pi}{2[f(\beta) + g(\beta)]}} = c^2(\beta)\frac{\sqrt{2\pi}}{2\sqrt{g(2\beta)}}.$$

The recursion relations for $f$ and $g$ imply

$$f(2\beta)g(2\beta) = f(\beta)g(\beta) = f(\beta/2)g(\beta/2) = \cdots = \frac{1}{4},$$

because of the high-temperature limit in eqn (3.34), and therefore

$$g(2\beta) = \frac{g(\beta) + (1/4)g^{-1}(\beta)}{2}. \qquad (3.36)$$

We can easily check that the only function satisfying eqn (3.36) with the limit in eqn (3.34) is

$$g(\beta) = \frac{1}{2} \coth \frac{\beta}{2} \implies f(\beta) = \frac{1}{2} \tanh \frac{\beta}{2}.$$

Knowing $g(\beta)$ and thus $g(2\beta)$, we can solve for $c(\beta)$ and arrive at

$$\rho^{\text{h.o.}}(x, x', \beta)$$
$$= \sqrt{\frac{1}{2\pi \sinh \beta}} \exp\left[ -\frac{(x + x')^2}{4} \tanh \frac{\beta}{2} - \frac{(x - x')^2}{4} \coth \frac{\beta}{2} \right], \qquad (3.37)$$

and the diagonal density matrix is

$$\rho^{\text{h.o.}}(x, x, \beta) = \sqrt{\frac{1}{2\pi \sinh \beta}} \exp\left( -x^2 \tanh \frac{\beta}{2} \right). \qquad (3.38)$$

To introduce physical units into these two equations, we must replace

$$x \rightarrow \sqrt{\frac{m\omega}{\hbar}} x,$$

$$\beta \rightarrow \hbar\omega\beta = \frac{\hbar\omega}{k_{\text{B}}T},$$

and also multiply the density matrix by a factor $\sqrt{m\omega/\hbar}$.

We used Alg. 3.2 (harmonic-density) earlier to compute the diagonal density matrix $\rho^{\text{h.o.}}(x, x, \beta)$ from the wave functions and energy eigenvalues. We now see that the resulting plots, shown in Fig. 3.3, are simply Gaussians of variance

$$\sigma^2 = \frac{1}{2 \tanh (\beta/2)}. \qquad (3.39)$$

For a classical harmonic oscillator, the analogous probabilities are obtained from the Boltzmann distribution

$$\pi^{\text{class.}}(x) \propto e^{-\beta E(x)} = \exp\left( -\beta x^2/2 \right).$$

This is also a Gaussian, but its variance ($\sigma^2 = 1/\beta$) agrees with that in the quantum problem only in the high-temperature limit (see eqn (3.39) for $\beta \rightarrow 0$). Integrating the diagonal density matrix over space gives the partition function of the harmonic oscillator:

$$Z^{\text{h.o.}}(\beta) = \int dx \, \rho^{\text{h.o.}}(x, x, \beta) = \frac{1}{2 \sinh (\beta/2)}, \qquad (3.40)$$

where we have used the fact that

$$\tanh \frac{\beta}{2} \sinh \beta = 2 \left( \sinh \frac{\beta}{2} \right)^2.$$

This way of computing the partition function agrees with what we obtained from the sum of energies in eqn (3.3). Matrix squaring also allows us to compute the ground-state wave function without solving the Schrödinger equation because, in the limit of zero temperature, eqn (3.38) becomes $\rho^{\text{h.o.}}(x, x, \beta) \propto \exp\left(-x^2\right) \propto \psi_0^{\text{h.o.}}(x)^2$ (see Alg. 3.1 (`harmonic-wavefunction`)).

In conclusion, we have obtained in this subsection an analytic expression for the density matrix of a harmonic oscillator, not from the energy eigenvalues and eigenfunctions, but using matrix squaring down from high temperature. We shall need this expression several times in this chapter and in Chapter 4.

### 3.2.3   Infinitesimal matrix products

The density matrix at low temperature (inverse temperature $\beta_1 + \beta_2$) $\rho(x, x', \beta_1 + \beta_2)$ is already contained in the density matrices at $\beta_1$ and $\beta_2$. We can also formulate this relation between density matrices at two different temperatures in terms of a differential relation, by taking one of the temperatures to be infinitesimally small:

$$\rho(\Delta\beta)\rho(\beta) = \rho(\beta + \Delta\beta).$$

Because of $\rho(\Delta\beta) = \mathrm{e}^{-\Delta\beta H} \simeq 1 - \Delta\beta H$, this is equivalent to

$$-H\rho \simeq \frac{\rho(\beta + \Delta\beta) - \rho(\beta)}{\Delta\beta} = \frac{\partial}{\partial\beta}\rho. \tag{3.41}$$

The effect of the hamiltonian on the density matrix is best clarified in the position representation, where one finds, either by inserting complete sets of states into eqn (3.41) or from the definition of the density matrix:

$$\frac{\partial}{\partial\beta}\rho(x, x', \beta) = -\sum_n \underbrace{E_n\psi_n(x)}_{H\psi_n(x)}\,\mathrm{e}^{-\beta E_n}\psi_n^*(x') = -H_x\rho(x, x', \beta), \tag{3.42}$$

where $H_x$ means that the Hamiltonian acts on $x$, not on $x'$, in the density matrix $\rho(x, x', \beta)$. Equation (3.42) is the differential version of matrix squaring, and has the same initial condition at infinite temperature: $\rho(x, x', \beta \to 0) \to \delta(x - x')$ (see eqn (3.14)).

We shall not need the differential equation (3.42) in the further course of this book and shall simply check in passing that it is satisfied by $\rho^{\text{free}}(x, x', \beta)$. We have

$$\frac{\partial}{\partial\beta}\rho^{\text{free}}(x, x', \beta) = \frac{\partial}{\partial\beta}\left\{\frac{1}{\sqrt{2\pi\beta}}\exp\left[-\frac{(x - x')^2}{2\beta}\right]\right\}$$

$$= \frac{-\beta + (x - x')^2}{2\beta^2}\rho^{\text{free}}(x, x', \beta).$$

On the other hand, we can explicitly check that:

$$\frac{\partial^2}{\partial x^2}\rho^{\text{free}}(x, x', \beta) = \rho^{\text{free}}(x, x', \beta)\frac{(x - x')^2}{\beta^2} - \rho^{\text{free}}(x, x', \beta)/\beta,$$

so that the free density matrix solves the differential equation (3.42).

In this book, we shall be interested only in the density matrix in statistical mechanics, shown in eqn (3.41), which is related to the evolution operator in real time, $e^{-itH}$. Formally, we can pass from real time $t$ to inverse temperature $\beta$ through the replacement

$$\beta = it,$$

and $\beta$ is often referred to as an "imaginary" time. The quantum Monte Carlo methods in this chapter and in Chapter 4 usually do not carry over to real-time quantum dynamics, because the weights would become complex numbers, and could then no longer be interpreted as probabilities.

## 3.3 The Feynman path integral

In matrix squaring, one of the subjects of Section 3.2, we convolute two density matrices at temperature $T$ to produce the density matrix at temperature $T/2$. By iterating this process, we can obtain the density matrix at any temperature from the quasi-classical high-temperature limit. Most often, however, it is impossible to convolute two density matrices analytically. With increasing numbers of particles and in high dimensionality, the available computer memory soon becomes insufficient even to store a reasonably discretized matrix $\rho(\mathbf{x}, \mathbf{x}', \beta)$, so that one cannot run Alg. 3.3 (matrix-square) on a discretized approximation of the density matrix. Monte Carlo methods are able to resolve this problem. They naturally lead to the Feynman path integral for quantum systems and to the idea of path sampling, as we shall see in the present section.

Instead of evaluating the convolution integrals one after the other, as is done in matrix squaring, we can write them out all together:

$$\rho(x, x', \beta) = \int dx''\, \rho(x, x'', \beta/2)\, \rho(x'', x', \beta/2)$$

$$= \iiint dx''dx'''dx''''\rho(x, x''', \tfrac{\beta}{4})\, \rho(x''', x'', \tfrac{\beta}{4})\, \rho(x'', x'''', \tfrac{\beta}{4})\, \rho(x'''', x', \tfrac{\beta}{4})$$

$$= \dots.$$

This equation continues to increasingly deeper levels, with the $k$th applications of the matrix-squaring algorithm corresponding to $\simeq 2^k$ integrations. Writing $\{x_0, x_1, \dots\}$ instead of the cumbersome $\{x, x', x'', \dots\}$, this gives

$$\rho(x_0, x_N, \beta) = \int \cdots \int dx_1 \dots dx_{N-1}$$
$$\times \rho\left(x_0, x_1, \frac{\beta}{N}\right) \dots \rho\left(x_{N-1}, x_N, \frac{\beta}{N}\right), \quad (3.43)$$

where we note that, for the density matrix $\rho(x_0, x_N, \beta)$, the variables $x_0$ and $x_N$ are fixed on both sides of eqn (3.43). For the partition function,

150 *Density matrices and path integrals*

there is one more integration, over the variable $x_0$, which is identified with $x_N$:

$$Z = \int \mathrm{d}x_0 \, \rho(x_0, x_0, \beta) = \int \cdots \int \mathrm{d}x_0 \ldots \mathrm{d}x_{N-1}$$
$$\times \rho\left(x_0, x_1, \frac{\beta}{N}\right) \cdots \rho\left(x_{N-1}, x_0, \frac{\beta}{N}\right). \quad (3.44)$$

The sequence $\{x_0, \ldots, x_N\}$ in eqns (3.43) and (3.44) is called a path, and we can imagine the variable $x_k$ at the value $k\beta/N$ of the imaginary-time variable $\tau$, which goes from 0 to $\beta$ in steps of $\Delta_\tau = \beta/N$ (see Feynman (1972)). Density matrices and partition functions are thus represented as multiple integrals over paths, called path integrals, both at finite $N$ and in the limit $N \to \infty$. The motivation for this representation is again that for large $N$, the density matrices under the multiple integral signs are at small $\Delta_\tau = \beta/N$ (high temperature) and can thus be replaced by their quasi-classical high-temperature approximations. To distinguish between the density matrix with fixed positions $x_0$ and $x_N$ and the partition function, where one integrates over $x_0 = x_N$, we shall refer to the paths in eqn (3.43) as contributing to the density matrix $\rho(x_0, x_N, \beta)$, and to the paths in eqn (3.44) as contributing to the partition function.

After presenting a naive sampling approach in Subsection 3.3.1, we discuss direct path sampling using the Lévy construction, in Subsection 3.3.2. The closely related later Subsection 3.5.1 introduces path sampling in Fourier space.

### 3.3.1 Naive path sampling

The Feynman path integral describes a single quantum particle in terms of paths $\{x_0, \ldots, x_N\}$ (often referred to as world lines), with weights given by the high-temperature density matrix or another suitable approximation:

$$Z = \underbrace{\int \int \mathrm{d}x_0, \ldots, \mathrm{d}x_{N-1}}_{\text{sum of paths}} \underbrace{\rho(x_0, x_1, \Delta_\tau) \cdots \rho(x_{N-1}, x_0, \Delta_\tau)}_{\text{weight } \pi \text{ of path}}.$$

More generally, any variable $x_k$ can represent a $d$-dimensional quantum system. The full path then lies in $d + 1$ dimensions.

Let us first sample the paths contributing to the partition function of a harmonic oscillator using a local Markov-chain algorithm (see Fig. 3.9). We implement the Trotter formula, as we would for an arbitrary potential. Each path comes with a weight containing terms as the following:

$$\underbrace{\cdots \rho^{\text{free}}(x_{k-1}, x_k, \Delta_\tau) \, \mathrm{e}^{-\frac{1}{2}\Delta_\tau V(x_k)}}_{\rho(x_{k-1}, x_k, \Delta_\tau) \text{ in Trotter formula}} \underbrace{\mathrm{e}^{-\frac{1}{2}\Delta_\tau V(x_k)} \rho^{\text{free}}(x_k, x_{k+1}, \Delta_\tau) \cdots}_{\rho(x_k, x_{k+1}, \Delta_\tau) \text{ in Trotter formula}}.$$

As shown, each argument $x_k$ appears twice, and any two contributions $\exp\left[-\frac{1}{2}\Delta_\tau V(x_k)\right]$, where $V(x) = \frac{1}{2}x^2$, can be combined into a single

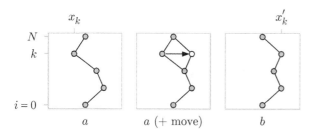

**Fig. 3.9** Naive path-sampling move. The ratio $\pi_b/\pi_a$ is computed from $\{x_{k-1}, x_k, x_{k+1}\}$ and from the new position $x'_k$.

term $\exp\left[-\Delta_\tau V(x_k)\right]$. To move from one position of the path to the next, we choose a random element $k$ and accept the move $x_k \rightarrow x_k + \delta_x$ using the Metropolis algorithm. The ratio of the weights of the new and the old path involves only two segments of the path and one interaction potential (see Alg. 3.4 (`naive-harmonic-path`)). A move of $x_k$, for $k \neq 0$, involves segments $\{x_{k-1}, x_k\}$ and $\{x_k, x_{k+1}\}$. Periodic boundary conditions in the $\tau$-domain have been worked in: for $k = 0$, we consider the density matrices between $\{x_{N-1}, x_0\}$ and $\{x_0, x_1\}$. Such a move across the horizon $k = 0$ changes $x_0$ and $x_N$, but they are the same (see the iteration $i = 10$ in Fig. 3.11).

**procedure** naive-harmonic-path
**input** $\{x_0, \ldots, x_{N-1}\}$
$\Delta_\tau \leftarrow \beta/N$
$k \leftarrow \mathbf{nran}\,(0, N-1)$
$k_\pm \leftarrow k \pm 1$
**if** $(k_- = -1)k_- \leftarrow N$
$x'_k \leftarrow x_k + \mathbf{ran}\,(-\delta, \delta)$
$\pi_a \leftarrow \rho^{\mathrm{free}}\left(x_{k_-}, x_k, \Delta_\tau\right) \rho^{\mathrm{free}}\left(x_k, x_{k_+}, \Delta_\tau\right) \exp\left(-\frac{1}{2}\Delta_\tau x_k^2\right)$
$\pi_b \leftarrow \rho^{\mathrm{free}}\left(x_{k_-}, x'_k, \Delta_\tau\right) \rho^{\mathrm{free}}\left(x'_k, x_{k_+}, \Delta_\tau\right) \exp\left(-\frac{1}{2}\Delta_\tau x'^2_k\right)$
$\Upsilon \leftarrow \pi_b/\pi_a$
**if** $(\mathbf{ran}\,(0,1) < \Upsilon)x_k \leftarrow x'_k$
**output** $\{x_0, \ldots, x_{N-1}\}$
———

**Algorithm 3.4** naive-harmonic-path. Markov-chain sampling of paths contributing to $Z^{\mathrm{h.o.}} = \int \mathrm{d}x_0\, \rho^{\mathrm{h.o.}}(x_0, x_0, \beta)$.

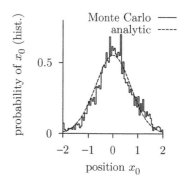

**Fig. 3.10** Histogram of positions $x_0$ (from Alg. 3.4 (`naive-harmonic-path`), with $\beta = 4$, $N = 8$, and $1 \times 10^6$ iterations).

Algorithm 3.4 (`naive-harmonic-path`) is an elementary path-integral Monte Carlo program. To test it, we can generate a histogram of positions for any of the $x_k$ (see Fig. 3.10). For large $N$, the error in the Trotter formula is negligible. The histogram must then agree with the analytical result for the probability $\pi(x) = \rho^{\mathrm{h.o.}}(x, x, \beta)/Z$, which we can also calculate from eqns (3.38) and (3.40). This simple path-integral

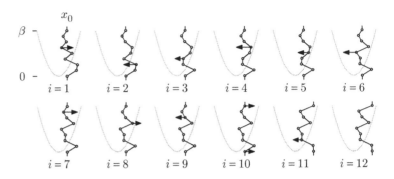

**Fig. 3.11** Markov-chain path sampling for a harmonic potential (from Alg. 3.4 (`naive-harmonic-path`)).

Monte Carlo program can in principle, but rarely in practice, solve problems in equilibrium quantum statistical physics.

Algorithm 3.4 (`naive-harmonic-path`), like local path sampling in general, is exceedingly slow. This can be seen from the fluctuations in the histogram in Fig. 3.10 or in the fact that the path positions in Fig. 3.11 are all on the positive side ($x_k > 0$), just as in the initial configuration: Evidently, a position $x_k$ cannot get too far away from $x_{k-1}$ and $x_{k+1}$, because the free density matrix then quickly becomes very small. Local path sampling is unfit for complicated problems.

### 3.3.2   Direct path sampling and the Lévy construction

To overcome the limitations of local path sampling, we must analyze the origin of the high rejection rate. As discussed in several other places in this book, a high rejection rate signals an inefficient Monte Carlo algorithm, because it forces us to use small displacements $\delta$. This is what happens in Alg. 3.4 (`naive-harmonic-path`). We cannot move $x_k$ very far away from its neighbors, and are also prevented from moving larger parts of the path, consisting of positions $\{x_k, \dots, x_{k+l}\}$. For concreteness, we first consider paths contributing to the density matrix of a free particle, and later on the paths for the harmonic oscillator. Let us sample the integral

$$\rho^{\text{free}}(x_0, x_N, \beta) = \int \cdots \int dx_1 \dots dx_{N-1}$$
$$\underbrace{\rho^{\text{free}}(x_0, x_1, \Delta_\tau) \, \rho^{\text{free}}(x_1, x_2, \Delta_\tau) \dots \rho^{\text{free}}(x_{N-1}, x_N, \Delta_\tau)}_{\pi(x_1, \dots, x_{N-1})}. \quad (3.45)$$

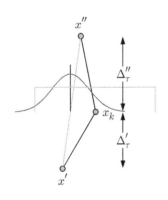

**Fig. 3.12** Proposed and accepted moves in Alg. 3.4 (`naive-harmonic-path`). The position $x_k$ is restrained by $x'$ and $x''$.

We focus for a moment on Monte Carlo moves where all positions except $x_k$ are frozen, in particular $x_{k-1}$ and $x_{k+1}$. Slightly generalizing the problem, we focus on a position $x_k$ sandwiched in between fixed positions

$x'$ and $x''$, with two intervals in $\tau$, $\Delta'_\tau$ and $\Delta''_\tau$ (see Fig. 3.12). In the naive path-sampling algorithm, the move is drawn randomly between $-\delta$ and $+\delta$, around the current position $x_k$ (see Fig. 3.12). The distribution of the accepted moves in Fig. 3.12 is given by

$$\pi^{\text{free}}(x_k|x',x'') \propto \rho^{\text{free}}(x',x_k,\Delta'_\tau)\,\rho^{\text{free}}(x_k,x'',\Delta''_\tau),$$

where

$$\rho^{\text{free}}(x',x_k,\Delta'_\tau) \propto \exp\left[-\frac{(x'-x_k)^2}{2\Delta'_\tau}\right],$$

$$\rho^{\text{free}}(x_k,x'',\Delta''_\tau) \propto \exp\left[-\frac{(x_k-x'')^2}{2\Delta''_\tau}\right].$$

Expanding the squares and dropping all multiplicative terms independent of $x_k$, we find the following for the probability of $x_k$:

$$\pi^{\text{free}}(x_k|x',x'') \propto \exp\left(-\frac{\cancel{x'^2}-2x'x_k+x_k^2}{2\Delta'_\tau} - \frac{x_k^2-2x_kx''+\cancel{x''^2}}{2\Delta''_\tau}\right)$$

$$\propto \exp\left[-\frac{(x_k-\langle x_k\rangle)^2}{2\sigma^2}\right], \quad (3.46)$$

where

$$\langle x_k\rangle = \frac{\Delta''_\tau x'+\Delta'_\tau x''}{\Delta'_\tau+\Delta''_\tau}$$

and

$$\sigma^2 = (1/\Delta''_\tau+1/\Delta'_\tau)^{-1}.$$

The mismatch between the proposed moves and the accepted moves generates the rejections in the Metropolis algorithm. We could modify the naive path-sampling algorithm by choosing $x_k$ from a Gaussian distribution with the appropriate parameters (taking $x' \equiv x_{k-1}$ (unless $k=0$), $x'' \equiv x_{k+1}$, and $\Delta'_\tau = \Delta''_\tau = \beta/N$). In this way, no rejections would be generated.

The conditional probability in eqn (3.46) can be put to much better use than just to suppress a few rejected moves in a Markov-chain algorithm. In fact, $\pi^{\text{free}}(x_k|x',x'')$ gives the weight of all paths which, in Fig. 3.12, start at $x'$, pass through $x_k$ and end up at $x''$. We can sample this distribution to obtain $x_1$ (using $x' = x_0$ and $x'' = x_N$). Between the freshly sampled $x_1$ and $x_N$, we may then pick $x_2$, and thereafter $x_3$ between $x_2$ and $x_N$ and, eventually, the whole path $\{x_1,\dots,x_N\}$ (see Fig. 3.14 and Alg. 3.5 (`levy-free-path`)). A directly sampled path with $N = 50\,000$ is shown in Fig. 3.13; it can be generated in a split second, has no correlations with previous paths, and its construction has caused no rejections. In the limit $N \to \infty$, $x(\tau)$ is a differentiable continuous function of $\tau$, which we shall further analyze in Section 3.5.

Direct path sampling—usually referred to as the Lévy construction—was introduced by Lévy (1940) as a stochastic interpolation between points $x_0$ and $x_N$. This generalizes interpolations using polynomials, trigonometric functions (see Subsection 3.5.1), splines, etc. The Lévy

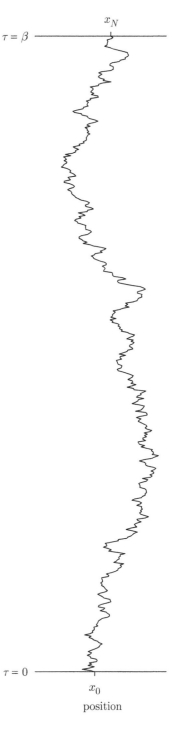

**Fig. 3.13** A path contributing to $\rho^{\text{free}}(x_0,x_N,\beta)$ (from Alg. 3.5 (`levy-free-path`), with $N = 50\,000$).

construction satisfies a local construction principle: the path $x(\tau)$, in any interval $[\tau_1, \tau_2]$, is the stochastic interpolation of its end points $x(\tau_1)$ and $x(\tau_2)$, but the behavior of the path outside the interval plays no role.

The Lévy construction is related to the theory of stable distributions, also largely influenced by Lévy (see Subsection 1.4.4), essentially because Gaussians, which are stable distributions, are encountered at each step. The Lévy construction can be generalized to other stable distributions, but it then would not generate a continuous curve in the limit $\Delta_\tau \to 0$.

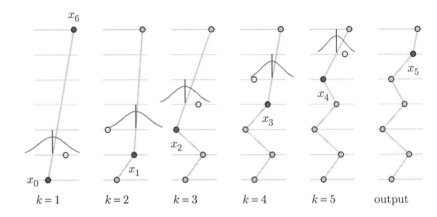

**Fig. 3.14** Lévy construction of a free-particle path from $x_0$ to $x_6$ (see Alg. 3.5 (`levy-free-path`)).

**procedure** `levy-free-path`
**input** $\{x_0, x_N\}$
$\Delta_\tau \leftarrow \beta/N$
**for** $k = 1, \ldots, N-1$ **do**
$\quad \begin{cases} \Delta'_\tau \leftarrow (N-k)\Delta_\tau \\ \langle x_k \rangle \leftarrow (\Delta'_\tau x_{k-1} + \Delta_\tau x_N)/(\Delta_\tau + \Delta'_\tau) \\ \sigma^{-2} \leftarrow \Delta_\tau^{-1} + \Delta_\tau'^{-1} \\ x_k \leftarrow \langle x_k \rangle + \mathbf{gauss}\,(\sigma) \end{cases}$
**output** $\{x_0, \ldots, x_N\}$
————

**Algorithm 3.5** `levy-free-path`. Sampling a path contributing to $\rho^{\text{free}}(x_0, x_N, \beta)$, using the Lévy construction (see Fig. 3.14).

We now consider the Lévy construction for a harmonic oscillator. The algorithm can be generalized to this case because the harmonic density matrix is a Gaussian (the exponential of a quadratic polynomial), and the convolution of two Gaussians is again a Gaussian:

$$\rho^{\text{h.o.}}(x', x'', \Delta'_\tau + \Delta''_\tau) = \int \mathrm{d}x_k \; \underbrace{\rho^{\text{h.o.}}(x', x_k, \Delta'_\tau)\,\rho^{\text{h.o.}}(x_k, x'', \Delta''_\tau)}_{\pi^{\text{h.o.}}(x_k | x', x'')}.$$

Because of the external potential, the mean value $\langle x_k \rangle$ no longer lies on the straight line between $x'$ and $x''$. From the nondiagonal harmonic density matrix in eqn (3.37), and proceeding as in eqn (3.46), we find the following:

$$\pi^{\text{h.o.}}(x_k | x', x'') \propto \exp\left[-\frac{1}{2\sigma^2}(x_k - \langle x_k \rangle)^2\right],$$

with parameters

$$\langle x_k \rangle = \frac{\Upsilon_2}{\Upsilon_1},$$
$$\sigma = \Upsilon_1^{-1/2},$$
$$\Upsilon_1 = \coth \Delta'_\tau + \coth \Delta''_\tau,$$
$$\Upsilon_2 = \frac{x'}{\sinh \Delta'_\tau} + \frac{x''}{\sinh \Delta''_\tau},$$

as already used in the analytic matrix squaring for the harmonic oscillator. We can thus directly sample paths contributing to the harmonic density matrix $\rho^{\text{h.o.}}(x_0, x_N, \beta)$ (see Alg. 3.6 (levy-harmonic-path)), and also paths contributing to $Z^{\text{h.o.}} = \int \mathrm{d}x_0\, \rho^{\text{h.o.}}(x_0, x_0, \beta)$, if we first sample $x_0$ from the Gaussian diagonal density matrix in eqn (3.38).

**procedure** levy-harmonic-path
**input** $\{x_0, x_N\}$
$\Delta_\tau \leftarrow \beta/N$
**for** $k = 1, \ldots, N-1$ **do**
$\left\{ \begin{array}{l} \Upsilon_1 \leftarrow \coth \Delta_\tau + \coth\left[(N-k)\Delta_\tau\right] \\ \Upsilon_2 \leftarrow x_{k-1}/\sinh \Delta_\tau + x_N/\sinh\left[(N-k)\Delta_\tau\right] \\ \langle x_k \rangle \leftarrow \Upsilon_2/\Upsilon_1 \\ \sigma \leftarrow 1/\sqrt{\Upsilon_1} \\ x_k \leftarrow \langle x_k \rangle + \mathbf{gauss}\,(\sigma) \end{array} \right.$
**output** $\{x_0, \ldots, x_N\}$
———

**Algorithm 3.6** levy-harmonic-path. Sampling a path contributing to $\rho^{\text{h.o.}}(x_0, x_N, \beta)$, using the Lévy construction (see Fig. 3.15).

In Alg. 3.5 (levy-free-path), we were not obliged to sample the path in chronological order (first $x_0$, then $x_1$, then $x_2$, etc.). After fixing $x_0$ and $x_N$, we could have chosen to sample the midpoint $x_{N/2}$, then the midpoint between $x_0$ and $x_{N/2}$ and between $x_{N/2}$ and $x_N$, etc. (see Alg. 3.8 (naive-box-path) and Fig. 3.19 later). We finally note that free paths can also be directly sampled using Fourier-transformation methods (see Subsection 3.5.1).

### 3.3.3 Periodic boundary conditions, paths in a box

We now turn to free paths in a box, first with periodic boundary conditions, and then with hard walls. A path contributing to the density

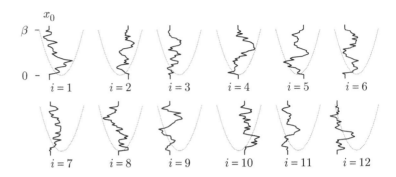

**Fig. 3.15** Paths contributing to $Z^{\text{h.o.}} = \int \mathrm{d}x_0\, \rho^{\text{h.o.}}(x_0, x_0, \beta)$ (from Alg. 3.6 (`levy-harmonic-path`), with $x_0$ first sampled from eqn (3.38)).

matrix $\rho^{\text{per},L}(x, x', \beta)$ may wind around the box, that is, go from $x$ to a periodic image $x' + wL$ rather than straight to $x'$ (see Fig. 3.16, where the path $i = 1$ has zero winding number, the path $i = 2$ has a winding number $w = 1$, etc.).

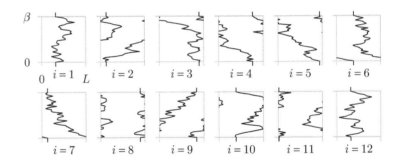

**Fig. 3.16** Paths for a free particle in a periodic box (from Alg. 3.7 (`levy-periodic-path`) with $x_0 = \mathtt{ran}\,(0, L)$).

The paths contributing to $\rho^{\text{per},L}(x, x', \beta)$ with a given winding number $w$ are exactly the same as the paths contributing to $\rho^{\text{free}}(x, x' + wL, \beta)$. The total weight of all the paths contributing to $\rho^{\text{per},L}(x, x', \beta)$ is therefore $\sum_w \rho^{\text{free}}(x, x' + wL, \beta)$, and this is the expression for the density matrix in a periodic box that we already obtained in eqn (3.18), from the Poisson sum formula.

We can sample the paths contributing to the periodic density matrix by a two-step procedure: as each of the images $x_N + wL$ carries a statistical weight of $\rho^{\text{free}}(x_0, x_N + wL, \beta)$, we may first pick the winding number $w$ by tower sampling (see Subsection 1.2.3). In the second step, the path between $x_0$ and $x_N + wL$ is filled in using the Lévy construction (see Alg. 3.7 (`levy-periodic-path`)).

The paths contributing to $\rho^{\text{box},[0,\,L]}(x, x', \beta)$ are the same as the free

```
procedure levy-periodic-path
input {x₀, xₙ}
for w' = ..., −1, 0, 1, ... do
   { πw' ← ρfree(x₀, xₙ + w'L, β)
w ← tower sampling (..., π₋₁, π₀, π₁, ...)
{x₁, ..., xₙ₋₁} ← levy-free-path(x₀, xₙ + wL, β)
output {x₀, ..., xₙ}
```

**Algorithm 3.7** `levy-periodic-path`. Sampling a free path between $x_0$ and $x_N$ in a box of length $L$ with periodic boundary conditions.

paths, with the obvious restriction that they should not cross the boundaries. We have already shown that the density matrix of a free particle in a box satisfies

$$\rho^{\mathrm{box},L}(x, x', \beta)$$
$$= \sum_{w=-\infty}^{\infty} \left[ \rho^{\mathrm{free}}(x, x' + 2wL, \beta) - \rho^{\mathrm{free}}(x, -x' + 2wL, \beta) \right]. \quad (3.47)$$

This is eqn (3.23) again. We shall now rederive it using a graphic method, rather than the Poisson sum formula of Subsection 3.1.3. Let us imagine boxes around an interval $[wL, (w+1)L]$, as in Fig. 3.17. For $x$ and $x'$ inside the interval $[0, L]$ (the original box), either the paths contributing to the free density matrix $\rho^{\mathrm{free}}(x, x', \beta)$ never leave the box, or they reenter the box from the left or right boundary before connecting with $x'$ at $\tau = \beta$:

$$\underbrace{\rho^{\mathrm{free}}(x, x', \beta)}_{\substack{x \text{ and } x' \\ \text{in same box}}} = \underbrace{\rho^{\mathrm{box}}(x, x', \beta)}_{\substack{\text{path does not} \\ \text{leave box}}} + \underbrace{\rho^{\mathrm{right}}(x, x', \beta)}_{\substack{\text{path reenters} \\ \text{from right}}} + \underbrace{\rho^{\mathrm{left}}(x, x', \beta)}_{\substack{\text{path reenters} \\ \text{from left}}} \quad (3.48)$$

(see Fig. 3.17).

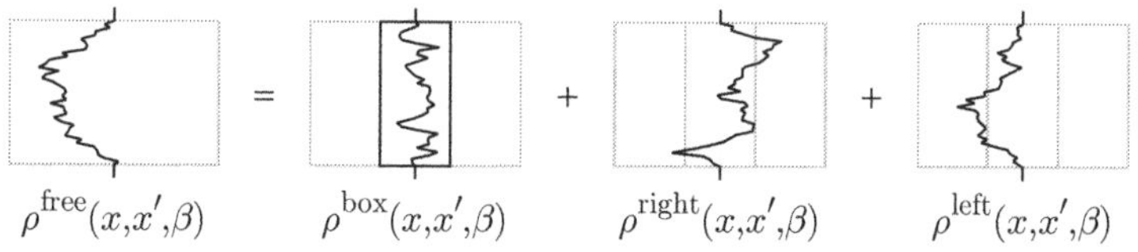

$$\rho^{\mathrm{free}}(x,x',\beta) \quad = \quad \rho^{\mathrm{box}}(x,x',\beta) \quad + \quad \rho^{\mathrm{right}}(x,x',\beta) \quad + \quad \rho^{\mathrm{left}}(x,x',\beta)$$

**Fig. 3.17** Free density matrix as a sum of three classes of paths.

When $x$ and $x'$ are not in the same box, the path connects back to $x'$ from either the right or the left border:

$$\underbrace{\rho^{\mathrm{free}}(x, x', \beta)}_{\substack{x \text{ and } x' \\ \text{not in same box}}} = \underbrace{\rho^{\mathrm{right}}(x, x', \beta)}_{\substack{\text{path enters box of } x' \\ \text{from right}}} + \underbrace{\rho^{\mathrm{left}}(x, x', \beta)}_{\substack{\text{path enters box of } x' \\ \text{from left}}}.$$

By flipping the final leg of a path, we can identify $\rho^{\text{right}}(x, x', \beta)$ with $\rho^{\text{left}}(x, 2L - x', \beta)$ etc., (see Fig. 3.18):

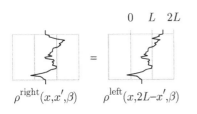

0    L    2L

$$\rho^{\text{right}}(x, x', \beta) = \rho^{\text{left}}(x, 2L - x', \beta)$$
$$= \rho^{\text{free}}(x, 2L - x', \beta) - \rho^{\text{right}}(x, 2L - x', \beta)$$
$$= \rho^{\text{free}}(x, 2L - x', \beta) - \rho^{\text{left}}(x, 2L + x', \beta)$$
$$= \rho^{\text{free}}(x, 2L - x', \beta) - \rho^{\text{free}}(x, 2L + x', \beta) + \rho^{\text{right}}(x, 2L + x', \beta).$$

$\rho^{\text{right}}(x,x',\beta)$          $\rho^{\text{left}}(x,2L-x',\beta)$

**Fig. 3.18** A flip relation between left and right density matrices.

The density matrix $\rho^{\text{right}}(x, 2L + x', \beta)$ can itself be expressed through free density matrices around $4L$, so that we arrive at

$$\rho^{\text{right}}(x, x', \beta) = \sum_{w=1}^{\infty} \left[ \rho^{\text{free}}(x, 2wL - x', \beta) - \rho^{\text{free}}(x, 2wL + x', \beta) \right].$$

Analogously, we obtain

$$\rho^{\text{left}}(x, x', \beta) = \rho^{\text{free}}(x, -x', \beta)$$
$$+ \sum_{w=-1,-2,\dots} \left[ \rho^{\text{free}}(x, 2wL - x', \beta) - \rho^{\text{free}}(x, 2wL + x', \beta) \right].$$

Entered into eqn (3.48), the last two equations yield the box density matrix.

We now sample the paths contributing to the density matrix in a box. Naively, we might start a Lévy construction, and abandon it after the first trespass over box boundaries. Within the naive approach, it is better to sample the path on large scales first, and then to work one's way to small scales (see Fig. 3.19 and Alg. 3.8 (`naive-box-path`)). Big moves, which carry a high risk of rejection, are made early. Bad paths are abandoned quickly.

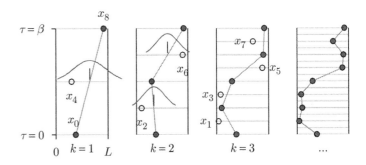

**Fig. 3.19** Path construction in a box, on increasingly finer scales (in Alg. 3.8 (`naive-box-path`)).

Each path of the free density matrix that hits the boundary of the box contributes to the rejection rate of Alg. 3.8 (`naive-box-path`), given simply by

$$p_{\text{reject}} = 1 - \frac{\rho^{\text{box},[0, L]}(x, x', \beta)}{\rho^{\text{free}}(x, x', \beta)}.$$

**procedure** `naive-box-path`
**input** $\{x_0, x_N\}$ ($N$ is power of two: $N = 2^K$)
1  **for** $k = 1, \dots, K$ **do**

$$\left\{ \begin{array}{l} \Delta_k \leftarrow 2^k \\ \Delta_\tau \leftarrow \beta/\Delta_k \\ \left\{ \begin{array}{l} \textbf{for } k' = 1, N/\Delta_k \textbf{ do} \\ \left\{ \begin{array}{l} k \leftarrow k'\Delta_k \\ k_\pm \leftarrow k \pm \Delta_k \\ \langle x_{k'\Delta} \rangle \leftarrow \frac{1}{2}\left(x_{k_-} + x_{k_+}\right) \\ x_k \leftarrow \langle x_{k_-} \rangle + \textbf{gauss}\left(\sqrt{2/\Delta_\tau}\right) \\ \textbf{if } (x_k < 0 \textbf{ or } x_k > L) \textbf{ goto } 1 \end{array} \right. \end{array} \right. \end{array} \right.$$

**output** $\{x_0, \dots, x_N\}$

---

**Algorithm 3.8** `naive-box-path`. Sampling a path contributing to $\rho^{\text{box},[0,\,L]}(x_0, x_N, \beta)$ from large scales down to small ones (see Fig. 3.19).

When $L \gg \sqrt{\beta}$, this rate is so close to one that the naive algorithm becomes impractical. We can then directly sample the positions $x_k$ noting that, for example, the position $x_4$ in Fig. 3.19 is distributed as

$$\pi(x_4|x_0, x_8) = \underbrace{\rho^{\text{box}}(x_0, x_4, \beta/2)\, \rho^{\text{box}}(x_4, x_8, \beta/2)}_{\text{explicitly known, see eqn (3.47)}}.$$

This one-dimensional distribution can be sampled without rejections using the methods of Subsection 1.2.3, even though we must recompute it anew for each value of $x_0$ and $x_8$. We shall not pursue the discussion of this algorithm, but again see that elegant analytic solutions lead to superior sampling algorithms.

## 3.4 Pair density matrices

Using wave functions and energy eigenvalues on the one side, and density matrices on the other, we have so far studied single quantum particles in external potentials. It is time to move closer to the core of modern quantum statistical mechanics and to consider the many-body problem—mutually interacting quantum particles, described by many degrees of freedom. Path integral methods are important conceptual and computational tools to study these systems.

For concreteness, and also for consistency with other chapters, we illustrate the quantum many-body problem for a pair of three-dimensional hard spheres of radius $\sigma$. The density matrix for a pair of hard spheres can be computed with paths, but also in the old way, as in Section 3.1, with wave functions and eigenvalues. Both approaches have subtleties. The naive path integral for hard spheres is conceptually simple. It illustrates how the path integral approach maps a quantum system onto a classical system of interacting paths. However, it is computationally awkward, even for two spheres. On the other hand, the wave functions and energy eigenvalues are not completely straightforward to obtain for a pair of hard spheres, but they then lead to a computationally fast, exact numerical solution for the pair density matrix. The two approaches come together in modern perfect action algorithms for the many-body problem, where analytical calculations are used to describe an $N$-body density matrix as a product of two-body terms, and where Monte Carlo calculations correct for the small error made in neglecting three-body and higher terms (see Subsection 3.4.3).

### 3.4.1   Two quantum hard spheres

We first consider noninteracting distinguishable particles, whose density matrix is the product of the individual density matrices because the wave function of pairs of noninteracting particles is the product of the single-particle wave functions. In the simplest terms, the density matrix is a sum of paths, and the paths for two or more distinguishable free particles in $d$ dimensions can be put together from the individual uncorrelated $d$-dimensional paths, one for each particle. In the following, we shall imagine that these three-dimensional paths are generated by Algorithm levy-free-path-3d, which simply executes Alg. 3.5 (levy-free-path) three times: once for each Cartesian coordinate.

To sample paths contributing to the density matrix for a pair of hard spheres, we naively generate individual paths as if the particles were noninteracting. We reject them if particles approach to closer than twice the sphere radius $\sigma$, anywhere on their way from $\tau = 0$ to $\tau = \beta$:

$$\rho^{\text{pair}}(\{\mathbf{x}_0, \mathbf{x}_0'\}, \{\mathbf{x}_N, \mathbf{x}_N'\}, \beta) = \{\text{sum of paths}\}$$

$$= \sum_{\text{paths } 1,\, 2} \left\{ \begin{array}{c} \text{path 1:} \\ \mathbf{x}_0 \text{ to } \mathbf{x}_N \end{array} \right\} \left\{ \begin{array}{c} \text{path 2:} \\ \mathbf{x}_0' \text{ to } \mathbf{x}_N' \end{array} \right\} \left\{ \begin{array}{c} \text{nowhere} \\ \text{closer than } 2\sigma \end{array} \right\}$$

$$= [1 - \underbrace{p^{\text{reject}}(\{\mathbf{x}_0, \mathbf{x}_0'\}, \{\mathbf{x}_N, \mathbf{x}_N'\}, \beta)}_{\text{rejection rate of Alg. 3.9}}]\, \rho^{\text{free}}(\mathbf{x}_0, \mathbf{x}_N, \beta)\, \rho^{\text{free}}(\mathbf{x}_0', \mathbf{x}_N', \beta)$$

(see Fig. 3.20 and Alg. 3.9 (naive-sphere-path)). As discussed throughout this book, rejection rates of naive sampling algorithms often have a profound physical interpretation. The naive algorithm for a pair of quantum hard spheres is no exception to this rule.

$$\begin{array}{ll} & \textbf{procedure naive-sphere-path} \\ 1 & \text{call levy-free-path-3d}\,(\mathbf{x}_0, \mathbf{x}_N, \beta, N) \\ & \text{call levy-free-path-3d}\,(\mathbf{x}_0', \mathbf{x}_N', \beta, N) \\ & \textbf{for } k = 1, \dots, N-1 \textbf{ do} \\ & \quad \{ \textbf{ if } (|\mathbf{x}_k - \mathbf{x}_k'| < 2\sigma)\ \textbf{goto } 1 \text{ (reject path—tabula rasa)} \\ & \textbf{output } \{\{\mathbf{x}_0, \mathbf{x}_0'\}, \dots, \{\mathbf{x}_N, \mathbf{x}_N'\}\} \end{array}$$

**Algorithm 3.9 naive-sphere-path.** Sampling a path contributing to $\rho^{\text{pair}}(\{\mathbf{x}_0, \mathbf{x}_0'\}, \{\mathbf{x}_N, \mathbf{x}_N'\}, \beta)$ (see also Alg. 3.10).

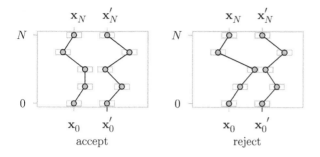

$\mathbf{x}_N$ $\mathbf{x}'_N$ $\qquad$ $\mathbf{x}_N$ $\mathbf{x}'_N$

$N$ $\qquad\qquad\qquad\qquad$ $N$

$0$ $\qquad\qquad\qquad\qquad$ $0$

$\mathbf{x}_0$ $\mathbf{x}'_0$ $\qquad\qquad$ $\mathbf{x}_0$ $\mathbf{x}'_0$

accept $\qquad\qquad\qquad$ reject

**Fig. 3.20** Accepted and rejected configurations in Alg. 3.9 (`naive-sphere-path`) (schematic reduction from three dimensions).

When running Alg. 3.9 (`naive-sphere-path`), for example with $\mathbf{x}_0 = \mathbf{x}_N$ and $\mathbf{x}'_0 = \mathbf{x}'_N$, we notice a relatively high rejection rate (see Fig. 3.21). This can be handled for a single pair of hard spheres, but becomes prohibitive for the $N$-particle case. An analogous problem affected the direct sampling of classical hard spheres, in Subsection 2.2.1, and was essentially overcome by replacing the direct-sampling algorithm by a Markov-chain program. However, it is an even more serious issue that our approximation of the pair density matrix, the free density matrix multiplied by the acceptance rate of Alg. 3.9 (`naive-sphere-path`), strongly depends on the number of time slices. This means that we have to use very large values of $N$, that is, can describe two quantum hard spheres as a system of classical spheres connected by lines (as in Fig. 3.20), but only if there are many thousands of them.

As a first step towards better algorithms, we write the pair density matrix as a product of two density matrices, one for the relative motion and the other for the center-of-mass displacement. For a pair of free particles, we have

The various terms in the second line of eqn (3.49) are rearranged as

$$\exp\left(-\frac{1}{2}A^2 - \frac{1}{2}B^2\right) =$$
$$\exp\left[-\frac{(A+B)^2}{4} - \frac{(A-B)^2}{4}\right]$$

$$\rho^{\text{free},m}(x_1, x'_1, \beta)\,\rho^{\text{free},m}(x_2, x'_2, \beta)$$
$$= \frac{m}{2\pi\beta}\exp\left[-\frac{m(x'_1 - x_1)^2}{2\beta} - \frac{m(x'_2 - x_2)^2}{2\beta}\right]$$
$$= \sqrt{\frac{2m}{2\pi\beta}}\exp\left[-\frac{2m(X' - X)^2}{2\beta}\right]\sqrt{\frac{m/2}{2\pi\beta}}\exp\left[-\frac{m(\Delta'_x - \Delta_x)^2}{4\beta}\right]$$
$$= \underbrace{\rho^{\text{free},2m}(X, X', \beta)}_{\text{center of mass motion}}\ \underbrace{\rho^{\text{free},\frac{1}{2}m}(\Delta_x, \Delta'_x, \beta)}_{\text{relative motion}}, \quad (3.49)$$

where $X = \frac{1}{2}(x_1 + x_2)$ and $X' = \frac{1}{2}(x'_1 + x'_2)$ describe the center of mass and $\Delta_x = x_1 - x_2$ and $\Delta'_x = x'_1 - x'_2$ the relative distance. Clearly, interactions influence only the relative motion, and it suffices to generate single-particle paths corresponding to the relative coordinate describing a particle of reduced mass $\mu = m/2$ (or equivalently a particle of mass $m$ at twice the inverse temperature $2\beta$, see Alg. 3.10

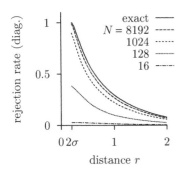

**Fig. 3.21** Rejection rate of Alg. 3.9 (`naive-sphere-path`) (from Alg. 3.10, with $\beta = 4$, $2\sigma = 0.2$, $\Delta_{\mathbf{x},0} = \Delta_{\mathbf{x},N}$, and $r = |\Delta_{\mathbf{x},0}|$; for the exact solution see Subsection 3.4.2).

(`naive-sphere-path(patch)`)). (In the following, we set $m = 1$.) The new program runs twice as fast as Alg. 3.9 (`naive-sphere-path`), if we only compute the rejection rate. To recover the paths of the original algorithm, we must sample an independent free path for the center-of-mass variable $\mathbf{X}$ (a path for a particle of mass $2m$), and reconstruct the original variables as $\mathbf{x}_{1,2} = \mathbf{X} \pm \frac{1}{2}\Delta_{\mathbf{x}}$.

**procedure** `naive-sphere-path(patch)`
$\Delta_{\mathbf{x},0} \leftarrow \mathbf{x}'_0 - \mathbf{x}_0$
$\Delta_{\mathbf{x},N} \leftarrow \mathbf{x}'_N - \mathbf{x}_N$
**call** `levy-free-path-3d`$(\Delta_{\mathbf{x},0}, \Delta_{\mathbf{x},N}, 2\beta, N)$
**for** $k = 1, \ldots, N-1$ **do**
$\left\{\begin{array}{l} \textbf{if } (|\Delta_{\mathbf{x},k}| < 2\sigma) \textbf{ then} \\ \quad \left\{\begin{array}{l} \textbf{output "reject"} \\ \textbf{exit} \end{array}\right. \end{array}\right.$
**output** "accept"

———

**Algorithm 3.10** `naive-sphere-path(patch)`. Computing the rejection rate of Alg. 3.9 from a single-particle simulation.

In conclusion, we have made first contact in this subsection with the path-integral Monte Carlo approach to interacting quantum systems. For concreteness, we considered the case of hard spheres, but other interaction potentials can be handled analogously, using the Trotter formula. We noted that the problem of two quantum hard spheres could be transformed into a problem involving a large number of interacting classical hard spheres. In Subsection 3.4.2, we shall find a more economical path-integral representation for quantum particles, which uses a much smaller number of time slices.

### 3.4.2 Perfect pair action

In Subsection 3.4.1, we tested a naive approach to the many-particle density matrix, one of the central objects of quantum statistical mechanics. The density matrix $\rho^{\text{pair}}(\{\mathbf{x}_0, \mathbf{x}'_0\}, \{\mathbf{x}_N, \mathbf{x}'_N\}, \beta)$ was computed by sending pairs of free paths from $\mathbf{x}_0$ to $\mathbf{x}_N$ and from $\mathbf{x}'_0$ to $\mathbf{x}'_N$. Any pair of paths that got too close was eliminated. All others contributed to the density matrix for a pair of hard spheres. Algorithm 3.9 (`naive-sphere-path`) can in principle be extended to more than two particles, and modified for arbitrary interactions. However, we must go to extremely large values of $N$ in order for the discrete paths to really describe quantum hard spheres (see Fig. 3.21).

In Alg. 3.10 (`naive-sphere-path(patch)`), we separated the center-of-mass motion, which is unaffected by interactions, from the relative motion, which represents a single free particle of reduced mass $\mu = \frac{1}{2}$ that cannot penetrate into a region $r < 2\sigma$. Let us suppose, for a moment, that in addition the particle cannot escape to radii beyond a cutoff $L$ (see Fig. 3.22; the cutoff will afterwards be pushed to infinity). This three-dimensional free particle in a shell $r \in [2\sigma, L]$ has wave functions and eigenvalues just as the harmonic oscillator from the first pages of this chapter.

In the present subsection, we shall first compute these wave functions, and the energy eigenvalues, and then construct the hard-sphere pair density matrix much like we did in Alg. 3.2 (harmonic-density). We shall also see how to treat directly the $L = \infty$ limit. The calculation will need some basic mathematical concepts common to electrodynamics and quantum mechanics: the Laplace operator in spherical coordinates, the spherical Bessel functions, the spherical harmonics, and the Legendre polynomials. To simplify notation, we shall suppose that this particle of reduced mass $\mu = \frac{1}{2}$ is described by variables $\{x, y, z\}$, and replace them by the relative variables $\{\Delta_x, \Delta_y, \Delta_z\}$ in the final equations only.

In Subsection 3.4.3, we show how the analytical calculation of a pair density matrix can be integrated into a perfect-action Monte Carlo program, similar to those that have been much used in statistical mechanics and in field theory.

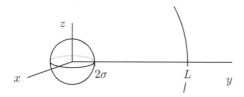

**Fig. 3.22** Solving the Schrödinger equation for a free particle in a shell $r \in [2\sigma, L]$.

The three-dimensional Hamiltonian for a free particle of mass $\mu$ is

$$-\frac{\hbar^2}{2\mu} \underbrace{\left( \frac{\partial^2}{\partial x^2} + \frac{\partial^2}{\partial y^2} + \frac{\partial^2}{\partial z^2} \right)}_{\text{Laplace operator, } \nabla^2} \psi(x, y, z) = E\psi(x, y, z). \qquad (3.50)$$

To satisfy the boundary conditions at $r = 2\sigma$ and $r = L$, we write the wave function $\psi(x, y, z) = \psi(r, \theta, \phi)$ and the Laplace operator $\nabla^2$ in spherical coordinates:

$$\nabla^2 \psi = \frac{1}{r} \frac{\partial^2}{\partial r^2} (r\psi) + \frac{1}{r^2 \sin\theta} \frac{\partial}{\partial \theta} \left( \sin\theta \frac{\partial \psi}{\partial \theta} \right) + \frac{1}{r^2 \sin^2\theta} \frac{\partial^2 \psi}{\partial \phi^2}.$$

The wave function must vanish at $r = 2\sigma$ and for $r = L$. The differential equation (3.50) can be solved by the separation of variables:

$$\psi_{klm}(r, \theta, \phi) = Y_{lm}(\theta, \phi) R_{kl}(r),$$

where $Y_{lm}$ are the spherical harmonic wave functions. For each value of $l$, the function $R_{kl}(r)$ must solve the radial Schrödinger equation

$$\left[ -\frac{\hbar^2}{2\mu r^2} \frac{\partial}{\partial r} r^2 \frac{\partial}{\partial r} + \frac{\hbar^2 l(l+1)}{2\mu r^2} \right] R_{kl}(r) = E R_{kl}(r).$$

The spherical Bessel functions $j_l(r)$ and $y_l(r)$ satisfy this differential equation. For small $l$, they are explicitly given by

$$j_0(r) = \frac{\sin r}{r}, \qquad\qquad y_0(r) = -\frac{\cos r}{r},$$

$$j_1(r) = \frac{\sin r}{r^2} - \frac{\cos r}{r}, \qquad y_1(r) = -\frac{\cos r}{r^2} - \frac{\sin r}{r}.$$

Spherical Bessel functions of higher order $l$ are obtained by the recursion relation

$$f_{l+1}(r) = \frac{2l+1}{r} f_l(r) - f_{l-1}(r), \qquad (3.51)$$

where $f$ stands either for the functions $j_l(r)$ or for $y_l(r)$. We find, for example, that

$$j_2(r) = \left(\frac{3}{r^3} - \frac{1}{r}\right)\sin r - \frac{3}{r^2}\cos r,$$

$$y_2(r) = \left(-\frac{3}{r^3} + \frac{1}{r}\right)\cos r - \frac{3}{r^2}\sin r,$$

etc. (The above recursion relation is unstable numerically for large $l$ and small $r$, but we only need it for $l \lesssim 3$.) For example, we can check that the function $j_0(r) = \sin r/r$, as all the other ones, is an eigenfunction of the radial Laplace operator with an eigenvalue equal to 1:

$$-\frac{1}{r^2}\frac{\partial}{\partial r}\underbrace{r^2\frac{\partial}{\partial r}\left(\frac{\sin r}{r}\right)}_{r\cos r - \sin r} = -\frac{1}{r^2}\frac{\partial}{\partial r}(r\cos r - \sin r) = \overbrace{\left(\frac{\sin r}{r}\right)}^{j_0(r)}.$$

It follows that, analogously, all the functions $j_l(kr)$ and $y_l(kr)$ are solutions of the radial Schrödinger equation, with an eigenvalue

$$k^2 = 2\mu E_k \Leftrightarrow E_k = \frac{k^2}{2\mu}. \qquad (3.52)$$

Besides satisfying the radial Schrödinger equation, the radial wave functions must vanish at $r = 2\sigma$ and $r = L$. The first condition, at $r = 2\sigma$, can be met by appropriately mixing $j_l(kr)$ and $y_l(kr)$, as follows:

$$R^\delta_{kl}(r) = \text{const} \cdot [j_l(kr)\cos\delta - y_l(kr)\sin\delta], \qquad (3.53)$$

where the mixing angle $\delta$ satisfies

$$\delta = \arctan\frac{j_l(2k\sigma)}{y_l(2k\sigma)} \implies \cos\delta = \frac{y_l(2k\sigma)}{j_l(2k\sigma)}\sin\delta, \qquad (3.54)$$

so that $R^\delta_{kl}(2\sigma) = 0$. The function $R^\delta_{kl}(r)$ vanishes at $r = L$ only for special values $\{k_0, k_1, \dots\}$. To find them, we can simply scan through the positive values of $k$ using a small step size $\Delta_k$. A change of sign between $R_{kl}(L)$ and $R_{(k+\Delta_k)l}(L)$ brackets a zero in the interval $[k, k+\Delta_k]$ (see Fig. 3.23 and Alg. 3.11 (naive-rad-wavefunction)). The three-

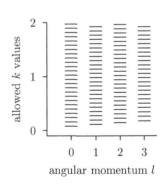

**Fig.  3.23**  Allowed  $k$-values $\{k_0, k_1, \dots\}$  for  small  $k$  (from Alg. 3.11 (naive-rad-wavefunction), with $2\sigma = 0.2$, and $L = 40$).

**procedure** `naive-rad-wavefunction`
**input** $\{r, \sigma, L\}$
$n \leftarrow 0$
**for** $k = 0, \Delta_k, \ldots$ **do**
$\left\{ \begin{array}{l} \delta \leftarrow \arctan \left[ j_l(2k\sigma)/y_l(2k\sigma) \right] \\ \textbf{if } (R^\delta_{(k+\Delta_k)l}(L)R^\delta_{kl}(L) < 0) \textbf{ then } \text{(function } R^\delta_{kl}(r) \text{ from eqn (3.53))} \\ \quad \left\{ \begin{array}{l} k_n \leftarrow k \\ \Upsilon \leftarrow 0 \text{ (squared norm)} \\ \textbf{for } r = 2\sigma, 2\sigma + \Delta_r, \ldots, L \textbf{ do} \\ \quad \{ \; \Upsilon \leftarrow \Upsilon + \Delta_r r^2 [R^\delta_{k_n l}(r)]^2 \\ \textbf{output } \{k_n, \{R_{k_n l}(2\sigma)/\sqrt{\Upsilon}, \ldots, R_{k_n l}(L)/\sqrt{\Upsilon}\}\} \\ n \leftarrow n + 1 \end{array} \right. \end{array} \right.$

**Algorithm 3.11** `naive-rad-wavefunction`. Computing normalized radial wave functions that vanish at $r = 2\sigma$ and at $r = L$.

dimensional wave functions must be normalized, i.e. must satisfy the relation

$$\int d^3\mathbf{x} \; |\psi_{klm}(\mathbf{x})|^2 = \underbrace{\int d\Omega \; Y_{lm}(\theta, \phi) Y^*_{lm}(\theta, \phi)}_{=1} \underbrace{\int_{2\sigma}^{L} dr \; r^2 R^2_{kl}(r)}_{\text{must be 1}} = 1.$$

The normalization condition on the radial wave functions is taken into account in Alg. 3.11 (`naive-rad-wavefunction`).

With the normalized radial wave functions $R^\delta_{k_n l}(r)$, which vanish at $r = 2\sigma$ and $r = L$, and which have eigenvalues as shown in eqn (3.52), we have all that it takes to compute the density matrix

$$\rho^{\text{rel}}(\Delta_{\mathbf{x}_0}, \Delta_{\mathbf{x}_N}, \beta) = \sum_{l=0}^{\infty} \underbrace{\sum_{m=-l}^{l} Y^*_{lm}(\theta_0, \phi_0) Y_{lm}(\theta_N, \phi_N)}_{\frac{2l+1}{4\pi} P_l(\cos\gamma)}$$

$$\times \sum_{n=0,1,\ldots} \exp\left(-\beta \frac{k_n^2}{2\mu}\right) R_{k_n l}(r_0) R_{k_n l}(r_N). \quad (3.55)$$

Here, the relative coordinates are written in polar coordinates $\Delta_{\mathbf{x}_0} = \{r_0, \theta_0, \phi_0\}$ and $\Delta_{\mathbf{x}_N} = \{r_N, \theta_N, \phi_N\}$. Furthermore, we have expressed in eqn (3.55) the sum over products of the spherical harmonics through the Legendre polynomials $P_l(\cos\gamma)$, using a standard relation that is familiar from classical electrodynamics and quantum mechanics. The argument of the Legendre polynomial involves the opening angle $\gamma$ between the vectors $\Delta_{\mathbf{x}_0}$ and $\Delta_{\mathbf{x}_N}$, defined by the scalar product $(\Delta_{\mathbf{x}_0} \cdot \Delta_{\mathbf{x}_N}) = r_0 r_N \cos\gamma$.

The Legendre polynomials $P_l$ could be computed easily, but for concreteness, we consider here only the diagonal density matrix, where $\Delta_{\mathbf{x}_0} = \Delta_{\mathbf{x}_N}$ (so that $\gamma = 0$). In this case, the Legendre polynomials

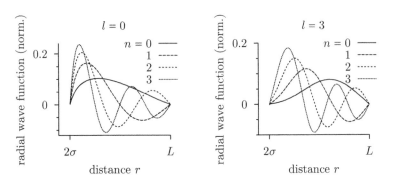

**Fig. 3.24** Normalized radial wave functions $R^{\delta}_{k_n l}$ (from Alg. 3.11 (`naive-rad-wavefunction`), with $2\sigma = 1$ and $L = 10$).

are all equal to 1 $(P_l(1) = 1)$, so that we do not have to provide a subroutine.

The relative-motion density matrix in eqn (3.55) is related to the density matrix for a pair of hard spheres as follows:

$$\rho^{\mathrm{pair}}(\{\mathbf{x}_0, \mathbf{x}'_0\}, \{\mathbf{x}_N, \mathbf{x}'_N\}, \beta)$$

$$= \underbrace{\left[\frac{\rho^{\mathrm{pair}}(\{\mathbf{x}_0, \mathbf{x}'_0\}, \{\mathbf{x}_N, \mathbf{x}'_N\}, \beta)}{\rho^{\mathrm{free}}(\mathbf{x}_0, \mathbf{x}_N, \beta)\, \rho^{\mathrm{free}}(\mathbf{x}'_0, \mathbf{x}'_N, \beta)}\right]}_{\text{depends on } \Delta_\mathbf{x} \text{ and } \Delta'_\mathbf{x}, \text{ only}}$$

$$\times\, \rho^{\mathrm{free}}(\mathbf{x}_0, \mathbf{x}_N, \beta)\, \rho^{\mathrm{free}}(\mathbf{x}'_0, \mathbf{x}'_N, \beta)\,, \quad (3.56)$$

where it is crucial that the piece in square brackets can be written as a product of center-of-mass density matrices and of relative-motion density matrices. The latter cancel, and we find

$$\left[\frac{\rho^{\mathrm{pair}}(\{\mathbf{x}_0, \mathbf{x}'_0\}, \{\mathbf{x}_N, \mathbf{x}'_N\}, \beta)}{\rho^{\mathrm{free}}(\mathbf{x}_0, \mathbf{x}_N, \beta)\, \rho^{\mathrm{free}}(\mathbf{x}'_0, \mathbf{x}'_N, \beta)}\right] = \left[\frac{\rho^{\mathrm{rel},\mu = \frac{1}{2}}(\Delta_{\mathbf{x}_0}, \Delta_{\mathbf{x}_N}, \beta)}{\rho^{\mathrm{free},\mu = \frac{1}{2}}(\Delta_{\mathbf{x}_0}, \Delta_{\mathbf{x}_N}, \beta)}\right].$$

**Table 3.2** Relative density matrix $\rho^{\mathrm{rel}}(\Delta_{\mathbf{x},0}, \Delta_{\mathbf{x},N}, \beta)$, and rejection rate of Alg. 3.9 (`naive-sphere-path`) (from Alg. 3.11 (`naive-rad-wavefunction`) and eqn (3.55), with $\beta = 4$, $2\sigma = 0.2$, $\Delta_{\mathbf{x},0} = \Delta_{\mathbf{x},N}$ ($r = |\Delta_{\mathbf{x},0}|$, compare with Fig. 3.21)).

| $r$ | $\rho^{\mathrm{rel}}$ | Rejection rate |
|-----|-----------------------|----------------|
| 0.2 | 0.00000 | 1.00 |
| 0.4 | 0.00074 | 0.74 |
| 0.6 | 0.00134 | 0.52 |
| 0.8 | 0.00172 | 0.39 |
| 1.0 | 0.00198 | 0.30 |

To obtain the virtually exact results shown in Fig. 3.21, it suffices to compute, for $l = \{0, \ldots, 3\}$, the first 25 wave numbers $\{k_0, \ldots, k_{24}\}$ for which the wave functions satisfy the boundary conditions (the first four of them are shown in Fig. 3.24, for $l = 0$ and $l = 3$) (see Table 3.2; the diagonal free density matrix is $(1/\sqrt{4\pi\beta})^3 = 0.0028$).

The density matrix for the relative motion can be computed directly in the limit $L \to \infty$ because the functions $j_l(r)$ and $y_l(r)$ behave as $\pm\sin(r)/r$ or $\pm\cos(r)/r$ for large $r$. (This follows from the recursion relation in eqn (3.51).) The normalizing integral becomes

$$\Upsilon = \int_a^L \mathrm{d}r\; r^2 [j_l(kr)]^2 \xrightarrow{L \to \infty} \frac{1}{k^2} \int_a^L \mathrm{d}r\; \sin^2(kr) = \frac{L}{2k^2}.$$

The asymptotic behavior of the spherical Bessel functions also fixes the separation between subsequent $k$-values to $\Delta_k = \pi/L$. (Under this condition, subsequent functions satisfy $\sin(kL) = 0$ and $\sin[(k + \Delta_k)L] = 0$, etc., explaining why there are about 25 states in an interval of length $\pi25/40 \simeq 2.0$, in Fig. 3.23). The sum over discrete eigenvalues $n$ can then be replaced by an integral:

$$\sum_k \cdots = \frac{1}{\Delta_k}\sum_k \Delta_k \cdots \simeq \frac{1}{\Delta_k}\int \mathrm{d}k \ \cdots,$$

and we arrive at the following pair density matrix in the limit $L \to \infty$:

$$\rho^{\mathrm{rel}}(\mathbf{x}, \mathbf{x}', \beta) = \sum_{l=0}^{\infty} P_l(\cos\gamma)\frac{2l+1}{4\pi}$$
$$\times \int_{k=0}^{\infty} \mathrm{d}k \ \exp\left(-\beta\frac{k^2}{2\mu}\right)\hat{R}_{kl}^{\delta}(r)\hat{R}_{kl}^{\delta}(r'). \quad (3.57)$$

In this equation, we have incorporated the constant stemming from the normalization and from the level spacing $\Delta_k$ into the radial wave function:

$$\hat{R}^{\delta}(r) = \sqrt{\frac{2}{\pi}}k\left[j_l(kr)\cos\delta - y_l(kr)\sin\delta\right].$$

The integrals in eqn (3.57) are done numerically (except for $l = 0$). The mixing angles $\delta(k, \sigma)$ ensure that $\tilde{R}_{kl}^{\delta}(2\sigma) = 0$ for all $k$ (see eqn (3.54)).

In conclusion, we have computed in this subsection the exact statistical weight for all continuous hard-sphere paths going through a discretized set of position $\{\mathbf{x}_0, \dots, \mathbf{x}_N\}$ and $\{\mathbf{x}_0', \dots, \mathbf{x}_N'\}$ (see Fig. 3.25). For clarity, let us collect variables on one slice $k$ into a single symbol $\mathcal{X}_k \equiv \{\mathbf{x}_k, \mathbf{x}_k'\}$. The weight of a discretized path, the exponential of the action $\mathcal{S}$, was determined as

$$\left\{\begin{array}{c}\text{weight}\\\text{of path}\end{array}\right\} \propto \exp\left[-\mathcal{S}(\{\mathcal{X}_0, \dots, \mathcal{X}_N\}, \Delta_\tau)\right]$$
$$= \rho^{\mathrm{pair}}(\mathcal{X}_0, \mathcal{X}_1, \Delta_\tau) \times \cdots \times \rho^{\mathrm{pair}}(\mathcal{X}_{N-1}, \mathcal{X}_N, \Delta_\tau). \quad (3.58)$$

Previously, the naive action $\mathcal{S}$ was either zero or infinite, and it described a pair of hard spheres badly, unless $\Delta_\tau$ was very small. In contrast, the weight of a path, in eqn (3.58), is assembled from the pair density matrices. It describes a pair of hard spheres exactly, at any $\Delta_\tau$, and it corresponds to the "perfect pair action" $\mathcal{S}(\{\mathcal{X}_0, \dots, \mathcal{X}_N\}, \Delta_\tau)$.

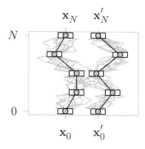

Fig. 3.25 A pair of discretized paths, representing continuous paths of hard spheres.

### 3.4.3 Many-particle density matrix

The pair density matrix from Subsection 3.4.2 links up with the full quantum $N$-body problem (for concreteness, we continue with the example of quantum hard spheres). It is easily generalized from two to $N$

particles:

$$\rho^{N\text{-part}}(\{\mathbf{x}_1, \ldots, \mathbf{x}_N\}, \{\mathbf{x}'_1, \ldots, \mathbf{x}'_N\}, \Delta_\tau) \simeq$$

$$\left\{ \prod_{k=1}^{N} \rho^{\text{free}}(\mathbf{x}_k, \mathbf{x}'_k, \Delta_\tau) \right\} \prod_{k<l} \underbrace{\frac{\rho^{\text{pair}}(\{\mathbf{x}_k, \mathbf{x}_l\}, \{\mathbf{x}'_k, \mathbf{x}'_l\}, \Delta_\tau)}{\rho^{\text{free}}(\mathbf{x}_k, \mathbf{x}'_k, \Delta_\tau)\, \rho^{\text{free}}(\mathbf{x}_l, \mathbf{x}'_l, \Delta_\tau)}}_{\text{prob. that paths } k \text{ and } l \text{ do not collide}}. \quad (3.59)$$

For two particles, this is the same as eqn (3.56), and it is exact. For $N$ particles, eqn (3.59) remains correct under the condition that we can treat the collision probabilities for any pair of particles as independent of those for other pairs. This condition was already discussed in the context of the virial expansion for classical hard spheres (see Subsection 2.2.2). It is justified at low density or at high temperature. In the first case (low density) paths rarely collide, so that the paths interfere very little. In the second case ($\Delta_\tau$ corresponding to high temperature), the path of particle $k$ does not move away far from the position $\mathbf{x}_k \simeq \mathbf{x}'_k$, and the interference of paths is again limited. Because of the relation $\Delta_\tau = \beta/N$, we can always find an appropriate value of $N$ for which the $N$-density matrix in eqn (3.59) is essentially exact. The representation of the density matrix in eqn (3.59) combines elements of a high-temperature expansion and of a low-density expansion. It is sometimes called a Wigner–Kirkwood expansion.

In all practical cases, the values of $N$ that must be used are much smaller than the number of time slices needed in the naive approach of Subsection 3.4.1 (see Pollock and Ceperley (1984), Krauth (1996)).

## 3.5    Geometry of paths

Quantum statistical mechanics can be formulated in terms of random paths in space and imaginary time. This is the path-integral approach that we started to discuss in Sections 3.3 and 3.4, and for which we have barely scratched the surface. Rather than continue with quantum statistics as it is shaped by path integrals, we analyze in this section the shapes of the paths themselves. This will lead us to new sampling algorithms using Fourier transformation methods. The geometry of paths also provides an example of the profound connections between classical statistical mechanics and quantum physics, because random paths do not appear in quantum physics alone. They can describe cracks in homogeneous media (such as a wall), interfaces between different media (such as between air and oil in a suspension) or else between different phases of the same medium (such as the regions of a magnet with different magnetizations). These interfaces are often very rough. They then resemble the paths of quantum physics, and can be described by very similar methods. This will be the subject of Subsection 3.5.3.

## 3.5.1   Paths in Fourier space

In the following, we describe paths by use of Fourier variables, the co-efficients of trigonometric functions. For computer implementation, we remain primarily interested in discrete paths $\{x_0, \ldots, x_N\}$, but we treat continuous paths first because they are mathematically simpler. As a further simplification, we consider here paths which start at zero and return back to zero ($x(0) = x(\beta) = 0$). Other cases will be considered in Subsection 3.5.3.

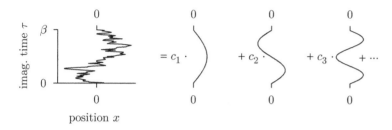

**Fig. 3.26** Representation of a continuous path $x(\tau)$ as an infinite sum over Fourier modes.

Any path $x(\tau)$ with $x(0) = x(\beta) = 0$ can be decomposed into an infinite set of sine functions:

$$x(\tau) = \sum_{n=1}^{\infty} c_n \sin\left(n\pi \frac{\tau}{\beta}\right) \quad \tau \in [0, \beta] \tag{3.60}$$

(see Fig. 3.26). The sine functions in eqn (3.60) are analogous to the wave functions $\psi_n^{\mathrm{box}}(x)$, the solutions of the Schrödinger equation in a box with walls (see Subsection 3.1.3). Now, however, we consider a path, a function of $\tau$ from 0 to $\beta$, rather than a wave function, extending in $x$, from 0 to $L$. The difference between pure sine functions and the combined series of sines and cosines mirrors the one between wave functions with hard walls and with periodic boundary conditions.

Each path $x(\tau)$ contributing to the density matrix $\rho^{\mathrm{free}}(0, 0, \beta)$ is described by Fourier coefficients $\{c_0, c_1, \ldots\}$. We first determine these coefficients for a given path and then express the weight of each path, and the density matrix, directly in Fourier space. The coefficients are obtained from the orthonormality relation of Fourier modes:

$$\int_0^{\beta} d\tau \underbrace{\sin\left(n\pi \frac{\tau}{\beta}\right) \sin\left(l\pi \frac{\tau}{\beta}\right)}_{\frac{1}{2}\{\cos[(n-l)\pi\tau/\beta] - \cos[(n+l)\pi\tau/\beta]\}} = \frac{\beta}{2}\delta_{nl}. \tag{3.61}$$

We can project out the coefficient $c_l$ of mode $l$ by multiplying the Fourier representation of a path, in eqn (3.60), on both sides by $\sin(l\pi\tau/\beta)$ and integrating over $\tau$ from 0 to $\beta$,

$$
\underline{\frac{2}{\beta} \int_0^\beta \mathrm{d}\tau \; \sin\left(l\pi\frac{\tau}{\beta}\right) x(\tau)} = \frac{2}{\beta} \int_0^\beta \mathrm{d}\tau \; \sin\left(l\pi\frac{\tau}{\beta}\right) \sum_{n=1}^\infty c_n \sin\left(n\pi\frac{\tau}{\beta}\right)
$$

$$
= \frac{2}{\beta} \sum_{n=1}^\infty c_n \underbrace{\int_0^\beta \mathrm{d}\tau \; \sin\left(l\pi\frac{\tau}{\beta}\right) \sin\left(n\pi\frac{\tau}{\beta}\right)}_{(\beta/2)\delta_{ln};\; \text{see eqn (3.61)}} = \underline{c_l}. \quad (3.62)
$$

We can thus determine the Fourier coefficients $\{c_1, c_2, \ldots\}$ for a given function $x(\tau)$, whereas eqn (3.60) allowed us to compute the function $x(\tau)$ for given Fourier coefficients.

We now express the statistical weight of the path $\{x_0, \ldots, x_N\}$ directly in Fourier variables. With $\Delta_\tau = \beta/N$, we find

$$
\left\{\begin{matrix} \text{weight} \\ \text{of path} \end{matrix}\right\} = \exp\left[-\mathcal{S}(\{x_0, \ldots, x_N\})\right]
$$

$$
= \rho^{\mathrm{free}}(x_0, x_1, \Delta_\tau)\, \rho^{\mathrm{free}}(x_1, x_2, \Delta_\tau) \times \cdots \times \rho^{\mathrm{free}}(x_{N-1}, x_N, \Delta_\tau).
$$

In the small-$\Delta_\tau$ limit, each term in the action can be written as

$$
\frac{1}{2}\frac{(x-x')^2}{\Delta_\tau} = \frac{1}{2}\Delta_\tau \frac{(x-x')^2}{\Delta_\tau^2} \to \frac{1}{2}\mathrm{d}\tau \left[\frac{\partial x(\tau)}{\partial \tau}\right]^2 \quad (3.63)
$$

and summing over all terms, in the limit $\Delta_\tau \to 0$, corresponds to an integration from 0 to $\beta$. In this limit, the action becomes

$$
\mathcal{S} = \frac{1}{2}\int_0^\beta \mathrm{d}\tau \left[\frac{\partial x(\tau)}{\partial \tau}\right]^2. \quad (3.64)
$$

We use this formula to express the action in Fourier space, using the Fourier representation of the path given in eqn (3.60). The derivative with respect to $\tau$ gives

$$
\frac{\partial}{\partial \tau} x(\tau) = \sum_{n=1}^\infty c_n \frac{n\pi}{\beta} \cos\left(n\pi\frac{\tau}{\beta}\right).
$$

The action in eqn (3.64) leads to a double sum of terms $\propto c_n c_m$ that is generated by the squared derivative. However, the nondiagonal terms again vanish. We arrive at

$$
\frac{1}{2}\int_0^\beta \mathrm{d}\tau \left(\frac{\partial x}{\partial \tau}\right)^2 = \frac{1}{2}\sum_{n=1}^\infty c_n^2 \frac{n^2\pi^2}{\beta^2} \underbrace{\int_0^\beta \mathrm{d}\tau \; \cos^2\left(n\pi\frac{\tau}{\beta}\right)}_{\beta/2} = \frac{1}{\beta}\sum_{n=1}^\infty \frac{c_n^2 n^2\pi^2}{4}.
$$

We should note that the derivative of $x(\tau)$ and the above exchange of differentiation and integration supposes that the function $x(\tau)$ is sufficiently smooth. We simply assume that the above operations are well defined. The statistical weight of a path is then given by

$$
\left\{\begin{matrix} \text{weight of} \\ \text{path} \end{matrix}\right\} \propto \exp\left(-\frac{1}{\beta}\sum_{n=1}^\infty c_n^2 \frac{n^2\pi^2}{4}\right),
$$

and the density matrix, written in terms of Fourier variables, is an infinite product of integrals:

$$\rho^{\text{free}}(0,0,\beta) \propto \prod_{n=1}^{\infty} \left[ \int_{-\infty}^{\infty} \frac{dc_n n\pi}{\sqrt{4\pi\beta}} \exp\left( -\frac{1}{\beta} \sum_{n=1}^{\infty} c_n^2 \frac{n^2\pi^2}{4} \right) \right].$$

The Fourier transform of a continuous path is operationally quite simple but, as mentioned, hides mathematical subtleties. These difficulties are absent for the Fourier transformation of discrete paths $\{x_0,\ldots,x_N\}$, a subject we now turn to.

A discrete function $\{x_0,\ldots,x_N\}$ can be represented by a finite number $N$ of Fourier modes:

$$x_k = \sum_{n=1}^{N-1} c_n \sin\left( n\pi\frac{k}{N} \right) \quad k = 0,\ldots,N \tag{3.65}$$

(see Fig. 3.27). Remarkably, the discrete sine functions remain mutually orthogonal if we simply replace the integral over $\tau$ in the orthogonality condition in eqn (3.61) by the sum over a discrete index $k$:

In the following, $k$ and $j$ are discretized $\tau$ indices, and $n$ and $l$ describe Fourier modes.

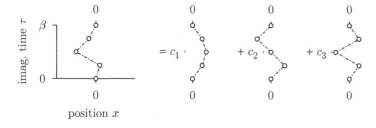

**Fig. 3.27** Representation of a discrete path $\{x_0,\ldots,x_4\}$ as a finite sum over Fourier modes.

$$\underbrace{\sum_{k=1}^{N-1} \sin\left( n\pi\frac{k}{N} \right) \sin\left( l\pi\frac{k}{N} \right)}_{\frac{1}{2}\{\cos[(n-l)\pi k/N]-\cos[(n+l)\pi k/N]\}} = \frac{N}{2}\delta_{nl}. \tag{3.66}$$

Equation (3.66) can be checked in exponential notation ($\cos x = \text{Re}(e^{ix})$) by summing the geometric series for noninteger $M/N$:

$$\sum_{k=0}^{N-1} \cos M\pi\frac{k}{N} = \text{Re} \sum_{k=0}^{N-1} \exp\left( i\pi\frac{M}{N} \right)^k = \text{Re}\left[ \frac{1 - e^{iM\pi}}{1 - \exp\left(i\frac{M}{N}\pi\right)} \right]. \tag{3.67}$$

In eqn (3.66), $(n-l)$ and $(n+l)$ are either both even or odd. In the first case, the sum in eqn (3.67) is zero. In the second, the two sums are easily seen to be equal. They thus cancel.

Again multiplying the discrete function $x_k$ of eqn (3.65) on both sides by $\sin(l\pi k/N)$ and summing over $l$, we find

$$\frac{2}{N}\sum_{k=1}^{N-1}\sin\left(l\pi\frac{k}{N}\right)x_k = \frac{2}{N}\sum_{k=1}^{N-1}\sin\left(l\pi\frac{k}{N}\right)\sum_{n=1}^{N-1}c_n\sin\left(n\pi\frac{k}{N}\right)$$

$$= \frac{2}{N}\sum_{n=1}^{N-1}c_n\underbrace{\sum_{k=1}^{N-1}\sin\left(l\pi\frac{k}{N}\right)\sin\left(n\pi\frac{k}{N}\right)}_{\frac{1}{2}N\delta_{ln},\text{ see eqn (3.66)}} = \underline{c_l}, \quad (3.68)$$

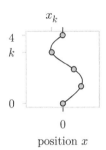

$x_k$

4
$k$

0

0

position $x$

**Fig. 3.28** Example path $\{x_0,\ldots,x_N\}$ of eqn (3.69). The trigonometric polynomial defined in eqn (3.70) passes through all the points.

in analogy with eqn (3.62). The $N-1$ Fourier coefficients $\{c_1,\ldots,c_{N-1}\}$ define a trigonometric interpolating polynomial $x(\tau)$ which passes exactly through the points $\{x_0,\ldots,x_N\}$ and which contains the same information as the Fourier coefficients. To illustrate this point, let us consider an example path for $N=4$ described by the two sets of variables:

$$\underbrace{\begin{bmatrix}x_0\\x_1\\x_2\\x_3\\x_4\end{bmatrix}=\begin{bmatrix}0.0\\0.25\\0.15\\-0.15\\0.0\end{bmatrix}}_{\text{real-space variables (in Fig. 3.28)}}\equiv\underbrace{\begin{bmatrix}c_0\\c_1\\c_2\\c_3\\c_4\end{bmatrix}=\begin{bmatrix}0.0\\0.1104\\0.2\\-0.039\\0.0\end{bmatrix}}_{\text{Fourier variables (in Fig. 3.28, from eqn (3.68))}}. \quad (3.69)$$

The trigonometric polynomial interpolating the points $\{x_0,\ldots,x_4\}$ is

$$x(\tau) = \underbrace{0.1104}_{c_1}\cdot\sin\left(\pi\frac{\tau}{\beta}\right)+\underbrace{0.2}_{c_2}\cdot\sin\left(2\pi\frac{\tau}{\beta}\right)\underbrace{-0.039}_{c_3}\cdot\sin\left(3\pi\frac{\tau}{\beta}\right), \quad (3.70)$$

and we easily check that $x(\frac{3}{4}\beta)=-0.15$, etc.

The weight of a path, a product of factors $\rho^{\text{free}}(x_k,x_{k+1},\Delta_\tau)$, can be expressed through Fourier variables. To do so, we write out the weight as before, but without taking the $\Delta_\tau\to 0$ limit:

$$\begin{Bmatrix}\text{weight}\\\text{of path}\end{Bmatrix}=\exp\left[-\mathcal{S}(\{x_0,\ldots,x_N\})\right]$$

$$=\exp\left[-\frac{(x_1-x_0)^2}{2\Delta_\tau}-\frac{(x_2-x_1)^2}{2\Delta_\tau}-\frac{(x_N-x_{N-1})^2}{2\Delta_\tau}\right]. \quad (3.71)$$

The action, $\mathcal{S}$, is transformed as

$$\sum_{k=1}^{N}\frac{(x_k-x_{k-1})^2}{2\Delta_\tau}=\frac{1}{2\Delta_\tau}\sum_{k=1}^{N}\sum_{n,l=1}^{N-1}c_nc_l$$

$$\times\left[\sin\left(n\pi\frac{k}{N}\right)-\sin\left(n\pi\frac{k-1}{N}\right)\right]\left[\sin\left(l\pi\frac{k}{N}\right)-\sin\left(l\pi\frac{k-1}{N}\right)\right].$$

Terms with $n \neq l$ vanish after summation over $k$, and we end up with:

$$S = \frac{1}{2\Delta_\tau} \sum_{j=1}^{N-1} c_j^2 \underbrace{\sum_{k=1}^{N} \left[ \sin\left(j\pi \frac{k}{N}\right) - \sin\left(j\pi \frac{k-1}{N}\right) \right]^2}_{4\cos^2\left[j\pi(k-\frac{1}{2})/N\right]\sin^2\left[j\pi/(2N)\right]}$$

$$= \frac{2}{\Delta_\tau} \sum_{n=1}^{N-1} c_n^2 \sin^2\left(\frac{n\pi}{2N}\right) \underbrace{\sum_{k=1}^{N} \cos^2\left[\frac{n\pi}{N}\left(k - \frac{1}{2}\right)\right]}_{N/2},$$

so that

$$\left\{ \begin{array}{c} \text{weight} \\ \text{of path} \end{array} \right\} = \exp\left[ -\frac{N}{\Delta_\tau} \sum_{n=1}^{N-1} c_n^2 \sin^2\left(\frac{n\pi}{2N}\right) \right]. \qquad (3.72)$$

It is instructive to check that the weight of our example path from eqn (3.69) comes out the same no matter whether it is computed it from $\{x_0, \ldots, x_N\}$, using eqn (3.71), or from the $\{c_0, \ldots, c_N\}$, using eqn (3.72).

We have arrived at the representation of the path integral as

$$\rho^{\text{free}}(0, 0, \beta) = \left\{ \begin{array}{c} \text{sum of paths} \\ \text{from } 0 \to 0 \end{array} \right\}$$

$$\propto \int_{\infty}^{\infty} \frac{dc_1}{\sqrt{2\pi}\sigma_1} \cdots \frac{dc_{N-1}}{\sqrt{2\pi}\sigma_{N-1}} \exp\left(-\frac{c_1^2}{2\sigma_1^2}\right) \cdots \exp\left(-\frac{c_{N-1}^2}{2\sigma_{N-1}^2}\right)$$

$$= \left[ \int_{\infty}^{\infty} \frac{dc_1}{\sqrt{2\pi}\sigma_1} \exp\left(-\frac{c_1^2}{2\sigma_1^2}\right) \right] \cdots \left[ \int_{\infty}^{\infty} \frac{dc_{N-1}}{\sqrt{2\pi}\sigma_{N-1}} \exp\left(-\frac{c_{N-1}^2}{2\sigma_{N-1}^2}\right) \right],$$

where

$$\sigma_n^2 = \frac{\beta}{2N^2 \sin^2\left(\frac{n\pi}{2N}\right)} \simeq \frac{2\beta}{\pi^2 n^2} + \cdots. \qquad (3.73)$$

In the above representation of the path integral in Fourier space, the integrals are independent of each other. The Fourier modes are thus uncorrelated ($\langle c_k c_l \rangle \propto \delta_{kl}$), and the autocorrelations $\langle c_k c_k \rangle$, the variance of mode $k$, are given by eqn (3.73). This can be checked by generating paths with Alg. 3.5 (levy-free-path) and by Fourier-transforming them (see Fig. 3.29). Most simply, free paths are described as independent Gaussian modes $n$ with zero mean and variance as $\propto 1/n^2$.

Paths $\{x_0, \ldots, x_N\}$ can not only be described but also sampled as independent Gaussian Fourier modes, that is, as Gaussian random numbers $\{c_0, \ldots, c_N\}$ which can be transformed back to real space (see Alg. 3.12 (fourier-free-path)). This algorithm is statistically identical to the Lévy construction. Using fast Fourier methods, it can be implemented in $\propto N \log N$ operations.

In this subsection, we have passed back and forth between the real-space and the Fourier representation of paths, using classic formulas for expressing the $\{c_1, \ldots, c_N\}$ in terms of the $\{x_1, \ldots, x_N\}$ and vice versa. We saw how to transform the single path, but also the statistical weight

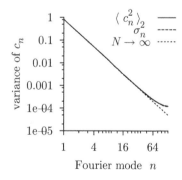

**Fig. 3.29** Correlation $\langle c_n c_n \rangle$ of Fourier-transformed Lévy paths, compared to eqn (3.73) (from Alg. 3.5 (levy-free-path), with $N = 128$, $\beta = 4$).

**procedure** `fourier-free-path`
**for** $n = 1, \ldots, N-1$ **do**
$\left\{ \begin{array}{l} \Upsilon_n \leftarrow 2N^2 \sin^2\left[n\pi/(2N)\right] \\ c_n \leftarrow \text{gauss}\left(\sqrt{\beta/\Upsilon_n}\right) \end{array} \right.$
**for** $k = 0, \ldots, N$ **do**
$\left\{ x_k \leftarrow \sum_{n=1}^{N-1} c_n \sin\left(n\pi\frac{k}{N}\right) \right.$
**output** $\{x_0, \ldots, x_N\}$

———

**Algorithm 3.12** `fourier-free-path`. Sampling a path contributing to $\rho^{\text{free}}(0, 0, \beta)$ in Fourier space, and then transforming to real variables.

of a path, and the path integral itself. The path integral decoupled in Fourier space, because the real-space action is translation invariant.

We thus have two direct sampling algorithms: one in real space—the Lévy construction, and one in Fourier space—the independent sampling of modes. However, these algorithms exist for completely different reasons: the real-space algorithm relies on a local construction property of the sequence $\{x_1, \ldots, x_N\}$, which allows us to assemble pieces of the path independently of the other pieces. In contrast, the Fourier transformation decouples the real-space action because the latter is invariant under translations. In the case of the free path integral, Fourier transformation offered new insights, but did not really improve the performance of the Monte Carlo algorithms. In many other systems, however, simulations can be extremely difficult when done with one set of coordinates, and much easier after a coordinate transformations, because the variables might be less coupled. A simple example of such a system will be shown in Subsection 3.5.3, where a real-space Monte Carlo simulation would necessarily be very slow, but a Fourier-space calculation can proceed by direct sampling, that is, at maximum speed.

In this subsection we did not touch on the subject of fast Fourier transformation methods, which would allow us to pass between the $\{x_1, \ldots, x_N\}$ and the $\{c_1, \ldots, c_N\}$ in about $N \log N$ operations, rather than $\propto N^2$ (see for example Alg. 3.12 (`fourier-free-path`)). For heavy use of Fourier transformation, the fast algorithms must be implemented, using widely available routines. However, the naive versions provided by eqns (3.62) and (3.68) must always be kept handy, as alternative subroutines. They help us avoid problems with numerical factors of two and of $\pi$, and with subtle shifts of indices. As mentioned in many other places throughout this book, there is great virtue in getting to run naive algorithms before embarking on more elaborate, and less transparent programming endeavors.

## 3.5.2 Path maxima, correlation functions

We continue to explore the geometry of free paths, which start and end at $x_0 = x_N = 0$. Let us compute first the probability distribution of the maximum of all $x$-values (see Fig. 3.30) i.e. the probability $\Pi_{\max}(x)$ for

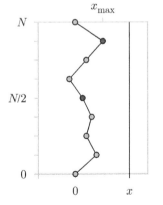

**Fig. 3.30** Geometry of a path. We compute the probability distribution of the midpoint $x_{N/2}$, and the probability of staying to the left of $x$.

the path to remain to the left of $x$. This path is contained in a box with the left wall at $-\infty$ and a right wall at $x$. It thus contributes to the density matrix $\rho^{\text{box}[-\infty,\,x]}(0,0,\beta)$:

$$\rho^{\text{box}[-\infty,\,x]}(0,0,\beta) = \lim_{L\to\infty} \rho^{\text{box}[0,\,L]}(L-x,L-x,\beta)$$

$$\xrightarrow[\text{see eqn (3.23)}]{L\to\infty} \underbrace{\rho^{\text{free}}(L-x,L-x,\beta)}_{\text{independent of } L \text{ and } x} - \underbrace{\rho^{\text{free}}(L-x,L+x,\beta)}_{\text{from eqn (3.23)}},$$

because in an infinite box the sum over windings is dominated by a single flip operation. We find

$$\Pi_{\text{max}}(x) = \left\{\begin{array}{c}\text{prob. that}\\ \text{max. position} < x\end{array}\right\} = \frac{\rho^{\text{box}[-\infty,\,x]}(0,0,\beta)}{\rho^{\text{free}}(0,0,\beta)}$$

$$= \frac{\rho^{\text{free}}(x,x,\beta) - \rho^{\text{free}}(-x,x,\beta)}{\rho^{\text{free}}(x,x,\beta)} = 1 - \exp\left(-\frac{2x^2}{\beta}\right).$$

The probability $\Pi_{\text{max}}(x+dx)$ counts all paths whose maximum is to the left of $x+dx$, and $\Pi_{\text{max}}(x)$ counts all those paths whose maximum position is smaller than $x$. The difference between the two amounts to all paths whose maximum falls between $x$ and $x+dx$. Therefore,

$$\pi_{\text{max}}(x) = \left\{\begin{array}{c}\text{prob. that}\\ x_{\text{max}} = x\end{array}\right\} = \frac{d\Pi_{\text{max}}(x)}{dx} = \frac{4x}{\beta}\exp\left(-\frac{2x^2}{\beta}\right) \qquad (3.74)$$

(see Fig. 3.31).

After the maximum positions, we now consider correlations $\langle x_k x_l\rangle$ between the different components of a path $\{x_0,\ldots,x_N\}$. The autocorrelation of $x_{N/2}$ (see Fig. 3.30) follows, for $N \to \infty$, from the Fourier representation (3.60) of $x_{N/2}$:

$$\left\langle x_{\frac{N}{2}} x_{\frac{N}{2}}\right\rangle = \sum_{k=1}^{\infty} \sigma_k^2 \underbrace{\sin^2\frac{\pi k}{2}}_{\substack{0 \text{ for } k=2,4,\ldots\\1 \text{ for } k=1,3,\ldots}} \to \frac{2\beta}{\pi^2}\underbrace{\left(1 + \frac{1}{3^2} + \frac{1}{5^2} + \cdots\right)}_{\pi^2/8} = \frac{\beta}{4}. \qquad (3.75)$$

This correlation is in fact independent of $N$, as we can see as follows. The probability distribution of the midpoint (corresponding to slice $N/2$ or, equivalently, to imaginary time $\tau = \beta/2$) is

$$\pi_{\beta/2}(x) = \frac{\rho^{\text{free}}(0,x,\frac{\beta}{2})\,\rho^{\text{free}}(x,0,\frac{\beta}{2})}{\rho^{\text{free}}(0,0,\beta)} = \sqrt{\frac{2}{\pi\beta}}\exp\left(-\frac{2x^2}{\beta}\right),$$

a Gaussian with zero mean and variance $\sigma^2 = \langle x^2\rangle = \beta/4\ \langle x^2\rangle = \beta/4$, in agreement with eqn (3.75). The root mean square width of the path grows with the square root of the length $\beta$ of the path, that is, with $\sqrt{\beta}$. This relation, (width) $\propto \sqrt{\text{(length)}}$, is the hallmark of diffusive processes, and of random walks.

We now determine all the path correlations $\langle x_k x_l\rangle$ from the path integral action. Using for concreteness $\Delta_\tau = 1$ so that $N = \beta$, the action is

$$S = \frac{1}{2}\left[(x_1 - x_0)^2 + \cdots + (x_N - x_{N-1})^2\right].$$

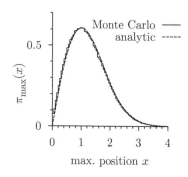

Fig. **3.31** Distribution of the path maximum (with $\beta = 4$, from modified Alg. 3.5 (levy-free-path), compared with eqn (3.74)).

With $x_0 = x_N = 0$, the action $\mathcal{S}$ can be written in matrix form:

$$\left\{\begin{array}{c}\text{weight of}\\ \text{path}\end{array}\right\} \propto \exp\left(-\frac{1}{2}\sum_{k,l=1}^{N-1} x_k \mathcal{M}_{kl} x_l\right).$$

Let us look at this $(N-1) \times (N-1)$ matrix, and its inverse, for $N = 8$:

$$\underbrace{\begin{bmatrix} 2 & -1 & \cdot & \cdot & \cdot & \cdot & \cdot \\ -1 & 2 & -1 & \cdot & \cdot & \cdot & \cdot \\ \cdot & -1 & 2 & -1 & \cdot & \cdot & \cdot \\ \cdot & \cdot & -1 & 2 & -1 & \cdot & \cdot \\ \cdot & \cdot & \cdot & -1 & 2 & -1 & \cdot \\ \cdot & \cdot & \cdot & \cdot & -1 & 2 & -1 \\ \cdot & \cdot & \cdot & \cdot & \cdot & -1 & 2 \end{bmatrix}^{-1}}_{\mathcal{M} \text{ (in action } \mathcal{S})} = \underbrace{\frac{1}{8}\begin{bmatrix} 7 & 6 & 5 & 4 & 3 & 2 & 1 \\ 6 & 12 & 10 & 8 & 6 & 4 & 2 \\ 5 & 10 & 15 & 12 & 9 & 6 & 3 \\ 4 & 8 & 12 & 16 & 12 & 8 & 4 \\ 3 & 6 & 9 & 12 & 15 & 10 & 5 \\ 2 & 4 & 6 & 8 & 10 & 12 & 6 \\ 1 & 2 & 3 & 4 & 5 & 6 & 7 \end{bmatrix}}_{\mathcal{M}^{-1} \text{ (correlation matrix } (\langle x_k x_l\rangle))}, \quad (3.76)$$

as is easily checked. In general, the inverse of the $(N-1) \times (N-1)$ matrix $\mathcal{M}$ of eqn (3.76) is

$$\mathcal{M}_{kl}^{-1} = \frac{1}{N}\min\left[(N-k)l, (N-l)k\right].$$

For the free path integral, the correlation functions are given by the inverse of the matrix $\mathcal{M}$:

$$\langle x_k x_l\rangle = (\mathcal{M}^{-1})_{kl} = \frac{\int dx_1\dots dx_{N-1} x_k x_l \exp\left(-\frac{1}{2}\sum x_n \mathcal{M}_{nm} x_m\right)}{\int dx_1\dots dx_{N-1}\exp\left(-\frac{1}{2}\sum x_n \mathcal{M}_{nm} x_m\right)}. \quad (3.77)$$

The general correlation formula in eqn (3.77) agrees with the mid-path correlation (eqn (3.75)), because

$$(\mathcal{M}^{-1})_{44} = \frac{16}{8} = \frac{\beta}{4}$$

(we note that in eqn (3.76), we supposed $\beta = N = 8$). Equation (3.77) has many applications. We use it here to prove the correctness of a trivial sampling algorithm for free paths, which generates a first path $\{\Upsilon_0, \dots, \Upsilon_N\}$ from a sum of uncorrelated Gaussian random numbers, without taking into account that the path should eventually return to $x = 0$. We define

$$\Upsilon_k = \begin{cases} 0 & \text{for } k = 0 \\ \Upsilon_{k-1} + \xi_k & \text{for } k = 1,\dots,N \end{cases},$$

where $\xi_k$ are uncorrelated Gaussian random numbers with variance 1. After the construction of this first path, we "pull back" $\Upsilon_N$ to zero (by an amount $\Upsilon_N$). For all $k$, $\Upsilon_k$ is pulled back by $\Upsilon_N k/N$ (see Alg. 3.13 (trivial-free-path) and Fig. 3.32). The pulled-back random variables $\eta_k$ (which are correlated) are also sums of the uncorrelated Gaussians $\xi_k$, and their variances depend on $k$:

$$\eta_k = \underbrace{\xi_1 + \dots + \xi_k}_{\Upsilon_k} - \frac{k}{N}\underbrace{(\xi_1 + \dots + \xi_N)}_{\Upsilon_N} = \sum_{l=1}^N a_{kl}\xi_l,$$

where

$$a_{kl} = \begin{cases} 1 - \frac{k}{N} & \text{if } l \le k \\ -\frac{k}{N} & \text{if } l > k \end{cases}.$$

> **procedure** trivial-free-path
> $x_0 \leftarrow 0, \Upsilon_0 \leftarrow 0$
> **for** $k = 1, \dots, N$ **do**
> $\quad \{ \ \Upsilon_k \leftarrow \Upsilon_{k-1} + \text{gauss}\left(\sqrt{\beta/N}\right)$
> **for** $k = 1, \dots, N$ **do**
> $\quad \{ \ x_k \leftarrow \Upsilon_k - \Upsilon_N k/N$
> **output** $\{x_0, \dots, x_N\}$

---

**Algorithm 3.13** trivial-free-path. Sampling a path contributing to $\rho^{\text{free}}(0, 0, \beta)$ with a trivial, yet correct, algorithm (see Fig. 3.32).

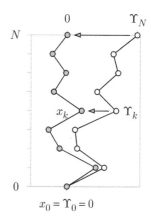

**Fig. 3.32** Direct sampling of a path contributing to $\rho^{\text{free}}(0, 0, \beta)$ by pulling back an unrestricted path.

The Gaussians $\eta_k$ are characterized by their means, which are all zero, and by their correlations which, for $k \le l$, are given by

$$\langle \eta_k \eta_l \rangle = \frac{\beta}{N} \sum_{j=1}^{N} a_{kj} a_{lj}$$

$$= \frac{1}{N^2} [\underbrace{k(N-k)(N-l)}_{j \le k} - \underbrace{(l-k)k(N-l)}_{k < j \le l} + \underbrace{(N-l)kl}_{l \le j}] = \underbrace{\frac{1}{N} k(N-l)}_{\text{see eqn (3.77)}}.$$

This agrees with the correlation matrix of the path integral. We see that the random variables $\{\eta_0, \dots, \eta_N\}$ from Alg. 3.13 (trivial-free-path) indeed sample paths contributing to the free density matrix.

### 3.5.3  Classical random paths

In this chapter, we have discussed a number of direct path-sampling algorithms for the free density matrix. All these algorithms produced statistically identical output, both for discrete paths and in the continuum limit. Performances were also roughly equivalent. What really differentiates these algorithms is how they generalize from the free paths (classical random walks). We saw, for example, that the Lévy construction can be made to sample harmonic-oscillator paths (see Subsection 3.3.2).

In the present subsection, we consider generalized Fourier sampling methods. For concreteness, we restrict our attention to the continuous paths with $\Delta_\tau = \beta/N \to 0$. In Fourier space, continuous paths are generated by independent Gaussian random variables with variance $\propto 1/n^2$, for all $n = \{1, 2, \dots\}$ (see Alg. 3.12 (fourier-free-path)). We now analyze the paths that arise from a scaling $\propto 1/n^\alpha$ of the variances. We again pass between the real-space and the Fourier representations, and adopt the most general Fourier transform, containing sines and cosines:

$$x(t) = \sum_{n=1}^{\infty} \left\{ a_n \cos\left(2n\pi \frac{t}{L}\right) + b_n \sin\left(2n\pi \frac{t}{L}\right) \right\}. \tag{3.78}$$

The paths described by eqn (3.78) have zero mean ($\int_0^L dt\, x(t)/L = 0$), but need not start at the origin $x = 0$. The Fourier-space action in eqn (3.64) can be written for the transform in eqn (3.78) and generalized to arbitrary values of $\alpha$:

$$
\mathcal{S} = \frac{1}{2}\sum_{n=1}^{\infty}\left(\frac{2\pi n}{L}\right)^{\alpha}\int_0^L dt\,\left[a_n^2\cos^2\left(2n\pi\frac{t}{L}\right) + b_n^2\sin^2\left(2n\pi\frac{t}{L}\right)\right]
$$

$$
= \frac{1}{2}\sum_n \underbrace{\left(\frac{2\pi n}{L}\right)^{\alpha}\frac{L}{2}}_{\sigma_n^{-2}}(a_n^2 + b_n^2). \quad (3.79)
$$

The $a_n$ and $b_n$ are Gaussian random variables with standard deviation

$$
\sigma_n = \frac{1}{(\pi n)^{1/2}}\left(\frac{L}{2\pi n}\right)^{\alpha/2-\frac{1}{2}}. \quad (3.80)
$$

The roughness exponent $\zeta = \alpha/2 - \frac{1}{2}$ in eqn (3.80) gives the scaling of the root mean square width of the path to its length so that we now have (width) $\propto$ (length)$^{\zeta}$ ($\zeta$ is pronounced "zeta"). All quantum paths, and all random walks have $\zeta = \frac{1}{2}$. However, many other paths appearing in nature are characterized by roughness exponents $\zeta$ different from $\frac{1}{2}$. Predicting these exponents for a given physical phenomenon is beyond the scope of this book. In this subsection, our goal is more restricted. We only aim at characterizing Gaussian paths, the simplest paths with nontrivial roughness exponents (with $0 < \zeta < 1$), and which are governed by the action in eqn (3.79).

**procedure fourier-gen-path**
**for** $n = 1, 2, \ldots$ **do**
$$
\begin{cases}
\sigma_n \leftarrow (\pi n)^{-\frac{1}{2}}\left(\frac{L}{2\pi n}\right)^{\zeta} \\
a_n \leftarrow \mathbf{gauss}\,(\sigma_n) \\
b_n \leftarrow \mathbf{gauss}\,(\sigma_n)
\end{cases}
$$
**for** $t = 0, \Delta_t, \ldots, L$ **do**
$\quad\{\ x(t) \leftarrow \sum_{n=1}^{\infty}\left[a_n\cos\left(2n\pi\frac{t}{L}\right) + b_n\sin\left(2n\pi\frac{t}{L}\right)\right]$
**output** $\{x(0), \ldots, x(L)\}$
——

**Algorithm 3.14 fourier-gen-path.** Sampling a periodic Gaussian path with roughness $\zeta$.

Periodic Gaussian paths can be easily sampled for various roughness exponents (see Fig. 3.33 and Alg. 3.14 (**fourier-gen-path**)). As discussed, the paths with larger $\zeta$ grow faster on large scales, but we see that they are also smoother, because the Fourier coefficients vanish faster as $n \to \infty$. Some paths appear wider than others (see Fig. 3.33). This is not a finite-size effect, as we can see as follows. A larger interval $L$ is generated by rescaling $L \to \Upsilon L$. Under this rescaling, the standard deviations of the Gaussian random numbers in Alg. 3.14 (**fourier-gen-path**) are

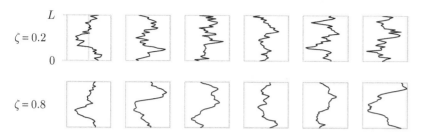

**Fig. 3.33** Periodic Gaussian paths with two different roughness exponents (from Alg. 3.14 (**fourier-gen-path**), with 40 Fourier modes).

uniformly rescaled as $\sigma_n \to \Upsilon^\zeta \sigma_n$. Under this transformation, each individual path is rescaled by factors $\Upsilon$ in $t$ (length) and $\Upsilon^\zeta$ in $x$ (width), but its shape remains unchanged. The same rescaling applies also to the ensemble of all paths; they are self-affine.

We can define the (mean square) width of a path as follows:

$$\omega_2 = \frac{1}{L} \int_0^L \mathrm{d}x \; x^2(t)$$

(we remember that the average position is zero). Wide paths have a larger value of $\omega_2$ than narrow paths. We can compute the probability distribution of the width, $\pi_\zeta(\omega_2)$, using Alg. 3.14 (**fourier-gen-path**) (see Fig. 3.36, later).

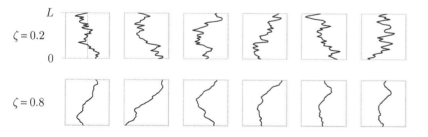

**Fig. 3.34** Free Gaussian paths with two different roughness exponents (from Alg. 3.15 (**fourier-cos-path**), with 40 Fourier modes).

However, Alg. 3.14 (**fourier-gen-path**) is not a unique prescription for generating Gaussian paths with roughness exponent $\zeta$. We can also generalize the Fourier sine transform of Subsection 3.5.1, or generate the paths by a Fourier cosine series:

$$x(t) = \sum_{n=1}^\infty c_n \cos\left(n\pi\frac{t}{L}\right),$$

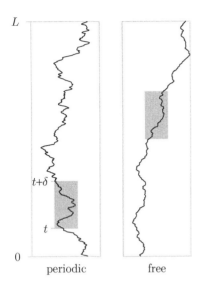

$L$

$t+\delta$

$t$

0

periodic        free

**Fig. 3.35** Periodic and free Gaussian paths with $\zeta = 0.8$. In small intervals $[t, t+\delta]$, with $\delta/L \to 0$, path fragments are statistically identical.

where the $c_n$ are again independent Gaussian random variables with a $\zeta$-dependent scaling (see Alg. 3.15 (fourier-cos-path) and Fig. 3.34). In these "free" paths, the boundary conditions no longer identify $x(0)$ and $x(L)$.

**procedure** fourier-cos-path
**for** $n = 1, 2, \ldots$ **do**
$$\begin{cases} \sigma_n \leftarrow \frac{2}{\pi n} \left(\frac{L}{\pi n}\right)^{\zeta} \\ c_n \leftarrow \text{gauss}\,(\sigma_n) \end{cases}$$
**for** $t = 0, \Delta_t, \ldots, L$ **do**
$\{\ x(t) \leftarrow \sum_{n=1}^{\infty} c_n \cos\left(n\pi \frac{t}{L}\right)$
**output** $\{x(0), \ldots, x(L)\}$

————

**Algorithm 3.15** fourier-cos-path. Sampling a free Gaussian path with roughness exponent $\zeta$.

The periodic paths in Fig. 3.33 differ from the free paths in Fig. 3.34 not only in the boundary conditions, but also in the width distributions. Nevertheless, the statistical properties of path fragments are equivalent in a small interval $[t, t+\delta]$, with $\delta/L \to 0$, for $\delta \ll t \ll L$ (see Fig. 3.35). To show this, we consider the mean value of the path fragment,

$$\langle x \rangle_{t, \delta} = \frac{1}{\delta} \int_t^{t+\delta} dt\ x(t'),$$

and the width of a path fragment,

$$w_2(t, \delta) = \int_t^{t+\delta} dt' \left[x(t') - \langle x \rangle_{t, \delta}\right]^2.$$

We rescale the distribution of $w_2$ such that its mean value is equal to 1. To obtain the width of a path fragment, we either generate the whole path from the explicit routines of this subsection or compute the width directly from the Fourier decomposition without generating $x(t)$. This is possible because a path fragment which is defined by Fourier coefficients $\{c_1, c_2, \ldots\}$ has width

$$w_2(t, \delta) = \sum_{n, m=1}^{\infty} c_n c_m D_{nm}(t, \delta), \qquad (3.81)$$

where the coefficients $D_{nm}(t, \delta)$ are given by

$$D_{nm}(t, \delta) = \frac{1}{\delta} \int_t^{t+\delta} dt'\ \overbrace{\cos\,(n\pi t') \cos\,(m\pi t')}^{\frac{1}{2}\{\cos\,[(m-n)\pi t'] + \cos\,[(m+n)\pi t']\}}$$
$$- \frac{1}{\delta^2} \left[\int_t^{t+\delta} dt'\ \cos\,(n\pi t')\right]\left[\int_t^{t+\delta} dt'\ \cos\,(m\pi t')\right].$$

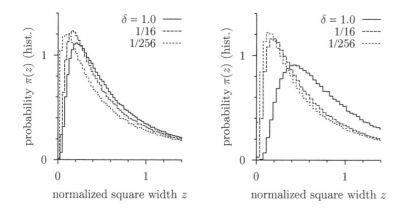

**Fig. 3.36** Width fluctuations (normalized) for free paths (*left*) and for periodic paths (*right*) with $\zeta = 0.75$ (from eqn (3.81), with $z = \omega_2/\langle\omega_2\rangle$).

These integrals can be computed analytically, once and for all. The width $\omega_2(t, \delta)$ of each path is then obtained directly from eqn (3.81), with coefficients $c_n$ taken from Alg. 3.15 (`fourier-cos-path`). The distribution of $\omega_2(t, \delta)$ can then be determined as an average over many paths. One can redo an analogous calculation for periodic paths (one finds three sets of coefficients, one for the sine–sine integrals, one for cosine–cosine integrals, and one for the mixed terms). For $\delta = 1$, the width distributions are quite different, but they converge to the same curve in the limit $\delta/L \to 0$ (see Fig. 3.36). For free paths, the width distribution depends on the length of the interval $\delta$ and also on the starting point $gt$ (for $\zeta \neq \frac{1}{2}$). For periodic paths, the distribution is independent of $t$ because of translational invariance.

In conclusion, this subsection introduced Gaussian paths, which allow us to describe real paths in nature, from the notorious random walk (with $\alpha = 2$, $\zeta = \frac{1}{2}$), to cracks, interfaces, and phase boundaries, among many others. A rough numerical simulation allowed us to show that the statistical properties of small path fragments become independent of the boundary condition. This statement can be made mathematically rigorous (see Rosso, Santachiara, and Krauth (2005)).

# Exercises

## (Section 3.1)

(3.1) Use Alg. 3.1 (`harmonic-wavefunction`) to generate the wave functions $\{\psi_0^{\text{h.o.}}(x), \ldots, \psi_{20}^{\text{h.o.}}(x)\}$ on a fine grid in the interval $x \in [-5, 5]$. Verify numerically that the wave functions are normalized and mutually orthogonal, and that they solve the Schrödinger equation (eqn (3.1)) with $\hbar = m = \omega = 1$. Analytically prove that the normalization is correct and that the wave functions are mutually orthogonal. Use the recursion relation to show analytically that the wave functions solve the Schrödinger equation.

NB: For the numerical checks, note that, on a grid with step size $\Delta_x$, the gradient is

$$\frac{\partial}{\partial x} \psi(x_k) \simeq \frac{\psi(x_{k+1}) - \psi(x_k)}{\Delta_x},$$

and the second derivative is approximated as

$$\frac{\partial^2}{\partial x^2} \psi(x_k) \simeq \frac{1}{\Delta_x}\left[\frac{\partial}{\partial x}\psi(x_k) - \frac{\partial}{\partial x}\psi(x_{k-1})\right]$$
$$\simeq \frac{\psi(x_{k+1}) - 2\psi(x_k) + \psi(x_{k-1})}{\Delta_x^2}.$$

(3.2) Determine the density matrix of the harmonic oscillator using Algs 3.2 (`harmonic-density`) and 3.1 (`harmonic-wavefunction`). Plot the diagonal density matrix $\rho^{\text{h.o.}}(x, x, \beta)$ for several temperatures. What is its relationship between the density matrix and the probability $\pi(x)$ of the quantum particle to be at position $x$? Compare this probability with the Boltzmann distribution $\pi(x)$ for a classical particle of mass $m = 1$ in a potential $V(x) = \frac{1}{2}x^2$.

(3.3) Familiarize yourself with the calculation, in eqn (3.10), of the free density matrix using plane-wave functions in an infinite box with periodic boundary conditions (pay attention to the use of the dummy parameter $\Delta_n$). Alternatively determine the free density matrix using the wave functions (eqn (3.20)) in a box with hard walls at positions $x = -L/2$ and $x = L/2$ in the limit $L \to \infty$. NB: First shift the functions in eqn (3.20) by $L/2$ to the left.

Finally, to illustrate the calculation of density matrices in arbitrary bases, expand the free density matrix in the harmonic oscillator basis:

$$\langle \psi_n^{\text{h.o.}} | H^{\text{free}} | \psi_m^{\text{h.o.}} \rangle = \langle \psi_n^{\text{h.o.}} | H^{\text{h.o.}} - \tfrac{1}{2}x^2 | \psi_m^{\text{h.o.}} \rangle.$$

Derive an explicit formula for $\langle \psi_n^{\text{h.o.}} | x^2 | \psi_m^{\text{h.o.}} \rangle$ from the recursion relation used in Alg. 3.1 (`harmonic-wavefunction`). Use the results of these calculations to compute numerically the density matrix as $\rho_{nm} = 1 - \beta H_{nm} + \frac{\beta^2}{2}\left(H^2\right)_{nm} - \cdots +$ and also

$$\rho^{\text{free}}(x, x', \beta) = \sum_{n,m=0}^{\infty} \psi_n^{\text{h.o.}}(x)\rho_{nm}\psi_m^{\text{h.o.}}(x').$$

Compare the density matrix obtained with the exact solution.

## (Section 3.2)

(3.4) Implement Alg. 3.3 (`matrix-square`) on a fine grid of equidistant points. Start from the high-temperature density matrix in eqn (3.30), and iterate several times, doubling $\beta$ at each time. Compare your results with the exact density matrix for the harmonic oscillator (eqn (3.37)).

(3.5) Consider the exactly solvable Pöschl–Teller potential

$$V(x) = \frac{1}{2}\left[\frac{\chi(\chi - 1)}{\sin^2 x} \frac{\lambda(\lambda - 1)}{\cos^2 x}\right].$$

Plot this potential for several choices of $\chi > 1$ and $\lambda > 1$, with $x$ in the interval $[0, \pi/2]$. The energy eigenvalues of a particle of mass $m = 1$ in this potential are

$$E_n^{\text{P–T}} = \frac{1}{2}(\chi + \lambda + 2n)^2 \quad \text{for } n = 0, \ldots, \infty.$$

All the wave functions are known analytically, and the ground state has the simple form:

$$\psi_0^{\text{P–T}}(x) = \text{const} \cdot \sin^{\chi} x \, \cos^{\lambda} x \quad x \in [0, \pi/2].$$

Use the Trotter formula, and matrix squaring (Alg. 3.3 (`matrix-square`)), to compute the density matrix $\rho^{\text{P–T}}(x, x', \beta)$ at arbitrary temperature. Plot its diagonal part at low temperatures, and show that

$$\rho^{\text{P–T}}(x, x, \beta) \xrightarrow[\beta \to \infty]{} \text{const} \cdot \left[\psi_0^{\text{P–T}}(x)\right]^2,$$

for various values of $\chi$ and $\lambda$. Can you deduce the value of $E_0^{\text{P–T}}$ from the output of the matrix-squaring routine? Compute the partition function

$Z^{\mathrm{P-T}}(\beta)$ using matrix squaring, and compare with the explicit solution given by the sum over eigenvalues $E_n^{\mathrm{P-T}}$. Check analytically that $\psi_0^{\mathrm{P-T}}(x)$ is indeed the ground-state wave function of the Pöschl–Teller potential.

(3.6) In Section 3.2.1, we derived the second-order Trotter formula for the density matrix at high temperature. Show that the expression

$$\rho\left(x, x', \Delta_\tau\right) \simeq \rho^{\mathrm{free}}\left(x, x', \Delta_\tau\right) e^{-\Delta_\tau V(x')}$$

is correct only to first order in $\Delta_\tau$. Study the becomings of this first-order approximation under the convolution in eqn (3.32), separately for diagonal matrix elements $\rho(x, x, \Delta_\tau)$ and for nondiagonal elements $\rho(x, x', \Delta_\tau)$, each at low temperature (large $\beta = N\Delta_\tau$).

(3.7) Consider a particle of mass $m = 1$ in a box of size $L$. Compute the probability $\pi^{\mathrm{box}}(x)$ to be at a position $x$, and inverse temperature $\beta$, in three different ways: first, sum explicitly over all states (adapt Alg. 3.2 (harmonic-density) to the box wave functions of eqn (3.20), with eigenvalues $E_n = \frac{1}{2}(n\pi/L)^2$). Second, use Alg. 3.3 (matrix-square) to compute $\rho^{\mathrm{box}}(x, x', \beta)$ from the high-temperature limit

$$\rho^{\mathrm{box}}\left(x, x', \beta\right) = \begin{cases} \rho^{\mathrm{free}}(x, x', \beta) & \text{if } 0 < x, x' < L \\ 0 & \text{otherwise} \end{cases}.$$

Finally, compute $\pi^{\mathrm{box},L}(x)$ from the density-matrix expression in eqn (3.23). Pay attention to the different normalizations of the density matrix and the probability $\pi^{\mathrm{box}}(x)$.

## (Section 3.3)

(3.8) Implement Alg. 3.4 (naive-harmonic-path). Check your program by plotting a histogram of the positions $x_k$ for $k = 0$ and for $k = N/2$. Verify that the distributions $\pi(x_0)$ and of $\pi(x_{N/2})$ agree with each other and with the analytic form of $\rho^{\mathrm{h.o.}}(x, x, \beta)/Z$ (see eqn (3.38)).

(3.9) Implement the Lévy construction for paths contributing to the free density matrix (Alg. 3.5 (levy-free-path)). Use this subroutine in an improved path-integral simulation of the harmonic oscillator (see Exerc. 3.8): cut out a connected piece of the path, between time slices $k$ and $k'$ (possible across the horizon), and thread in a new piece, generated with Alg. 3.5 (levy-free-path). Determine the acceptance probabilities in the Metropolis algorithm, taking into account that the free-particle

Hamiltonian is already incorporated in the Lévy-construction. Run your program for a sufficiently long time to allow careful comparison with the exact solution (see eqn (3.38)).

(3.10) Use Markov-chain methods to sample paths contributing to the partition function of a particle in the Pöschl–Teller potential of Exerc. 3.5. As in Exerc. 3.9, cut out a piece of the path, between time slices $k$ and $k'$ (possibly across the horizon), and thread in a new path, again generated with Alg. 3.5 (levy-free-path) (compare with Exerc. 3.9). Correct for effects of the potential using the Metropolis algorithm, again taking into account that the free Hamiltonian is already incorporated in the Lévy construction. If possible, check Monte Carlo output against the density matrix $\rho^{\mathrm{P-T}}(x, x, \beta)$ obtained in Exerc. 3.5. Otherwise, check consistency at low temperature with the ground-state wave function $\psi_0^{\mathrm{P-T}}(x)$ quoted in Exerc. 3.5.

(3.11) Use Alg. 3.8 (naive-box-path) in order to sample paths contributing to $\rho^{\mathrm{box}}(x, x', \beta)$. Generalize to sample paths contributing to $Z^{\mathrm{box}}(\beta)$ (sample $x_0 = x_N$ from the diagonal density matrix, as in Fig. 3.15, then use Alg. 3.8 (naive-box-path)). Sketch how this naive algorithm can be made into a rejection-free direct sampling algorithm, using the exact solution for $\rho^{\mathrm{box}}(x, x', \beta)$ from eqn (3.47). Implement this algorithm, using a fine grid of $x$-values in the interval $[0, L]$.
NB: In the last part, use tower sampling, from Alg. 1.14 (tower-sample), to generate $x$-values.

## (Section 3.5)

(3.12) Compare the three direct-sampling algorithms for paths contributing to $\rho^{\mathrm{free}}(0, 0, \beta)$, namely Alg. 3.5 (levy-free-path), Alg. 3.12 (fourier-free-path), and finally Alg. 3.13 (trivial-free-path). Implement them. To show that they lead to equivalent results, Compute the correlation functions $\langle x_k x_l \rangle$. Can each of these algorithms be generalized to sample paths contributing to $\rho^{\mathrm{h.o.}}(0, 0, \beta)$?

(3.13) Generate periodic random paths with various roughness exponents $0 < \zeta < 1.5$ using Alg. 3.14 (fourier-gen-path). Plot the mean square width $\omega_2$ as a function of $L$. For given $L$, determine the scaling of the mean square deviation $\langle |x(t) - x(0)|^2 \rangle$ as a function of $t$. Explain why these two quantities differ qualitatively for $\zeta > 1$ (see Leschhorn and Tang (1993)).

# References

Feynman R. P. (1972) *Statistical Mechanics: A Set of Lectures*, Benjamin/Cummings, Reading, Massachusetts

Hess G. B., Fairbank W. M. (1967) Measurement of angular momentum in superfluid helium, *Physical Review Letters* **19**, 216–218

Krauth W. (1996) Quantum Monte Carlo calculations for a large number of bosons in a harmonic trap, *Physical Review Letters* **77**, 3695–3699

Leggett A. J. (1973) Topics in the theory of helium, *Physica Fennica* **8**, 125–170

Leschhorn H., Tang L. H. (1993) Elastic string in a random potential–comment, *Physical Review Letters* **70**, 2973

Lévy P., (1940) Sur certains processus stochastiques homogènes [in French], *Composition Mathematica* **7**, 283–339

Pollock E. L., Ceperley D. M. (1984) Simulation of quantum many-body systems by path-integral methods, *Physical Review B* **30**, 2555–2568

Pollock E. L., Ceperley D. M. (1987) Path-integral computation of superfluid densities, *Physical Review B* **36**, 8343–8352

Rosso A., Santachiara R., Krauth W. (2005) Geometry of Gaussian signals, *Journal of Statistical Mechanics: Theory and Experiment*, L08001

Storer R. G. (1968) Path-integral calculation of quantum-statistical density matrix for attractive Coulomb forces, *Journal of Mathematical Physics* **9**, 964–970

# Bosons

The present chapter introduces the statistical mechanics and computational physics of identical bosons. Thus, after studying, in Chapter 3, the manifestation of Heisenberg's uncertainty principle at finite temperature for a single quantum particle or several distinguishable quantum particles, we now follow identical particles to low temperature, where they lose their individual characters and enter a collective state of matter characterized by Bose condensation and superfluidity.

We mostly focus in this chapter on the modeling of noninteracting (ideal) bosons. Our scope is thus restricted, by necessity, because quantum systems are more complex than the classical models of earlier chapters. However, ideal bosons are of considerably greater interest than other noninteracting systems because they can have a phase transition. In fact, a statistical interaction originates here from the indistinguishability of the particles, and can best be studied in the model of ideal bosons, where it is not masked by other interactions. This chapter's restriction to ideal bosons—leaving aside the case of interacting bosons—is thus, at least partly, a matter of choice.

Ideal bosons are looked at from two technically and conceptually different perspectives. First, we focus on the energy-level description of the ideal Bose gas. This means that particles are distributed among single-particle energy levels following the laws of bosonic statistics. As in other chapters, we stress concrete calculations with a finite number of particles.

The chapter's second viewpoint is density matrices and path integrals, which are important tools in computational quantum physics. We give a rather complete treatment of the ideal Bose gas in this framework, leading up to a direct-sampling algorithm for ideal bosons at finite temperature. We have already discussed in Chapter 3 the fact that interactions are very easily incorporated into the path-integral formalism. Our ideal-boson simulation lets us picture real simulations of interacting bosons in three-dimensional traps and homogeneous periodic boxes.

Our main example system, bosons in a harmonic trap, has provided the setting for groundbreaking experiments in atomic physics, where Bose–Einstein condensation has actually been achieved and studied in a way very close to what we shall do in our simulations using path-integral Monte Carlo methods. Path-integral Monte Carlo methods have become a standard approach to interacting quantum systems, from $^4$He, to interacting fermions and bosons in condensed matter physics, and to atomic gases.

We suppose that a three-dimensional harmonic trap, with harmonic potentials in all three space dimensions, is filled with bosons (see Fig. 4.1). Well below a critical temperature, most particles populate the state with the lowest energy. Above this temperature, they are spread out into many states, and over a wide range of positions in space. In the harmonic trap, the Bose–Einstein condensation temperature increases as the number of particles grows. We shall discuss bosonic statistics and calculate condensation temperatures, but also simulate thousands of (ideal) bosons in the trap, mimicking atomic gases, where Bose–Einstein condensation was first observed, in 1995, at microkelvin temperatures.

**Fig. 4.1** Energy levels $\{E_x, E_y, E_z\}$ of a quantum particle in a harmonic trap. The total energy is $E = E_x + E_y + E_z$.

# 4.1   Ideal bosons (energy levels)

In this section, we consider ideal bosons in a three-dimensional harmonic trap, as realized in experiments in atomic physics, and also bosons in a three-dimensional box with periodic boundary conditions, a situation more closely related to liquid and gaseous helium ($^4$He). Both systems will be described in the usual framework of energy levels. Many calculations will be redone in Section 4.2, in the less familiar but more powerful framework of density matrices.

## 4.1.1   Single-particle density of states

In this subsection, we review the concept of a single-particle state and compute the single-particle degeneracy $\mathcal{N}(E)$, that is, the number of these states with energy $E$. Let us start with the harmonic trap. For simplicity, we choose all of the spring constants $\{\omega_x, \omega_y, \omega_z\}$ equal to one,[1] so that the eigenvalues satisfy

$$\left.\begin{array}{c} E_x \\ E_y \\ E_z \end{array}\right\} = 0, 1, 2, \ldots$$

(see Fig. 4.1). To simplify the notation, in this chapter we subtract the zero-point energy, that is, we set the energy of the ground state equal to zero. The total energy of one particle in the potential is $E = E_x + E_y + E_z$. We need to compute the number of different choices for $\{E_x, E_y, E_z\}$ which give an energy $E$ (see Alg. 4.1 (`naive-degeneracy`), and Table 4.1).

**Table 4.1** Degeneracy $\mathcal{N}(E)$ for the harmonic trap

| $E$ | $\{E_x, E_y, E_z\}$ | $\mathcal{N}(E)$ |
|---|---|---|
| 0 | $\{0,0,0\}$ | 1 |
| 1 | $\{0,0,1\}$ $\{0,1,0\}$ $\{1,0,0\}$ | 3 |
| 2 | $\{0,0,2\}$ $\{0,1,1\}$ $\{0,2,0\}$ $\{1,0,1\}$ $\{1,1,0\}$ $\{2,0,0\}$ | 6 |
| 3 | ... | 10 |
| 4 | ... | 15 |
| ... | ... | ... |

---

**procedure** `naive-degeneracy`
$\{\mathcal{N}(0), \ldots, \mathcal{N}(E_{\max})\} \leftarrow \{0, \ldots, 0\}$
**for** $E_x = 0, \ldots, E_{\max}$ **do**
$\left\{\begin{array}{l} \textbf{for } E_y = 0, \ldots, E_{\max} \textbf{ do} \\ \quad\left\{\begin{array}{l} \textbf{for } E_z = 0, \ldots, E_{\max} \textbf{ do} \\ \quad\left\{\begin{array}{l} E \leftarrow E_x + E_y + E_z \\ \textbf{if } (E \leq E_{\max}) \textbf{ then} \\ \quad\{ \ \mathcal{N}(E) \leftarrow \mathcal{N}(E) + 1 \end{array}\right. \end{array}\right. \end{array}\right.$
**output** $\{\mathcal{N}(0), \ldots, \mathcal{N}(E_{\max})\}$

---

**Algorithm 4.1** `naive-degeneracy`. Single-particle degeneracy $\mathcal{N}(E)$ for the harmonic trap (see Table 4.1).

In the case of the harmonic trap, $\mathcal{N}(E)$ can be computed explicitly:

$$\big\{\text{total energy}\big\} = E = E_x + \underbrace{E_y}_{0 \leq E_y \leq E - E_x} + \underbrace{\text{remainder}}_{\geq 0}.$$

[1] Throughout this chapter, the word "harmonic trap" refers to an isotropic three-dimensional harmonic potential with $\omega_x = \omega_y = \omega_z = 1$.

For each choice of $E_x$, the energy $E_y$ may be any integer from 0 to $E - E_x$, but then, for given $E_x$ and $E_y$, the remainder, $E_z$, is fixed:

$$\mathcal{N}(E) = \sum_{E_x=0}^{E} \left\{ \begin{array}{c} \text{number of choices} \\ \text{for } E_y \text{ given } E_x \end{array} \right\} = \sum_{E_x=0}^{E} (E - E_x + 1)$$

$$= (E+1) + (E) + \cdots + (1) = \frac{(E+1)(E+2)}{2}. \quad (4.1)$$

Splitting the energy into $\{E_x, E_y\}$ and a remainder is a nice trick, but it is better to use a systematic method[2] for computing $\mathcal{N}(E)$ before using it in far more complicated contexts. The method consists in writing the density of states as a free sum over $\{E_x, E_y, E_z\}$

$$\mathcal{N}(E) = \sum_{E_x=0}^{E} \sum_{E_y=0}^{E} \sum_{E_z=0}^{E} \delta_{(E_x+E_y+E_z),E}, \quad (4.2)$$

where the Kronecker $\delta$-function is defined as

$$\delta_{j,k} = \begin{cases} 1 \text{ if } j = k \\ 0 \text{ if } j \neq k \end{cases}.$$

Because of the $\delta$-function, only combinations of $\{E_x, E_y, E_z\}$ with a sum equal to $E$ contribute to $\mathcal{N}(E)$ in eqn (4.2). The Kronecker $\delta$-function may be represented as an integral,

$$\delta_{j,k} = \int_{-\pi}^{\pi} \frac{d\lambda}{2\pi} e^{i(j-k)\lambda}. \quad (4.3)$$

This formula is evidently correct for $j = k$ (we integrate $1/(2\pi)$ from $-\pi$ to $\pi$), but it is also correct for integers $j \neq k$, because the oscillatory terms sum to zero.

We enter the integral representation of the Kronecker $\delta$-function into the density of states in eqn (4.2), exchange sums and integrals, and see that the three sums in the density of states have become independent:

$$\mathcal{N}(E) = \int_{-\pi}^{\pi} \frac{d\lambda}{2\pi} e^{-iE\lambda} \left( \sum_{E_x=0}^{E} e^{iE_x\lambda} \right) \left( \sum_{E_y=0}^{E} e^{iE_y\lambda} \right) \left( \sum_{E_z=0}^{E} e^{iE_z\lambda} \right).$$

The three geometric sums can be evaluated explicitly:

$$\mathcal{N}(E) = \int_{-\pi}^{\pi} \frac{d\lambda}{2\pi} e^{-iE\lambda} \underbrace{\left[ \frac{1 - e^{i(E+1)\lambda}}{1 - e^{i\lambda}} \right]^3}_{\mathcal{N}(E,\lambda)}. \quad (4.4)$$

This integral can be evaluated by Riemann integration (see Fig. 4.2). It reproduces eqn (4.1). It may also be evaluated exactly. The substitution $e^{i\lambda} = z$ gives the complex contour integral

$$\mathcal{N}(E) = \frac{1}{2\pi i} \oint_{|z|=1} \frac{dz}{z^{E+1}} \left( \frac{1 - z^{E+1}}{1 - z} \right)^3. \quad (4.5)$$

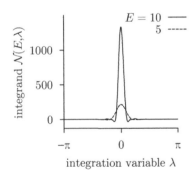

integrand $\mathcal{N}(E,\lambda)$

$E = 10$ ——
$5$ -----

1000

500

0

$-\pi$    0    $\pi$

integration variable $\lambda$

**Fig. 4.2** Real part of the integrand of eqn (4.4).

[2]What works once is a trick; what works twice is a method.

Using

$$\left(\frac{1}{1-z}\right)^3 = \frac{1}{2}\left(1\times 2 + 2\times 3z + 3\times 4z^2 + \cdots\right),$$

we expand the integrand into a Laurent (power) series around the singularity at $z = 0$:

$$\mathcal{N}(E) = \frac{1}{2\pi i}\oint \frac{dz}{z^{E+1}}\frac{1}{2}\left(1\times 2 + 2\times 3z + 3\times 4z^2 + \cdots\right)\left(1 - z^{E+1}\right)^3$$

$$= \frac{1}{2\pi i}\oint dz \left[\cdots + \frac{1}{2}(E+1)(E+2)z^{-1} + \cdots\right]. \qquad (4.6)$$

The residue theorem of complex analysis states that the coefficient of $z^{-1}$, namely $\frac{1}{2}(E+1)(E+2)$, is the value of the integral. Once more, we obtain eqn (4.1).

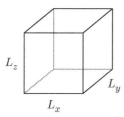

**Fig. 4.3** Three-dimensional cubic box with periodic boundary conditions and edge lengths $L_x = L_y = L_z = L$.

After dealing with the single-particle degeneracy for a harmonic trap, we now consider the same problem for a cubic box with periodic boundary conditions (see Fig. 4.3). The total energy is again a sum of the energies $E_x$, $E_y$, and $E_z$. As discussed in Subsection 3.1.2, the allowed energies for a one-dimensional line of length $L$ with periodic boundary conditions are

$$\left.\begin{array}{r}E_x\\E_y\\E_z\end{array}\right\} = \frac{2\pi^2}{L^2}\left[\ldots, (-2)^2, (-1)^1, 0, 1^2, 2^2, \ldots\right]$$

(see eqn (3.10)), For the moment, let us suppose that the cube has a side length $L = \sqrt{2}\pi$. Every integer lattice site $\{n_x, n_y, n_z\}$ in a three-dimensional lattice then contributes to the single-particle degeneracy of the energy $E = n_x^2 + n_y^2 + n_z^2$ (see Fig. 4.4 for a two-dimensional representation). The single-particle density of states is easily computed (see Alg. 4.2 (`naive-degeneracy-cube`)). The number of states with an energy below $E$ is roughly equal to the volume of a sphere of radius $\sqrt{E}$, because the integer lattice has one site per unit volume (see Table 4.2). The density of states $\mathcal{N}(E)$ is the key ingredient for calculating the thermodynamic properties of a homogeneous Bose gas.

**Table 4.2** Single-particle degeneracy for a cubic box (from Alg. 4.2 (`naive-degeneracy-cube`), $L = \sqrt{2}\pi$)

| $E$ | $\mathcal{N}(E)$ | $\sum_{E'\le E}\mathcal{N}(E')$ | $\frac{4}{3}\pi E^{3/2}$ |
|---|---|---|---|
| 0 | 1 | 1 | 0.00 |
| 1 | 6 | 7 | 4.19 |
| 2 | 12 | 19 | 11.85 |
| 3 | 8 | 27 | 21.77 |
| 4 | 6 | 33 | 33.51 |
| 5 | 24 | 57 | 46.83 |
| 6 | 24 | 81 | 61.56 |
| 7 | 0 | 81 | 77.58 |
| 8 | 12 | 93 | 94.78 |
| 9 | 30 | 123 | 113.10 |

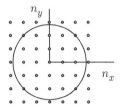

**Fig. 4.4** $\mathcal{N}(E)$ in a square box of edge length $\sqrt{2}\pi$ with periodic boundary conditions. Integer lattice point $\{n_x, n_y\}$ contribute to $\mathcal{N}\left(n_x^2 + n_y^2\right)$.

**procedure** `naive-degeneracy-cube`
$\{\mathcal{N}(0), \dots, \mathcal{N}(E_{\max})\} \leftarrow \{0, \dots, 0\}$
$n_{\max} \leftarrow \text{int}\sqrt{E_{\max}}$
**for** $n_x = -n_{\max}, \dots, n_{\max}$ **do**
$\left\lceil\quad \textbf{for } n_y = -n_{\max}, \dots, n_{\max} \textbf{ do}\right.$
$\quad\left\lceil\quad \textbf{for } n_z = -n_{\max}, \dots, n_{\max} \textbf{ do}\right.$
$\quad\quad\left\lceil\quad E \leftarrow n_x^2 + n_y^2 + n_z^2\right.$
$\quad\quad\quad \textbf{if } (|E| \leq E_{\max}) \textbf{ then}$
$\quad\quad\quad\quad \{\ \mathcal{N}(E) \leftarrow \mathcal{N}(E) + 1$
**output** $\{\mathcal{N}(0), \dots, \mathcal{N}(E_{\max})\}$

**Algorithm 4.2** `naive-degeneracy-cube`. Single-particle degeneracy $\mathcal{N}(E)$ for a periodic cubic box of edge length $\sqrt{2}\pi$.

### 4.1.2   Trapped bosons (canonical ensemble)

The preliminaries treated so far in this chapter have familiarized us with the concepts of single-particle states and the corresponding density of states. The integral representation of the Kronecker $\delta$-function was used to transform a constraint on the particle number into an integral. (The integration variable $\lambda$ will later be related to the chemical potential.) In this subsection, we apply this same method to $N$ trapped bosons. For concreteness, we shall first work on what we call the five-boson bounded trap model, which consists of five bosons in the harmonic trap, as in Fig. 4.5, but with a cutoff on the single-particle energies, namely $E_\sigma \leq 4$. For this model, naive enumeration still works, so we quickly get an initial result.

The five-boson bounded trap model keeps the first 35 single-particle states listed in Table 4.1:

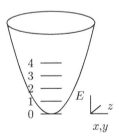

**Fig. 4.5** The five-boson bounded trap model with a cutoff on the single-particle spectrum ($E_{\max} = 4$). The degeneracies are as in eqn (4.7).

$$
\begin{aligned}
E &= 4 \quad (15 \text{ states:} \quad \sigma = 20, \dots, 34), \\
E &= 3 \quad (10 \text{ states:} \quad \sigma = 10, \dots, 19), \\
E &= 2 \quad (6 \text{ states:} \quad \sigma = 4, \dots, 9 \quad ), \\
E &= 1 \quad (3 \text{ states:} \quad \sigma = 1, 2, 3 \quad ), \\
E &= 0 \quad (1 \text{ state:} \quad \sigma = 0 \quad\quad ).
\end{aligned}
\tag{4.7}
$$

We construct the five-particle states by packing particle 1 into state $\sigma_1$, particle 2 into state $\sigma_2$, etc., each state being taken from the list of 35 states in eqn (4.7):

$$\left\{ \begin{array}{c} \text{five-particle} \\ \text{state} \end{array} \right\} = \{\sigma_1, \ldots, \sigma_5\}.$$

The quantum statistical mechanics of the five-boson bounded trap model derives from its partition function

$$Z_{\text{btm}} = \sum_{\substack{\text{all five-particle} \\ \text{states}}} e^{-\beta E_{\text{tot}}(\sigma_1, \ldots, \sigma_5)}, \qquad (4.8)$$

where the total five-particle energy is equal to

$$E_{\text{tot}} = E_{\text{tot}}(\sigma_1, \ldots, \sigma_5) = E_{\sigma_1} + E_{\sigma_2} + E_{\sigma_3} + E_{\sigma_4} + E_{\sigma_5}.$$

While we should be careful not to forget any five-particle states in the partition function in eqn (4.8), we are not allowed to overcount them either. The problem appears because bosons are identical particles, so that there is no way to tell them apart: the same physical five-particle state corresponds to particle 1 being in (single-particle) state 34, particle 2 in state 5, and particle 3 in state 8, or to particle 1 being in state 8, particle 2 in state 5, and particle 3 in state 34, etc.:

$$\begin{bmatrix} \sigma_1 \leftarrow 34 \\ \sigma_2 \leftarrow 5 \\ \sigma_3 \leftarrow 8 \\ \sigma_4 \leftarrow 0 \\ \sigma_5 \leftarrow 11 \end{bmatrix} \begin{array}{c} \text{same} \\ \text{as} \end{array} \begin{bmatrix} \sigma_1 \leftarrow 8 \\ \sigma_2 \leftarrow 5 \\ \sigma_3 \leftarrow 34 \\ \sigma_4 \leftarrow 0 \\ \sigma_5 \leftarrow 11 \end{bmatrix} \begin{array}{c} \text{same} \\ \text{as} \end{array} \begin{bmatrix} \sigma_1 \leftarrow 0 \\ \sigma_2 \leftarrow 5 \\ \sigma_3 \leftarrow 8 \\ \sigma_4 \leftarrow 11 \\ \sigma_5 \leftarrow 34 \end{bmatrix} \text{etc.} \qquad (4.9)$$

The groundbreaking insight of Bose (in 1923, for photons) and of Einstein (in 1924, for massive bosons) that the partition function should count only one of the states in eqn (4.9) has a simple combinatorial implementation. To avoid overcounting in eqn (4.8), we consider only those states which satisfy

$$0 \leq \sigma_1 \leq \sigma_2 \leq \sigma_3 \leq \sigma_4 \leq \sigma_5 \leq 34. \qquad (4.10)$$

Out of all the states in eqn (4.9), we thus pick the last one.

The ordering trick in eqn (4.10) lets us write the partition function and the mean energy as

$$Z_{\text{btm}} = \sum_{0 \leq \sigma_1 \leq \cdots \leq \sigma_5 \leq 34} e^{-\beta E_{\text{tot}}(\sigma_1, \ldots, \sigma_5)},$$

$$\langle E \rangle = \frac{1}{Z_{\text{btm}}} \sum_{0 \leq \sigma_1 \leq \cdots \leq \sigma_5 \leq 34} E_{\text{tot}}(\sigma_1, \ldots, \sigma_5) e^{-\beta E_{\text{tot}}(\sigma_1, \ldots, \sigma_5)}.$$

Some of the particles $k$ may be in the ground state ($\sigma_k = 0$). The number of these particles, $N_0(\sigma_1, \ldots, \sigma_5)$, is the ground-state occupation number of the five-particle state $\{\sigma_1, \ldots, \sigma_5\}$. (In the five-particle state in eqn (4.9), $N_0 = 1$.) The number $N_0$ can be averaged over the

**Table 4.3** Thermodynamics of the five-boson bounded trap model (from Alg. 4.3 (naive-bosons))

| $T$ | $Z_{\text{btm}}$ | $\langle E\rangle/N$ | $\langle N_0\rangle/N$ |
|-----|------|------|------|
| 0.1 | 1.000 | 0.000 | 1.000 |
| 0.2 | 1.021 | 0.004 | 0.996 |
| 0.3 | 1.124 | 0.026 | 0.976 |
| 0.4 | 1.355 | 0.074 | 0.937 |
| 0.5 | 1.780 | 0.157 | 0.878 |
| 0.6 | 2.536 | 0.282 | 0.801 |
| 0.7 | 3.873 | 0.444 | 0.711 |
| 0.8 | 6.237 | 0.634 | 0.616 |
| 0.9 | 10.359 | 0.835 | 0.526 |
| 1.0 | 17.373 | 1.031 | 0.447 |

Boltzmann distribution analogously to the way the mean energy was calculated. The mean number of particles in the ground state divided by the total number of particles is called the condensate fraction. Algorithm 4.3 (naive-bosons) evaluates the above sums and determines the partition function, the mean energy, and the condensate fraction of the five-boson bounded trap model. It thus produces our first numerically exact results in many-particle quantum statistical mechanics (see Table 4.3).

**procedure** naive-bosons
$\{E_0,\dots,E_{34}\} \leftarrow \{0,1,1,1,2,\dots,4\}$ (from eqn (4.7))
$Z_{\text{btm}} \leftarrow 0$
$\langle E\rangle \leftarrow 0$
$\langle N_0\rangle \leftarrow 0$
**for** $\sigma_1 = 0,\dots,34$ **do**
$\quad\begin{cases} \textbf{for } \sigma_2 = \sigma_1,\dots,34 \textbf{ do} \\ \quad\begin{cases} \cdots \\ \dots \textbf{for } \sigma_5 = \sigma_4,\dots,34 \textbf{ do} \\ \quad\begin{cases} E_{\text{tot}} \leftarrow E_{\sigma_1} + \cdots + E_{\sigma_5} \\ N_0 \leftarrow \#\text{ of zeros among } \{\sigma_1,\dots,\sigma_5\} \\ Z_{\text{btm}} \leftarrow Z_{\text{btm}} + e^{-\beta E_{\text{tot}}} \\ \langle E\rangle \leftarrow \langle E\rangle + E_{\text{tot}}e^{-\beta E_{\text{tot}}} \\ \langle N_0\rangle \leftarrow \langle N_0\rangle + N_0 e^{-\beta E_{\text{tot}}} \end{cases} \\ \cdots \end{cases} \end{cases}$
$\langle E\rangle \leftarrow \langle E\rangle/Z_{\text{btm}}$
$\langle N_0\rangle \leftarrow \langle N_0\rangle/Z_{\text{btm}}$
**output** $\{Z_{\text{btm}}, \langle E\rangle, \langle N_0\rangle\}$
—

**Algorithm 4.3** naive-bosons. Thermodynamics of the five-boson bounded trap model at temperature $1/\beta$.

At low temperature, almost all particles are in the ground state and the mean energy is close to zero, as is natural at temperatures $T \lesssim 1$. This is not yet (but soon will be) the phenomenon of Bose–Einstein condensation, which concerns the macroscopic population of the ground state for large systems at temperatures much higher than the difference in energy between the ground state and the excited states, in our case at temperatures $T \gg 1$.

Algorithm 4.3 (naive-bosons) goes 575 757 times through its inner loop. To check that this is indeed the number of five-particle states in the model, we can represent the 35 single-particle states as modular offices, put together from two fixed outer walls and 34 inner walls:

35 modular offices, 2 outer walls and 34 inner walls

Five identical particles are placed into these offices, numbered from 0 to

34, for example as in the following:

$$\underbrace{] \,|\, \bullet \,|| \,\bullet \,||| \,\bullet \,\bullet \,||||||||||| \,\bullet \,||||||||||||||||||[}_{\text{one particle in office (single-particle state) 1, one in 3, two in 6, one in 17}}. \quad (4.11)$$

Each bosonic five-particle state corresponds to one assignment of office walls as in eqn (4.11). It follows that the number of ways of distributing $N$ particles into $k$ offices is the same as that for distributing $N$ particles and $k-1$ inner walls, i.e. a total number of $N+k-1$ objects. The inner walls are identical and the bosons are also identical, and so we have to divide by combinatorial factors $N!$ and $(k-1)!$:

$$\left\{ \begin{array}{c} \text{number of } N\text{-particle} \\ \text{states from} \\ k \text{ single-particle states} \end{array} \right\} = \frac{(N+k-1)!}{N!(k-1)!} = \binom{N+k-1}{N}. \quad (4.12)$$

Equation (4.12) can be evaluated for $N = 5$ and $k = 35$. It gives 575 757 five-particle states for the five-boson bounded trap model, and allows Alg. 4.3 (**naive-bosons**) to pass an important first test. However, eqn (4.12) indicates that for larger particle numbers and more states, a combinatorial explosion will put a halt to naive enumeration, and this incites us to refine our methods.

Considerably larger particle numbers and states can be treated by characterizing any single-particle state $\sigma = 0, \ldots, 34$ by an occupation number $n_\sigma$. In the example of eqn (4.11), we have

$$\left\{ \begin{array}{c} \text{five-particle state} \\ \text{in config. (4.11)} \end{array} \right\} : \ \text{all } n_\sigma = 0 \quad \text{except} \quad \begin{bmatrix} n_1 = 1 \\ n_3 = 1 \\ n_6 = 2 \\ n_{17} = 1 \end{bmatrix}.$$

A five-particle state can thus be defined in terms of all the occupation numbers

$$\left\{ \begin{array}{c} \text{five-particle} \\ \text{state} \end{array} \right\} = \underbrace{\{n_0, \ldots, n_{34}\}}_{n_0 + \cdots + n_{34} = 5},$$

and its energy by

$$E_{\text{tot}} = n_0 E_0 + n_1 E_1 + \cdots + n_{34} E_{34}.$$

The statistical weight of this configuration is given by the usual Boltzmann factor $e^{-\beta E_{\text{tot}}}$. The partition function of the five-boson bounded trap model is

$$Z_{\text{btm}}(\beta) = \sum_{n_0=0}^{5} \cdots \sum_{n_{34}=0}^{5} e^{-\beta(n_0 E_0 + \cdots + n_{34} E_{34})} \delta_{(n_0 + \cdots + n_{34}),5}, \quad (4.13)$$

where the Kronecker $\delta$-function fixes the number of particles to five. Equations (4.8) and (4.13) are mathematically equivalent expressions for the partition function but eqn (4.13) is hardly the simpler one: instead of summing over $575\,757$ states, there is now an equivalent sum over $6^{35} \simeq 1.72 \times 10^{27}$ terms. However, eqn (4.13) has been put together from an unrestricted sum and a constraint (a Kronecker $\delta$-function), and the equation can again be simplified using the providential integral representation given in Subsection 4.1.1 (see eqn (4.3)):

$$Z_{\text{btm}}(\beta) = \int_{-\pi}^{\pi} \frac{d\lambda}{2\pi} e^{-iN\lambda}$$

$$\times \underbrace{\left( \sum_{n_0} e^{n_0(-\beta E_0 + i\lambda)} \right)}_{f_0(\beta,\lambda)} \cdots \underbrace{\left( \sum_{n_{34}} e^{n_{34}(-\beta E_{34} + i\lambda)} \right)}_{f_{34}(\beta,\lambda)}. \quad (4.14)$$

In this difficult problem, the summations have again become independent, and can be performed. As written in eqn (4.13), the sums should go from 0 to $N$ for all states. However, for the excited states ($E > 0$), the absolute value of $\Upsilon = e^{-\beta E + i\lambda}$ is smaller than 1:

$$|\Upsilon| = \left| e^{-\beta E} e^{i\lambda} \right| = \underbrace{\left| e^{-\beta E} \right|}_{\substack{<1 \text{ if} \\ E>0}} \underbrace{\left| e^{i\lambda} \right|}_{1},$$

which allows us to take the sum to infinity:

$$\sum_{n=0}^{N} \Upsilon^n = \frac{1 - \Upsilon^{N+1}}{1 - \Upsilon} \xrightarrow[|\Upsilon|<1]{N\to\infty} \frac{1}{1 - \Upsilon}.$$

The Kronecker $\delta$-function picks up the correct terms even from an infinite sum, and we thus take the sums in eqn (4.14) for the excited states to infinity, but treat the ground state differently, using a finite sum. We also take into account the fact that the $f_k$ depend only on the energy $E_k$, and not explicitly on the state number:

$$f_E(\beta, \lambda) = \frac{1 - \exp\left[i(N+1)\lambda\right]}{1 - \exp(i\lambda)}, \quad E = 0, \text{ (ground state),} \quad (4.15)$$

$$f_E(\beta, \lambda) = \frac{1}{1 - \exp(-\beta E + i\lambda)}, \quad E > 0, \text{ (excited state).} \quad (4.16)$$

(The special treatment of the ground state is "naive", that is, appropriate for a first try. In Subsection 4.1.3, we shall move the integration contour for $\lambda$ in the complex plane, and work with infinite sums for all energies.)

The partition function is finally written as

$$Z_N(\beta) = \int_{-\pi}^{\pi} \frac{d\lambda}{2\pi} \underbrace{e^{-iN\lambda} \prod_{E=0}^{E_{\max}} [f_E(\beta, \lambda)]^{\mathcal{N}(E)}}_{Z_N(\beta,\lambda)}. \quad (4.17)$$

This equation is generally useful for an $N$-particle problem. For the five-boson bounded trap model, we can input $\mathcal{N}(E) = \frac{1}{2}(E+1)(E+2)$

directly from eqn (4.1), and set $E_{max} = 4$. We may lack the courage to evaluate this one-dimensional integral analytically in the complex plane, as for $\mathcal{N}(E)$, but can easily integrate numerically, using Riemann sums and complex arithmetic (see Fig. 4.6). For the five-boson bounded trap model, the results of Table 4.3 are reproduced, but we can now go to larger particle numbers and also push $E_{max}$ to infinity.

From the partition function, it is possible to obtain the mean energy,

$$\langle E_N \rangle = -\frac{\partial \log Z_N(\beta)}{\partial \beta}, \tag{4.18}$$

by comparing $\log Z_N$ at two nearby temperatures. It is better to differentiate directly inside the sum of eqn (4.13), so that, in the five-boson bounded trap model:

$$\langle E_{btm} \rangle = \frac{1}{Z_{btm}(\beta)} \sum_{n_0} \cdots \sum_{n_{34}} (n_0 E_0 + \cdots + n_{34} E_{34})$$
$$\times \exp\left[ -\beta(n_0 E_0 + \cdots + n_{34} E_{34}) \right] \delta_{(n_0 + \cdots + n_{34}), 5}. \tag{4.19}$$

Each of the terms in parentheses in eqn (4.19) generates expressions of the general form

$$E(\beta, \lambda) = -\sum_k \frac{1}{f_k} \frac{\partial f_k}{\partial \beta} = \sum_{E \neq 0} \mathcal{N}(E) \, E \cdot \left( \frac{e^{-\beta E + i\lambda}}{1 - e^{-\beta E + i\lambda}} \right),$$

which give, with $E(\beta, \lambda)$ from the above expression,

$$\langle E_N(\beta) \rangle = \frac{1}{Z_N(\beta)} \int_{-\pi}^{\pi} \frac{d\lambda}{2\pi} e^{-iN\lambda} E(\beta, \lambda) \prod_{E=0}^{E_{max}} [f_E(\beta, \lambda)]^{\mathcal{N}(E)}.$$

The condensate corresponds to the mean number of particles in the ground state $\sigma = 0$. This number is obtained by differentiating $\log f_0$ with respect to $i\lambda$. Using the fact that the ground state $\sigma = 0$ has zero energy, we find

$$N_0(\beta, \lambda) = \frac{\partial}{i\partial\lambda} \log f_0 = \left[ -\frac{(N+1)\, e^{i\lambda(N+1)}}{1 - e^{i\lambda(N+1)}} + \frac{e^{i\lambda}}{1 - e^{i\lambda}} \right],$$

$$\langle N_0(\beta) \rangle = \frac{1}{Z_N(\beta)} \int_{-\pi}^{\pi} \frac{d\lambda}{2\pi} e^{-iN\lambda} N_0(\beta, \lambda) \prod_{E=0}^{E_{max}} [f_E(\beta, \lambda)]^{\mathcal{N}(E)}.$$

The calculation of the partition function, energy, and condensate fraction relies on the degeneracies $\mathcal{N}(E)$ in the form of a table or an explicit formula, and also on the temperature-dependent functions $f_E(\beta, \lambda)$ defined in eqn (4.16) (see Alg. 4.4 (`canonic-bosons`); the special treatment the ground state is naive, see eqn (4.15)). This algorithm reproduces Table 4.3 for $E_{max} = 4$ and $N = 5$. It allows us to push both the energy cutoff and the particle numbers to much larger values, and avoids any combinatorial explosion (see Table 4.4). In Subsection 4.2.3,

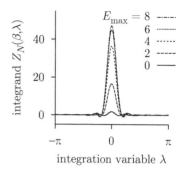

**Fig. 4.6** Real part of the integrand of eqn (4.17) for $N = 10$ and $T = 1$.

**procedure** `canonic-bosons`
**input** $\Delta_\mu$ (step size for Riemann sums)
$Z \leftarrow 0; \langle E \rangle \leftarrow 0; \langle N_0 \rangle \leftarrow 0$
**for** $\lambda = -\pi, -\pi + \Delta_\mu, \ldots, \pi$ **do**
$\left\{\begin{array}{l} Z_\lambda \leftarrow f_0(\lambda) \\ E_\lambda \leftarrow 0 \\ \textbf{for } E = 1, \ldots, E_{\max} \textbf{ do} \\ \quad \left\{\begin{array}{l} Z_\lambda \leftarrow Z_\lambda f_E(\lambda, \beta)^{\mathcal{N}(E)} \\ E_\lambda \leftarrow E_\lambda + \mathcal{N}(E)\, E \cdot \left(e^{-\beta E + i\lambda}\right) / \left(1 - e^{-\beta E + i\lambda}\right) \end{array}\right. \\ Z \leftarrow Z + Z_\lambda e^{-iN\lambda} \frac{\Delta_\mu}{2\pi} \\ \langle E \rangle \leftarrow \langle E \rangle + E_\lambda Z_\lambda e^{-iN\lambda} \frac{\Delta_\mu}{2\pi} \end{array}\right.$
$\langle E \rangle \leftarrow \langle E \rangle / Z$
**output** $Z, \langle E \rangle$

————

**Algorithm 4.4** `canonic-bosons`. Thermodynamics for $N$ bosons using the integral representation of the Kronecker $\delta$-function.

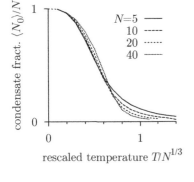

**Fig. 4.7** Condensate fraction $\langle N_0 \rangle / N$ of $N$ bosons in the harmonic trap.

an algorithm based on path integrals will greatly improve on Alg. 4.4 (`canonic-bosons`).

In Fig. 4.7, all particles populate the ground state in the limit of zero temperature. For temperatures much smaller than the energy gap between the ground state and the first excited level, this is a simple consequence of Boltzmann statistics. Bose condensation, in contrast, is what occurs when a finite fraction of particles populates the ground state at temperatures much larger than the energy gap (in the harmonic trap, the gap is equal to 1). This is illustrated for the harmonic trap by a plot of the condensate fraction $\langle N_0 \rangle / N$ against $T/N^{1/3}$; we shall discuss in Subsection 4.1.3 the reason for rescaling the temperature as $N^{1/3}$. On the basis of this plot, we can conjecture correctly that the condensate fraction falls to zero at a rescaled temperature $T_c/N^{1/3} \simeq 1$, so that the transition temperature $(T_c \simeq N^{1/3})$ is, for large $N$, indeed much larger than the difference $(= 1)$ between the ground state and the first excited state. Figure 4.7 does describe a system that undergoes Bose–Einstein condensation. Throughout this subsection, the number of particles was kept fixed; we considered what is called the canonical ensemble. In the canonical ensemble, calculations are sometimes more complicated than when we allow the total number of particles to fluctuate (see Subsection 4.1.3). However, the canonical ensemble is the preferred framework for many computational approaches, in classical and in quantum physics alike.

### 4.1.3   Trapped bosons (grand canonical ensemble)

So far, we have computed exact $N$-boson partition functions and related quantities. In the present subsection, we study the limit $N \to \infty$, where the integral in eqn (4.17) may be done by the saddle point method. This means that instead of adding up contributions to this integral over many

**Table 4.4** Thermodynamics for $N = 40$ bosons in the harmonic trap (from Alg. 4.4 (canonic-bosons) with $E_{\max} = 50$)

| $T/N^{1/3}$ | $Z$ | $\langle E \rangle /N$ | $\langle N_0 \rangle /N$ |
|---|---|---|---|
| 0.0 | 1.000 | 0.000 | 1.000 |
| 0.2 | 3.710 | 0.056 | 0.964 |
| 0.4 | 1083.427 | 0.572 | 0.797 |
| 0.6 | $0.128 \times 10^9$ | 2.355 | 0.449 |
| 0.8 | $0.127 \times 10^{17}$ | 5.530 | 0.104 |
| 1.0 | $0.521 \times 10^{25}$ | 8.132 | 0.031 |

values of the integration variable $\lambda$, we merely evaluate it at one specific point in the complex plane, which gives the dominant contribution. This point is called the chemical potential. It has a clear physical interpretation as the energy that it costs to introduce an additional particle into the system.

In Subsection 4.1.2, we integrated the function

$$Z_N(\beta, \lambda) = \prod_{E=0}^{E_{\max}} \left( \frac{1}{1 - e^{-\beta E + i\lambda}} \right)^{\mathcal{N}(E)} e^{-iN\lambda} \qquad (4.20)$$

over the variable $\lambda$, in the interval $\lambda \in [-\pi, \pi]$. This function can also be considered as a function of the complex variable $\lambda = \mathrm{Re}\,\lambda + i\mathrm{Im}\,\lambda$ (see Fig. 4.8, for a plot of $|Z_N(\beta, \lambda)|$ for $N = 5$ and $\beta = 1$).

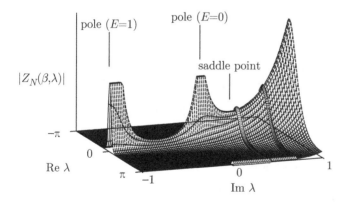

**Fig. 4.8** Absolute value of $Z_N(\beta, \lambda)$ for complex $\lambda$ (for $N = 5$ and $\beta = 1$). Two equivalent integration paths are shown.

For further study of this integral, we use that the integration contour of analytic functions can be moved in the complex plane, as long as we do not cross any singular points of the integrand. Instead of integrating the real $\lambda$, from $-\pi$ to $\pi$, we can integrate a complex $\lambda$ from the point $(-\pi, 0)$ up at constant real value to $(-\pi, \lambda_i)$, then from that point to $(+\pi, \lambda_i)$, and finally back to the real axis, at $(\pi, 0)$. We can confirm by elementary Riemann integration that the integral does not change. (This Riemann integration runs over a sequence of complex points $\{\lambda_1, \ldots, \lambda_K\}$. We put $\Delta_\lambda = \lambda_{k+1} - \lambda_k$ in the usual Riemann integration formula.) The first and third legs of the integration path have a negligible contribution to the integral, and the contributions come mostly from the part of the path for which $\mathrm{Re}\,\lambda \simeq 0$, as can be clearly seen in Fig. 4.8 (see also Fig. 4.9). Functions as that in eqn (4.20) are best integrated along paths passing through saddle points. As we can see, in this case, most of the oscillations of the integrand are avoided, and large positive contributions at some parts are not eliminated by large negative ones elsewhere. For large values of $N$, the value of the integral is dominated by the neighborhood of the saddle point; this means that the saddle point of $Z_N(\beta, \lambda)$ carries the entire contribution to the integral in eqn (4.17), in the limit $N \to \infty$. As we can see from Fig. 4.8, at the saddle point $\lambda$ is purely imaginary, and $Z_N$ is real valued. To find this value, we need to find the minimum of $Z_N$ or, more conveniently, the minimum of $\log Z_N$. This leads to the following:

$$\left\{\begin{array}{c} \text{saddle} \\ \text{point} \end{array}\right\} : \frac{\partial}{\mathrm{i}\partial\lambda} \left\{ -\mathrm{i}N\lambda + \sum_E \mathcal{N}(E) \log\left(1 - \mathrm{e}^{-\beta E + \mathrm{i}\lambda}\right) \right\} = 0.$$

We also let a new real variable $\mu$, the chemical potential, stand for essentially the imaginary part of $\lambda$:

$$\lambda = \mathrm{Re}\,\lambda + \mathrm{i}\underbrace{\mathrm{Im}\,\lambda}_{-\beta\mu}.$$

In terms of the chemical potential $\mu$, we find by differentiating the above saddle point equation:

$$\left\{\begin{array}{c} \text{saddle point } \mu \Leftrightarrow N \\ \text{(canonical ensemble)} \end{array}\right\} : \quad N = \sum_E \mathcal{N}(E) \frac{\mathrm{e}^{-\beta(E-\mu)}}{1 - \mathrm{e}^{-\beta(E-\mu)}}. \quad (4.21)$$

At the saddle point, the function $Z_N(\beta, \mu)$ takes the following value:

$$Z_N(\beta, \mu) = \prod_{\substack{\text{single part.} \\ \text{states } \sigma}} \left[1 + \mathrm{e}^{-\beta(E_\sigma - \mu)} + \mathrm{e}^{-2\cdot\beta(E_\sigma - \mu)} + \cdots\right]$$

$$= \prod_{E=0,1\ldots} \left(\frac{1}{1 - \mathrm{e}^{-\beta(E-\mu)}}\right)^{\mathcal{N}(E)} = \prod_{\substack{\text{single part.} \\ \text{states } \sigma}} \frac{1}{1 - \exp\left[-\beta(E_\sigma - \mu)\right]}. \quad (4.22)$$

This equation describes at the same time the saddle point of the canonical partition function (written as an integral over the variable $\lambda$) and a system with independent states $\sigma$ with energies $E_\sigma - \mu$. This system is called the grand canonical ensemble. Equation (4.22), the saddle point of the canonical partition function, defines the partition function in the grand canonical ensemble, with fluctuating total particle number.

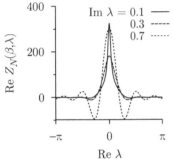

**Fig. 4.9** Values of the integrand along various integration contours in Fig. 4.8.

In the grand canonical ensemble, the probability of there being $k$ particles in state $\sigma$ is

$$\left\{\begin{array}{l} \text{probability of having} \\ k \text{ particles in state } \sigma \end{array}\right\} : \quad \pi(N_\sigma = k) = \frac{\exp\left[-\beta(E_\sigma - \mu)k\right]}{\sum_{N_\sigma=0}^{\infty} \exp\left[-\beta(E_\sigma - \mu)N_\sigma\right]}.$$

For the ground state, we find

$$\pi(N_0) = e^{\beta\mu N_0} - e^{\beta\mu(N_0+1)}. \tag{4.23}$$

The mean number of particles in state $\sigma$ is

$$\langle N_\sigma \rangle = \pi(N_\sigma = 1) \cdot 1 + \pi(N_\sigma = 2) \cdot 2 + \pi(N_\sigma = 3) \cdot 3 + \cdots$$
$$= \frac{\sum_{N_\sigma=0}^{\infty} N_\sigma \exp\left[-\beta(E_\sigma - \mu)N_\sigma\right]}{\sum_{N_\sigma=0}^{\infty} \exp\left[-\beta(E_\sigma - \mu)N_\sigma\right]} = \frac{\exp\left[-\beta(E_\sigma - \mu)\right]}{1 - \exp\left[-\beta(E_\sigma - \mu)\right]}. \tag{4.24}$$

The mean ground-state occupancy in the grand canonical ensemble is

$$\langle N_0 \rangle = \frac{e^{\beta\mu}}{1 - e^{\beta\mu}}. \tag{4.25}$$

The mean total number of particles in the harmonic trap is, in the grand canonical ensemble,

$$\left\{\begin{array}{l} \text{mean total number} \\ \text{(grand canonical)} \end{array}\right\} : \quad \langle N(\mu) \rangle = \sum_{E=0}^{\infty} \mathcal{N}(E) \frac{e^{-\beta(E-\mu)}}{1 - e^{-\beta(E-\mu)}}. \tag{4.26}$$

Equation (4.26) determines the mean particle number for a given chemical potential. We should note that eqns (4.26) and (4.21) are essentially the same expressions. The chemical potential denotes the saddle point of the canonical partition function for $N$ particles. It is also the point in the grand canonical system at which the mean particle number satisfies $\langle N \rangle = N$.

The inverse function of eqn (4.26), the chemical potential as a function of the particle number, is obtained by a basic bisection algorithm (see Alg. 4.5 (grandcan-bosons), and Table 4.5).

From the chemical potential corresponding to the mean particle number chosen, we can go on to compute condensate fractions, energies, and other observables. All these calculations are trivial to perform for the ideal gas, because in the grand canonical ensemble all states $\sigma$ are independent. As a most interesting example, the condensate fraction $\langle N_0 \rangle / \langle N \rangle$ follows directly from eqn (4.25), after calculating $\mu$ using Alg. 4.5 (grandcan-bosons) (see Fig. 4.10).

In conclusion, we have studied in this subsection the function $Z_N(\beta, \lambda)$ which, when integrated over $\lambda$ from $\pi$ to $\pi$, gives the exact canonical partition function for $N$ particles. In the complex $\lambda$-plane, this function has a saddle point defined by a relation of $N$ with the chemical potential $\mu$, essentially the rescaled imaginary part of the original integration variable $\lambda$. In the large-$N$ limit, this point dominates the integral for $Z_N$ (which gives an additional constant which we did not compute), and for

**Table 4.5** Mean particle number $\langle N \rangle$ vs. $\mu$ for bosons in the harmonic trap at temperature $T = 10$ (from Alg. 4.5 (grandcan-bosons))

| $\langle N \rangle$ | $\mu$ |
|---|---|
| 400 | −11.17332 |
| 800 | −4.82883 |
| 1000 | −2.92846 |
| 1200 | −1.48065 |
| 1400 | −0.42927 |
| 2000 | −0.01869 |
| 5000 | −0.00283 |
| 10 000 | −0.00117 |
| ... | ... |

```
procedure grandcan-bosons
input ⟨N⟩ (target mean number of particles)
input μ_min (with ⟨N(μ_min)⟩ < ⟨N⟩)
μ_max ← 0
for i = 1, 2, ... do
  ⎧  μ ← (μ_min + μ_max)/2
  ⎪  if ( ⟨N(μ)⟩ < N) then (evaluate eqn (4.26))
  ⎨    { μ_min ← μ
  ⎪  else
  ⎩    { μ_max ← μ
μ ← (μ_min + μ_max)/2
output μ
```

**Algorithm 4.5 grandcan-bosons.** Computing the chemical potential $\mu$ for a given mean number $\langle N \rangle$ of bosons.

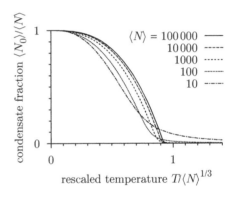

**Fig. 4.10** Condensate fraction in the harmonic trap, in the grand canonical ensemble (from Alg. 4.5 (**grandcan-bosons**), see also eqn (4.31)).

all other (extensive) observables, as the energy and the specific heat (for which the above constant cancels). The saddle point of the canonical ensemble also gives the partition function in the grand canonical ensemble. It is significant that at the saddle point the particle number $N$ in the canonical ensemble agrees with the mean number $\langle N \rangle$ in the grand canonical ensemble, that is, that eqns (4.21) and (4.26) agree even at finite $N$. Other extensive observables are equivalent only in the limit $N \to \infty$.

## 4.1.4   Large-$N$ limit in the grand canonical ensemble

Notwithstanding our excellent numerical control of the ideal Bose gas in the grand canonical ensemble, it is crucial, and relatively straightforward, to solve this problem analytically in the limit $N \to \infty$. The key to an analytic solution for the ideal Bose gas is that at low temperatures, the chemical potential is negative and very close to zero. For example, at temperature $T = 10$, we have

$$\left\{ \begin{array}{l} \mu = -0.00283 \\ \text{(see Table 4.5)} \end{array} \right\} \Rightarrow \begin{array}{l} \langle N_0 \rangle = 3533, \\ \langle N_{\sigma=1} \rangle = 9.48, \end{array}$$

$$\left\{ \begin{array}{l} \mu = -0.00117 \\ \text{(see Table 4.5)} \end{array} \right\} \Rightarrow \begin{array}{l} \langle N_0 \rangle = 8547, \\ \langle N_{\sigma=1} \rangle = 9.50. \end{array}$$

Thus, the occupation of the excited state changes very little; at a chemical potential $\mu = 0$, where the total number of particles diverges, it reaches $\langle N_{\sigma=1} \rangle = 9.51$. For $\mu$ around zero, only the ground-state occupation changes drastically; it diverges at $\mu = 0$. For all the excited states, we may thus replace the actual occupation numbers of states (at a chemical potential very close to 0) by the values they would have at $\mu = 0$. This brings us to the very useful concept of a temperature-dependent saturation number, the maximum mean number of particles that can be put into the excited states:

$$\langle N_{\text{sat}} \rangle = \sum_{\substack{E \geq 1 \\ \text{(excited)}}} \mathcal{N}(E) \underbrace{\frac{e^{-\beta E}}{1 - e^{-\beta E}}}_{\langle N_{\text{sat}}(E) \rangle}. \tag{4.27}$$

As $N$ increases (and $\mu$ approaches 0) the true occupation numbers of the excited states approach the saturation numbers $\langle N_{\text{sat}}(E) \rangle$ (see Fig. 4.11).

The saturation number gives a very transparent picture of Bose–Einstein condensation for the ideal gas. Let us imagine a trap getting filled with particles, for example the atoms of an experiment in atomic physics, or the bosons in a Monte Carlo simulation. Initially, particles go into excited states until they are all saturated, a point reached at the critical temperature $T_c$. After saturation, any additional particles have no choice but to populate the ground state and to contribute to the Bose–Einstein condensate.

We shall now compute analytically the saturation number for large temperatures $T$ (small $\beta$, large number of particles), where the above picture becomes exact. In the limit $\beta \to 0$, the sum in eqn (4.27) can be approximated by an integral. It is best to introduce the distance between two energy levels $\Delta_E = 1$ as a bookkeeping device:

$$\langle N_{\text{sat}} \rangle = \sum_{\substack{E \geq 1 \\ \text{(excited)}}} \Delta_E \frac{(E+1)(E+2)}{2} \frac{e^{-\beta E}}{1 - e^{-\beta E}}.$$

Changing the sum variable from $E$ to $\beta E = x$, this gives, with $\Delta_E = \Delta_x/\beta$, the term-by-term identical sum

$$\langle N_{\text{sat}} \rangle = \frac{1}{\beta} \sum_{x=\beta,2\beta\ldots} \Delta_x \frac{(x/\beta+1)(x/\beta+2)}{2} \frac{e^{-x}}{1 - e^{-x}} \tag{4.28}$$

$$\xrightarrow[\beta \to 0]{} \frac{1}{2\beta^3} \int_0^\infty dx\, x^2 \frac{e^{-x}}{1 - e^{-x}}. \tag{4.29}$$

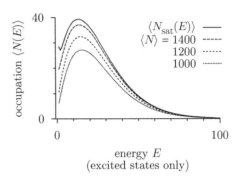

occupation $\langle N(E) \rangle$

energy $E$
(excited states only)

**Fig. 4.11** Occupation numbers and saturation numbers for excited states in the harmonic trap at $T = 10$ (from Alg. 4.5 (`grandcan-bosons`)).

The series

$$\frac{1}{1-e^{-x}} = 1 + e^{-x} + e^{-2x} + e^{-3x} + \cdots$$

leads to the integrals

$$\int_0^\infty dx\, x^2 e^{-nx}$$
$$= \frac{\partial^2}{\partial n^2} \underbrace{\int_0^\infty dx\, e^{-nx}}_{1/n \text{ for } n>0} = \frac{2}{n^3}. \quad (4.30)$$

In the last step, we have only kept the largest term, proportional to $1/\beta^3$. Depending on our taste and upbringing, we may look this integral up in a book, or on a computer, program it as a Riemann sum, or expand the denominator of eqn (4.29) into a geometric series, multiply by the numerator, evaluate the integrals in eqn (4.30), and sum the series. In any case, the result is

$$\langle N_{\text{sat}} \rangle \xrightarrow[\beta \to 0]{} \frac{1}{\beta^3} \underbrace{\sum_{n=1}^\infty \frac{1}{n^3}}_{\substack{\text{Riemann} \\ \text{zeta function } \zeta(3)}} = \frac{1.202}{\beta^3}.$$

The series $\sum_{n=1}^\infty 1/n^\alpha$ is called the Riemann zeta function of $\alpha$, $\zeta(\alpha)$, so that $\langle N_{\text{sat}} \rangle \to \zeta(3)/\beta^3$. The numerical value of $\zeta(3)$ can be looked up, or programmed in a few seconds.

The calculation of the saturation numbers in the limit $\langle N \rangle \to \infty$ allows us to determine the critical temperature of the ideal Bose gas and the dependence of the condensate fraction on the temperature. For large $\langle N \rangle$, the saturation number equals the particle number at the critical temperature, where

$$\langle N \rangle = \langle N_{\text{sat}} \rangle \qquad \Longleftrightarrow \qquad T = T_{\text{c}},$$
$$\langle N \rangle = 1.202 T_{\text{c}}^3 \qquad \Longleftrightarrow \qquad T_{\text{c}} = \frac{\langle N \rangle^{1/3}}{\sqrt[3]{1.202}}.$$

This is equivalent to

$$T_{\text{c}}/\langle N \rangle^{1/3} = 0.94.$$

Furthermore, below $T_{\text{c}}$, the difference between the particle number and the saturation number must come from particles in the ground state, so that $\langle N_0 \rangle = \langle N \rangle - \langle N_{\text{sat}} \rangle$. This simple reasoning allows us to compute the condensate fraction. We use one of the above equations ($\langle N \rangle =$

$1.202T_c^3$), and also the dependence of the saturation number on temperature ($\langle N_{\text{sat}}(T) \rangle = 1.202T^3$), to arrive at

$$\frac{\langle N_0 \rangle}{\langle N \rangle} = 1 - \frac{T^3}{T_c^3}. \tag{4.31}$$

The critical temperature $T_c$ increases with the cube root of the mean particle number. This justifies the scaling used for representing our data for the five-boson bounded trap model (see Fig. 4.7).

After the thermodynamics, we now turn to structural properties. We compute the density distribution in the gas, in the grand canonical ensemble. The grand canonical density is obtained by multiplying the squared wave functions in the harmonic trap by the occupation numbers $N_\sigma$ of eqn (4.24):

$$\left\{ \begin{array}{c} \text{density at point} \\ \mathbf{r} = \{x, y, z\} \end{array} \right\} \equiv \eta(x, y, z) = \sum_\sigma \psi_\sigma^2(x, y, z) \langle N_\sigma \rangle. \tag{4.32}$$

The states $\sigma$, indexed by the energies $\{E_x, E_y, E_z\}$, run through the list in Table 4.1, where

$$\psi_\sigma(x, y, z) = \psi_{E_x}^{\text{h.o.}}(x) \psi_{E_y}^{\text{h.o.}}(y) \psi_{E_z}^{\text{h.o.}}(z).$$

The wave functions of the harmonic oscillator can be computed by Alg. 3.1 (`harmonic-wavefunction`). The naive summation in eqn (4.32) is awkward because to implement it, we would have to program the harmonic-oscillator wave functions. It is better to reduce the density at point $\{x, y, z\}$ into a sum of one-particle terms. This allows us to express $\eta(x, y, z)$ in terms of the single-particle density matrix of the harmonic oscillator, which we determined in Subsection 3.2.2. We expand the denominator of the explicit expression for $\langle N_\sigma \rangle$ back into a geometric sum and multiply by the numerator:

$$\langle N_\sigma \rangle = \frac{\exp\left[-\beta(E_\sigma - \mu)\right]}{1 - \exp\left[-\beta(E_\sigma - \mu)\right]} = \sum_{k=1}^{\infty} \exp\left[-k\beta(E_\sigma - \mu)\right]$$

$$= \sum_{k=1}^{\infty} \exp\left[-k\beta(E_x + E_y + E_z - \mu)\right].$$

The density in eqn (4.32) at the point $\{x, y, z\}$ turns into

$$\eta(x, y, z) = \sum_{k=1}^{\infty} e^{k\beta\mu}$$

$$\times \underbrace{\sum_{E_x=0}^{\infty} e^{-k\beta E_x} \psi_{E_x}^2(x)}_{\rho^{\text{h.o.}}(x, x, k\beta)} \underbrace{\sum_{E_y=0}^{\infty} e^{-k\beta E_y} \psi_{E_y}^2(y)}_{\rho^{\text{h.o.}}(y, y, k\beta)} \underbrace{\sum_{E_z=0}^{\infty} e^{-k\beta E_z} \psi_{E_z}^2(z)}_{\rho^{\text{h.o.}}(z, z, k\beta)}.$$

Each of the sums in this equation contains a diagonal one-particle harmonic-oscillator density matrix at temperature $1/(k\beta)$, as determined analytically in Subsection 3.2.2, but for a different choice of the ground-state

204 *Bosons*

energy. With our present ground-state energy $E_0 = 0$, the diagonal density matrix of the one-dimensional harmonic oscillator is

$$\rho^{\text{h.o.}}(x,x,k\beta) = \sqrt{\frac{1}{2\pi \sinh(k\beta)}} \exp\left[-x^2 \tanh\left(\frac{k\beta}{2}\right) + \frac{k\beta}{2}\right].$$

Together with terms containing $\exp\left[-y^2 \ldots\right]$ and $\exp\left[-z^2 \ldots\right]$, this gives

$$\eta(r) = \sum_{k=1}^{\infty} e^{k\beta\mu} \left[\frac{1}{2\pi \sinh(k\beta)}\right]^{3/2} \exp\left[-r^2 \tanh\left(\frac{k\beta}{2}\right) + \frac{3}{2}k\beta\right]. \quad (4.33)$$

This formula can be programmed as it stands. It uses as input the values of the chemical potential in Table 4.5. However, convergence as $k$ increases is slow for $\mu \lesssim 0$ (it diverges for $\mu = 0$ because $\sinh^{3/2}(k\beta) \simeq \frac{1}{2}\exp\left(\frac{3}{2}k\beta\right)$ for large $\beta k$, so that each term in the sum is constant). It is better to compute only the excited-state density, i.e. to add to and then subtract from eqn (4.33) the ground-state term, put together from

$$\langle N_0 \rangle = \frac{e^{\beta\mu}}{1 - e^{\beta\mu}} = \sum_{k=1}^{\infty} e^{k\beta\mu},$$

and the ground-state wave function

$$\sum_{k=1}^{\infty} e^{k\beta\mu} \frac{1}{\pi^{3/2}} \exp\left(-r^2\right).$$

This yields

$$\eta^{\text{exc}}(r) = \sum_{k=1}^{\infty} e^{k\beta\mu}$$
$$\times \left\{\left[\frac{1}{2\pi \sinh(k\beta)}\right]^{3/2} \exp\left(-r^2 \tanh\frac{k\beta}{2} + \frac{3}{2}k\beta\right) - \frac{e^{-r^2}}{\pi^{3/2}}\right\}. \quad (4.34)$$

The density of the excited states at position $r$ saturates in the same way as the number of excited particles. Below the critical temperature, the density $\eta^{\text{exc}}(r)$ is very close to the saturation density at $\mu = 0$. The total density is finally given by

$$\eta(r) = \eta^{\text{exc}}(r) + \frac{e^{\beta\mu}}{1 - e^{\beta\mu}} \frac{1}{\pi^{3/2}} \exp\left(-r^2\right). \quad (4.35)$$

This equation, applied to the ground state $E = 0$, allows us to obtain the number of bosons at the point $\mathbf{x} = \{x,y,z\}$, with $|\mathbf{x}| = r$ per unit volume element in the harmonic trap (see Fig. 4.12). The sudden density increase in the center is a hallmark of Bose–Einstein condensation in a harmonic trap. It was first seen in the velocity distribution of atomic gases in the Nobel-prize-winning experiments of the groups of Cornell and Wieman (with rubidium atoms) and of Ketterle (with sodium atoms) in 1995.

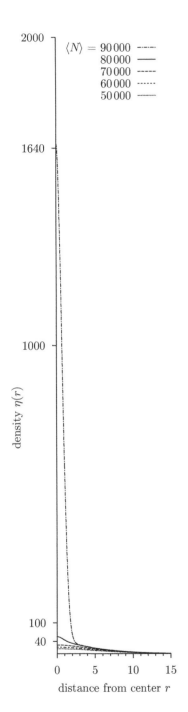

**Fig. 4.12** Density $\eta(r)$ in the harmonic trap at temperature $T = 40$ (grand canonical ensemble, from eqn (4.35)).

### 4.1.5   Differences between ensembles—fluctuations

The theory of saddle point integration guarantees that calculations performed in the canonical ensemble (with fixed particle number) and in the grand canonical ensemble (with fixed chemical potential) give identical results in the limit $N \to \infty$ for extensive quantities. The differences in the condensate fraction between the two ensembles are indeed quite minor, even for $N = 10$ or $40$ bosons, as we can see by comparing output of Alg. 4.4 (`canonic-bosons`) and of Alg. 4.5 (`grandcan-bosons`) with each other (see Fig. 4.13). The condensate fractions, in the two ensembles, are mean values of probability distributions which we can also compute. For the grand canonical ensemble, $\pi(N_0)$ is given explicitly by eqn (4.23); for the canonical ensemble, it can be obtained using Alg. 4.4 (`canonic-bosons`). The two probability distributions here are completely different, even though their mean values agree very well. The difference between the two distributions shown in Fig. 4.14 persists in the limit $N \to \infty$.

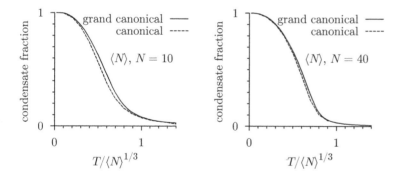

**Fig. 4.13** Condensate fraction in the harmonic trap (from Alg. 4.4 (`canonic-bosons`) and Alg. 4.5 (`grandcan-bosons`)).

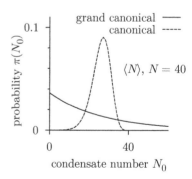

**Fig. 4.14** Probability $\pi(N_0)$ in the harmonic trap with $T/N^{1/3} = T/\langle N \rangle^{1/3} = 0.5$ (from modified Alg. 4.4 (`canonic-bosons`) and eqn (4.23)).

The huge fluctuations in the grand canonical ideal Bose gas are non-physical. They are caused by the accidental degeneracy of the chemical potential with the ground-state energy in the limit $N \to \infty$. Below the critical temperature, the chemical potential is asymptotically equal to the ground-state energy, so that the energy it costs to add a particle into the system, $E_0 - \mu$, vanishes. The fluctuations in $N_0$ are limited only because the mean particle number is fixed. In interacting systems, the chemical potential differs from the ground-state energy, and any interacting system will have normal fluctuations of particle number. We have seen in this subsection that the ideal canonical gas also has more "physical" fluctuations than the grand canonical gas (see Fig. 4.14).

### 4.1.6   Homogeneous Bose gas

In this subsection, we study Bose–Einstein condensation in a cubic box with periodic boundary conditions. The main difference with the trap is that the mean density in the homogeneous system of the periodic box cannot change with temperature. It must be the same all over the system at all temperatures, because it is protected by the translation symmetry in the same way as the classical density of hard disks in a periodic box, in Chapter 2, was necessarily constant throughout the system. This single-particle observable thus cannot signal the Bose–Einstein condensation transition as was possible in the trap (see Fig. 4.12). Another difference between the harmonic trap and the homogeneous gas lies in the way the thermodynamic limit is taken. Earlier, a trap with fixed values of $\{\omega_x, \omega_y, \omega_z\}$ was filled with more and more particles, and the thermodynamic limit took place at high temperature. Now we compare systems of the same density in boxes of increasing size. We shall again study the canonical and the grand canonical ensembles and interpret the transition in terms of saturation densities.

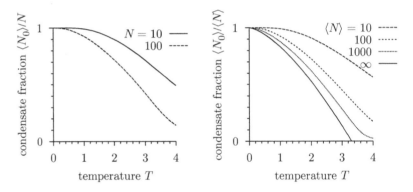

**Table 4.6** Thermodynamics for $N = 64$ bosons in a cube with periodic boundary conditions (from Alg. 4.4 (canonic-bosons) with $\eta = 1$)

| $T$ | $Z$ | $\langle E \rangle /N$ | $\langle N_0 \rangle /N$ |
|-----|-----|------------------------|--------------------------|
| 0.0 | 1.000 | 0.000 | 1.000 |
| 0.5 | 1.866 | 0.014 | 0.990 |
| 1.0 | 31.129 | 0.112 | 0.939 |
| 1.5 | 1992.864 | 0.329 | 0.857 |
| 2.0 | $3.32 \times 10^5$ | 0.691 | 0.749 |
| 2.5 | $1.22 \times 10^8$ | 1.224 | 0.619 |
| 3.0 | $8.92 \times 10^{10}$ | 1.944 | 0.470 |
| 3.5 | $1.18 \times 10^{14}$ | 2.842 | 0.313 |
| 4.0 | $2.25 \times 10^{17}$ | 3.812 | 0.188 |

**Fig. 4.15** Condensate fraction in a periodic cube ($\eta = 1$) in the canonical ensemble (*left*) and the grand canonical ensemble (*right*) (see eqn (4.40)).

In a cubic box with periodic boundary conditions, the energy eigenvalues depend on the edge length $L$ as

$$E(n_x, n_y, n_z) = \underbrace{\frac{2\pi^2}{L^2} \left( n_x^2 + n_y^2 + n_z^2 \right)}_{k} \tag{4.36}$$

(see eqn (3.10)). The following relation between the particle number, the volume, and the density,

$$N = \eta L^3,$$

allows us to write the energy levels in terms of the density and the dimensions of the box:

$$E(n_x, n_y, n_z) = \underbrace{\frac{2\pi^2}{L^2} (n_x^2 + n_y^2 + n_z^2)}_{\hat{E}} = \underbrace{\left[2\pi^2 \left(\frac{\eta}{N}\right)^{2/3}\right]}_{\Upsilon} \hat{E}. \quad (4.37)$$

Rescaling the temperature as $\hat{\beta} = \Upsilon\beta$ (with $\Upsilon$ from the above equation), we find the following, in the straightforward generalization of eqn (4.17):

$$Z_N(\beta) = \int_{-\pi}^{\pi} \frac{d\lambda}{2\pi} e^{-iN\lambda} \prod_{E=0}^{E_{max}} \left[f_E(\hat{\beta}, \lambda)\right]^{\mathcal{N}(E)}. \quad (4.38)$$

The single-particle degeneracies are given by

$$\{\mathcal{N}(0), \mathcal{N}(1), \dots\} = \underbrace{\{1, 6, 12, 8, 4, \dots\}}_{\text{from Table 4.2}},$$

directly obtained from Alg. 4.2 (`naive-degeneracy-cube`). The calculation of the mean energy and the condensate fraction is analogous to the calculation in the trap (see Table 4.6).

As in the harmonic trap, we can compute the saturation number and deduce from it the transition temperature and the condensate fraction in the thermodynamic limit (compare with Subsection 4.1.4). There is no explicit formula for the single-particle degeneracies, but we can simply sum over all sites of the integer lattice $\{n_x, n_y, n_z\}$ (whereas for the trap, we had a sum over energies). With dummy sum variables $\Delta_n = 1$, as in eqn (4.28), we find

$$\langle N_{\text{sat}} \rangle = \sum_{\substack{\{n_x,n_y,n_z\}=-n_{max} \\ \neq\{0,0,0\}}}^{n_{max}} \Delta_n^3 \frac{e^{-\beta E(n_x,n_y,n_z)}}{1 - e^{-\beta E(n_x,n_y,n_z)}},$$

where $E(n_x, n_y, n_z)$ is defined through eqn (4.37).

Using a change of variables $\sqrt{2\hat{\beta}}(\pi/L)n_x = x$, and analogously for $n_y$ and $n_z$, we again find a term-by-term equivalent Riemann sum (with $\Delta_x = \sqrt{2\hat{\beta}}\pi/L\Delta_n$):

$$\langle N_{\text{sat}} \rangle = \frac{L^3}{\pi^3(2\hat{\beta})^{3/2}} \sum_{\substack{\{x_i,y_i,z_i\} \\ \neq\{0,0,0\}}} \Delta_x\Delta_y\Delta_z \frac{e^{-(x_i^2+y_i^2+z_i^2)}}{1 - e^{-(x_i^2+y_i^2+z_i^2)}}.$$

For $\sqrt{\hat{\beta}}/L \to 0$, $\Delta_x$, etc., become differentials, and we find

$$\xrightarrow[\beta\to 0]{} \frac{L^3}{\pi^3(2\hat{\beta})^{3/2}} \int dV \frac{e^{-(x^2+y^2+z^2)}}{1 - e^{-(x^2+y^2+z^2)}}.$$

We again expand the denominator $\Upsilon/(1 - \Upsilon) = \Upsilon + \Upsilon^2 + \cdots$ (as for

If $r^2 = x^2 + y^2 + z^2$,

$$\int_{-\infty}^{\infty} dx\, dy\, dz\, f(r) = 4\pi \int_0^{\infty} dr\, r^2 f(r).$$

Also,

$$\int_{-\infty}^{\infty} d\mu \exp\left(-n\mu^2\right) = \sqrt{\frac{\pi}{n}}$$

(see eqn (3.12)) implies

$$\int_{-\infty}^{\infty} du\, u^2 \exp\left(-nu^2\right)$$

$$= -\frac{\partial}{\partial n}\sqrt{\frac{\pi}{n}} = \frac{\sqrt{\pi}}{2n^{3/2}}.$$

eqn (4.29)) and find

$$\underline{\langle N_{\mathrm{sat}}\rangle = \frac{2L^3}{(2\beta)^{3/2}\pi^2} \sum_{n=1}^{\infty} \int_{-\infty}^{\infty} du\ u^2 e^{-nu^2}}$$

$$= \frac{L^3}{(2\pi\beta)^{3/2}} \underbrace{\sum_{n=1}^{\infty} \frac{1}{n^{3/2}}}_{\zeta(3/2)=2.612\ldots} = \underline{\frac{L^3}{(2\pi\beta)^{3/2}} \cdot 2.612.}$$

Again, the total number of particles equals the saturation number at the critical temperature:

$$\langle N \rangle = \langle N_{\mathrm{sat}} \rangle \qquad \Longleftrightarrow \qquad T = T_{\mathrm{c}},$$

$$\langle N \rangle = \frac{2.612 L^3}{(2\pi\beta_{\mathrm{c}})^{3/2}} \qquad \Longleftrightarrow \qquad T_{\mathrm{c}} = 2\pi \left( \frac{1}{2.612} \frac{\langle N \rangle}{L^3} \right)^{2/3},$$

so that

$$\underline{T_{\mathrm{c}}(\eta) = 3.3149 \eta^{2/3}.} \tag{4.39}$$

Below $T_{\mathrm{c}}$, we can again obtain the condensate fraction as a function of temperature by noticing that particles are either saturated or in the condensate:

$$\langle N \rangle = \langle N_{\mathrm{sat}} \rangle + \langle N_0 \rangle,$$

which is equivalent to (using $\eta = 2.612/(2\pi\beta_{\mathrm{c}})^{3/2}$)

$$\frac{\langle N_0 \rangle}{\langle N \rangle} = 1 - \left( \frac{T}{T_{\mathrm{c}}} \right)^{3/2}. \tag{4.40}$$

This curve is valid in the thermodynamic limit of a box at constant density in the limit $L \to \infty$. It was already compared to data for finite cubes, and to the canonical ensemble (see Fig. 4.15).

Finally, we note that the critical temperature in eqn (4.39) can also be written as

$$\sqrt{2\pi\beta_{\mathrm{c}}} = 1.38 \eta^{-1/3}.$$

In Chapter 3, we discussed in detail (but in other terms) that the thermal extension of a quantum particle is given by the de Broglie wavelength $\lambda_{\mathrm{dB}} = \sqrt{2\pi\beta_{\mathrm{c}}}$ (see eqn (3.75)). On the other hand, $\eta^{-1/3}$ is the mean distance between particles. Bose–Einstein condensation takes place when the de Broglie wavelength is of the order of the inter-particle distance.

We can also write this same relation as

$$[\lambda_{\mathrm{dB}}]^3 \eta = \zeta(3/2) = 2.612. \tag{4.41}$$

It characterizes Bose–Einstein condensation in arbitrary three-dimensional geometries, if we replace the density $\eta$ by $\max_{\mathbf{x}}[\eta(\mathbf{x})]$. In the trap, Bose–Einstein condensation takes place when the density in the center satisfies eqn (4.41).

## 4.2   The ideal Bose gas (density matrices)

In Section 4.1, we studied the ideal Bose gas within the energy-level description. This approach does not extend well to the case of interactions, either for theory or for computation, as it builds on the concept of single-particle states, which ceases to be relevant for interacting particles. The most systematic approach for treating interactions employs density matrices and path integrals, which easily incorporate interactions. Interparticle potentials simply reweight the paths, as we saw for distinguishable quantum hard spheres in Subsection 3.4.3. In the present section, we treat the ideal Bose gas with density matrices. We concentrate first on thermodynamics and arrive at much better methods for computing $Z$, $\langle E \rangle$, $\langle N_0 \rangle$, etc., in the canonical ensemble. We then turn to the description of the condensate fraction and the superfluid density within the path integral framework. These concepts were introduced by Feynman in the 1950s but became quantitatively clear only much later.

### 4.2.1   Bosonic density matrix

In this subsection, we obtain the density matrix for ideal bosons as a sum of single-particle nondiagonal distinguishable density matrices. We derive this fundamental result only for two bosons and two single-particle energy levels, but can perceive the general case ($N$ bosons, arbitrary number of states, and interactions) through this example.

In Chapter 3, the density matrix of a single-particle quantum system was defined as

$$\rho(x, x', \beta) = \sum_n \psi_n(x) e^{-\beta E_n} \psi_n^*(x'),$$

with orthonormal wave functions $\psi_n$ ($\int dx \, |\psi_n(x)|^2 = 1$ etc.). The partition function is the trace of the density matrix:

$$Z = \sum_n e^{-\beta E_n} = \operatorname{Tr} \rho = \int dx \, \rho(x, x, \beta).$$

We may move from one to many particles without changing the framework simply by replacing single-particle states by $N$-particle states:

$$\left\{ \begin{array}{l} \text{many-particle} \\ \text{density matrix} \end{array} \right\} : \rho^{\text{dist}}(\{x_1, \dots, x_N\}, \{x'_1, \dots, x'_N\}, \beta)$$

$$= \underbrace{\sum}_{\substack{\text{orthonormal} \\ N\text{-particle} \\ \text{states } \psi_n}} \psi_n(x_1, \dots, x_N) e^{-\beta E_n} \psi_n(x'_1, \dots, x'_N). \quad (4.42)$$

For ideal distinguishable particles, the many-body wave functions are the products of the single-particle wave functions ($n \equiv \{\sigma_1, \dots, \sigma_N\}$):

$$\psi_{\{\sigma_1, \dots, \sigma_N\}}(x_1, \dots, x_N) = \psi_{\sigma_1}(x_1) \psi_{\sigma_2}(x_2) \dots \psi_{\sigma_N}(x_N).$$

The total energy is the sum of single-particle energies:

$$E_n = E_{\sigma_1} + \cdots + E_{\sigma_N}.$$

For two levels $\sigma_i = 1, 2$ and two particles, there are three states $\{\sigma_1, \sigma_2\}$ with
$\sigma_1 \leq \sigma_2$:

$$\{\sigma_1, \sigma_2\} = \{1, 1\},$$
$$\{\sigma_1, \sigma_2\} = \{1, 2\},$$
$$\{\sigma_1, \sigma_2\} = \{2, 2\}.$$

The last three equations give for the $N$-particle density matrix of distinguishable ideal particles:

$$\rho^{\text{dist}}(\{x_1, \ldots, x_N\}, \{x_1', \ldots, x_N'\}, \beta) = \prod_{k=1}^{N} \rho(x_k, x_k', \beta),$$

a product of single-particle density matrices. We see that the quantum
statistics of ideal (noninteracting) distinguishable quantum particles is
no more difficult than that of a single particle.

The bosonic density matrix is defined in analogy with eqn (4.42):

$$\rho^{\text{sym}}(\mathbf{x}, \mathbf{x}', \beta) = \sum_{\substack{\text{symmetric, orthonormal} \\ N\text{-particle wave functions } \psi_n^{\text{sym}}}} \psi_n^{\text{sym}}(\mathbf{x}) e^{-\beta E_n} \psi_n^{\text{sym}}(\mathbf{x}').$$

Here, "symmetric" refers to the interchange of particles $x_k \leftrightarrow x_l$. Again,
all wave functions are normalized. The symmetrized wave functions have
characteristic, nontrivial normalization factors. For concreteness, we go
on with two noninteracting particles in two single-particle states, $\sigma_1$ and
$\sigma_2$. The wave functions belonging to this system are the following:

$$\psi_{\{1,1\}}^{\text{sym}}(x_1, x_2) = \psi_1(x_1)\psi_1(x_2),$$
$$\psi_{\{1,2\}}^{\text{sym}}(x_1, x_2) = \frac{1}{\sqrt{2}} [\psi_1(x_1)\psi_2(x_2) + \psi_1(x_2)\psi_2(x_1)], \quad (4.43)$$
$$\psi_{\{2,2\}}^{\text{sym}}(x_1, x_2) = \psi_2(x_1)\psi_2(x_2).$$

These three wave functions are symmetric with respect to particle ex-
change (for example, $\psi_{12}^{\text{sym}}(x_1, x_2) = \psi_{12}^{\text{sym}}(x_2, x_1) = \psi_1(x_1)\psi_1(x_2)$).
They are also orthonormal ($\int dx_1 \, dx_2 \, \psi_{12}^{\text{sym}}(x_1, x_2)^2 = 1$, etc.). Hence
the ideal-boson density matrix is given by

$$\rho^{\text{sym}}(\{x_1, x_2\}, \{x_1', x_2'\}, \beta) = \psi_{11}^{\text{sym}}(x_1, x_2) e^{-\beta E_{11}} \psi_{11}(x_1', x_2')$$
$$+ \psi_{12}^{\text{sym}}(x_1, x_2) e^{-\beta E_{12}} \psi_{12}(x_1', x_2')$$
$$+ \psi_{22}^{\text{sym}}(x_1, x_2) e^{-\beta E_{22}} \psi_{22}(x_1', x_2').$$

The various terms in this unwieldy object carry different prefactors, if
we write the density matrix in terms of the symmetric wave functions in
eqn (4.43). We shall, however, now express the symmetric density matrix
through the many-particle density matrix of distinguishable particles
without symmetry requirements, and see that the different normalization
factors disappear.

We rearrange $\psi_{11}^{\text{sym}}(x_1, x_2)$ as $\frac{1}{2} [\psi_1(x_1)\psi_1(x_2) + \psi_1(x_2)\psi_1(x_1)]$, and
analogously for $\psi_{22}^{\text{sym}}$. We also write out the part of the density matrix
involving $\psi_{12}$ twice (with a prefactor $\frac{1}{2}$). This gives the first two lines
of the following expression; the third belongs to $\psi_{11}^{\text{sym}}$ and the fourth to

$\psi_{22}^{\text{sym}}$:

$\frac{1}{4} \left[\psi_1(x_1)\psi_1(x_2) + \psi_1(x_2)\psi_1(x_1)\right] \left[\psi_1(x_1')\psi_1(x_2') + \psi_1(x_2')\psi_1(x_1')\right] \mathrm{e}^{-\beta E_{11}}$

$\frac{1}{4} \left[\psi_1(x_1)\psi_2(x_2) + \psi_1(x_2)\psi_2(x_1)\right] \left[\psi_1(x_1')\psi_2(x_2') + \psi_1(x_2')\psi_2(x_1')\right] \mathrm{e}^{-\beta E_{12}}$

$\frac{1}{4} \left[\psi_2(x_1)\psi_1(x_2) + \psi_2(x_2)\psi_1(x_1)\right] \left[\psi_2(x_1')\psi_1(x_2') + \psi_2(x_2')\psi_1(x_1')\right] \mathrm{e}^{-\beta E_{21}}$

$\underbrace{\frac{1}{4} \left[\psi_2(x_1)\psi_2(x_2) + \psi_2(x_2)\psi_2(x_1)\right]}_{\text{all permutations } Q \text{ of } \{x_1, x_2\}} \underbrace{\left[\psi_2(x_1')\psi_2(x_2') + \psi_2(x_2')\psi_2(x_1')\right]}_{\text{all permutations } P \text{ of } \{x_1', x_2'\}} \mathrm{e}^{-\beta E_{22}}$

The rows of this expression correspond to a double sum over single-particle states. Each one has the same prefactor $1/4$ (more generally, for $N$ particles, we would obtain $(1/N!)^2$) and carries a double set ($P$ and $Q$) of permutations of $\{x_1, x_2\}$, and of $\{x_1', x_2'\}$. More generally, for $N$ particles and a sum over states $\sigma$, we have

$$\rho^{\text{sym}}(\{x_1, \ldots, x_N\}, \{x_1', \ldots, x_N'\}, \beta) = \sum_{\{\sigma_1, \ldots, \sigma_N\}} \sum_Q \sum_P \left(\frac{1}{N!}\right)^2$$

$$\times [\psi_{\sigma_1}(x_{Q_1}) \ldots \psi_{\sigma_N}(x_{Q_N})] [\psi_{\sigma_1}(x_{P_1}') \ldots \psi_{\sigma_N}(x_{P_N}')] \mathrm{e}^{-\beta(E_{\sigma_1} + \cdots + E_{\sigma_N})}.$$

This agrees with

$$\rho^{\text{sym}}(\{x_1, \ldots, x_N\}, \{x_1', \ldots, x_N'\}, \beta)$$

$$= \frac{1}{N!} \sum_P \rho(x_1, x_{P_1}', \beta) \ldots \rho(x_N, x_{P_N}', \beta), \quad (4.44)$$

where we were able to eliminate one set of permutations. We thus reach the bosonic density matrix in eqn (4.44) from the distinguishable density matrix by summing over permutations and dividing by $1/N!$, writing $\{x_{P_1}', \ldots, x_{P_N}'\}$ instead of $\{x_1', \ldots, x_N'\}$. The expression obtained for $N$ ideal bosons—even though we strictly derived it only for two bosons in two states—carries over to interacting systems, where we find

$$\overbrace{\rho^{\text{sym}}(\{x_1, \ldots, x_N\}, \{x_1', \ldots, x_N'\}, \beta)}^{\text{bosonic density matrix}} =$$

$$\frac{1}{N!} \sum_P \underbrace{\rho^{\text{dist}}(\{x_1, \ldots, x_N\}, \{x_{P_1}', \ldots, x_{P_N}'\}, \beta)}_{\text{distinguishable-particle density matrix}}. \quad (4.45)$$

In conclusion, in this subsection we have studied the density matrix of an $N$-particle system. For distinguishable particles, we can easily generalize the earlier definition by simply replacing normalized single-particle states with normalized $N$-particle states. For bosonic systems, these states have to be symmetrized. In the example of two particles in two states, the different normalization factors of the wave functions in eqn (4.43) gave rise to a simple final result, showing that the bosonic density matrix is the average of the distinguishable-particle density matrix with permuted indices. This result is generally valid for $N$ particles, with or without interactions. In the latter case, the density matrix of distinguishable particles becomes trivial. In contrast, in the bosonic density

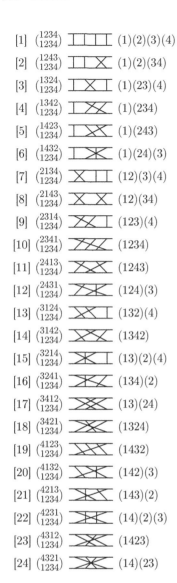

**Fig. 4.16** All 24 permutations of four elements.

matrix, the permutations connect the particles (make them interact), as we shall study in more detail in Subsection 4.2.3.

### 4.2.2 Recursive counting of permutations

Permutations play a pivotal role in the path-integral description of quantum systems, and we shall soon need to count permutations with weights, that is, compute general "partition functions" of permutations of $N$ particles

$$Y_N = \sum_{\text{permutations } P} \text{weight}(P).$$

If the weight of each permutation is 1, then $Y_N = N!$, the number of permutations of $N$ elements. For concreteness, we shall consider the permutations of four elements (see Fig. 4.16). Anticipating the later application, in Subsection 4.2.3, we allow arbitrary weights depending on the length of the cycles. For coherence with the later application, we denote the weight of a cycle of length $k$ by $z_k$, such that a permutation with one cycle of length 3 and another of length 1 has weight $z_3 z_1$, whereas a permutation with four cycles of length 1 has weight $z_1^4$, etc.

We now derive a crucial recursion formula for $Y_N$. In any permutation of $N$ elements, the last element (in our example the element $N = 4$) is in what may be called the last-element cycle. (In permutation [5] in Fig. 4.16, the last-element cycle, of length 3, contains $\{2, 3, 4\}$. In permutation [23], the last-element cycle, of length 4, contains all elements). Generally, this last-element cycle involves $k$ elements $\{n_1, \ldots, n_{k-1}, N\}$. Moreover, $N - k$ elements do not belong to the last-element cycle. The partition function of these elements is $Y_{N-k}$, because we know nothing about them, and they are unrestricted.

$Y_N$ is determined by the number of choices for $k$ and the cycle weight $z_k$, the number of different sets $\{n_1, \ldots, n_{k-1}\}$ given $k$, the number of different cycles given the set $\{n_1, \ldots, n_{k-1}, N\}$, and the partition function $Y_M$ of the elements not participating in the last-element cycle:

$$Y_N = \sum_{k=1}^{N} z_k \left\{ \begin{array}{c} \text{number of} \\ \text{choices for} \\ \{n_1, \ldots, n_{k-1}\} \end{array} \right\} \left\{ \begin{array}{c} \text{number of} \\ \text{cycles with} \\ \{n_1, \ldots, n_k\} \end{array} \right\} Y_{N-k}.$$

From Fig. 4.16, it follows that there are $(k-1)!$ cycles of length $k$ with the same $k$ elements. Likewise, the number of choices of different elements for $\{n_1, \ldots, n_{k-1}\}$ is $\binom{N-1}{k-1}$. We find

$$\underline{Y_N} = \sum_{k=1}^{N} z_k \binom{N-1}{k-1} (k-1)! \, Y_{N-k}$$

$$= \sum_{k=1}^{N} \frac{1}{N} z_k \frac{N!}{(N-k)!} Y_{N-k} \quad \text{(with } Y_0 = 1\text{).} \quad (4.46)$$

Equation (4.46) describes a recursion because it allows us to compute $Y_N$ from $\{Y_0, \ldots, Y_{N-1}\}$.

We now use the recursion formula in eqn (4.46) to count permutations of $N$ elements for various choices of cycle weights $\{z_1, \ldots, z_N\}$. Let us start with the simplest case, $\{z_1, \ldots, z_N\} = \{1, \ldots, 1\}$, where each cycle length has the same weight, and every permutation has unit weight. We expect to find that $Y_N = N!$, and this is indeed what results from the recursion relation, as we may prove by induction. It follows, for this case of equal cycle weights, that the weight of all permutations with a last-element cycle of length $k$ is the same for all $k$ (as is $z_k Y_{N-k}/(N-k)!$ in eqn (4.46)). This is a nontrivial theorem. To illustrate it, let us count last-element cycles in Fig. 4.16:

$$
\left\{ \begin{array}{l} \text{in Fig. 4.16,} \\ \text{element 4 is} \end{array} \right\}
\left\{ \begin{array}{l} \text{in 6 cycles of length 4} \\ \text{in 6 cycles of length 3} \\ \text{in 6 cycles of length 2} \\ \text{in 6 cycles of length 1} \end{array} \right. .
$$

The element 4 is in no way special, and this implies that any element among $\{1, \ldots, N\}$ is equally likely to be in a cycle of length $\{1, \ldots, N\}$. As a consequence, in a random permutation (for example generated with Alg. 1.11 (**ran-perm**)), the probability of having a cycle of length $k$ is $\propto 1/k$. We need more cycles of shorter length to come up with the same probability. Concretely, we find

$$
\left\{ \begin{array}{c} \text{in Fig. 4.16,} \\ \text{there are} \end{array} \right\}
\left\{ \begin{array}{l} 6 \text{ cycles of length 4} \\ 8 \text{ cycles of length 3} \\ 12 \text{ cycles of length 2} \\ 24 \text{ cycles of length 1} \end{array} \right. .
$$

The number of cycles of length $k$ is indeed inversely proportional to $k$. As a second application of the recursion relation, let us count permutations containing only cycles of length 1 and 2. Now $\{z_1, z_2, z_3, \ldots, z_N\} = \{1, 1, 0, \ldots, 0\}$ (every permutation has the same weight, under the condition that it contains no cycles of length 3 or longer). We find $Y_0 = 1$ and $Y_1 = 1$, and from eqn (4.46) the recursion relation

$$
Y_N = Y_{N-1} + (N-1)\, Y_{N-2},
$$

so that $\{Y_0, Y_1, Y_2, Y_3, Y_4, \ldots\} = \{1, 1, 2, 4, 10, \ldots\}$. Indeed, for $N = 4$, we find 10 such permutations in Fig. 4.16 ([1], [2], [3], [6], [7], [8], [15], [17], [22], and [24]).

In conclusion, we have described in this subsection a recursion formula for counting permutations that lets us handle arbitrary cycle weights. We shall apply it, in Subsection 4.2.3, to ideal bosons.

## 4.2.3   Canonical partition function of ideal bosons

In Subsection 4.2.1, we expressed the partition function of a bosonic system as a sum over diagonal and nondiagonal density matrices for distinguishable particles:

$$Z_N = \frac{1}{N!} \sum_P Z_P \tag{4.47}$$

$$= \frac{1}{N!} \sum_P \int d^N x \rho^{\text{dist}} \left( \{x_1, \ldots, x_N\}, \{x_{P(1)}, \ldots, x_{P(N)}\}, \beta \right). \tag{4.48}$$

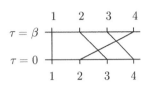

**Fig. 4.17** The permutation $\left(\begin{smallmatrix} 1 & 4 & 2 & 3 \\ 1 & 2 & 3 & 4 \end{smallmatrix}\right)$ represented as a path.

For ideal particles, the distinguishable-particle density matrix separates into a product of single-particle density matrices, but the presence of permutations implies that these single-particle density matrices are not necessarily diagonal. For concreteness, we consider, for $N = 4$ particles, the permutation $P = \left(\begin{smallmatrix} 1 & 4 & 2 & 3 \\ 1 & 2 & 3 & 4 \end{smallmatrix}\right)$, which in cycle representation is written as $P = (1)(243)$ (see Fig. 4.17). This permutation consists of one cycle of length 1 and one cycle of length 3. The permutation-dependent partition function $Z_{(1)(243)}$ is

$$Z_{(1)(243)} = \int dx_1 \, \rho(x_1, x_1, \beta) \int dx_2$$
$$\times \underbrace{\left[ \int dx_3 \int dx_4 \, \rho(x_2, x_4, \beta) \rho(x_4, x_3, \beta) \rho(x_3, x_2, \beta) \right]}_{\rho(x_2, x_2, 3\beta)}. \tag{4.49}$$

The last line of eqn (4.49) contains a double convolution and can be written as a diagonal single-particle density matrix at temperature $T = 1/(3\beta)$. This is an elementary application of the matrix squaring described in Chapter 3. After performing the last two remaining integrations, over $x_1$ and $x_2$, we find that the permutation-dependent partition function $Z_{(1)(243)}$ is the product of single-particle partition functions, one at temperature $1/\beta$ and the other at $1/(3\beta)$:

$$Z_{(1)(243)} = z(\beta) z(3\beta). \tag{4.50}$$

Here, and in the remainder of the present chapter, we denote the single-particle partition functions with the symbol $z(\beta)$:

$$\left\{ \begin{matrix} \text{single-particle} \\ \text{partition function} \end{matrix} \right\} : z(\beta) = \int dx \, \rho(x, x, \beta) = \sum_\sigma e^{-\beta E_\sigma}. \tag{4.51}$$

Equation (4.50) carries the essential message that—for ideal bosons—the $N$-particle partition function $Z(\beta)$ can be expressed as a sum of products of single-particle partition functions. However, this sum of $N!$ terms is nontrivial, unlike the one for the gas of ideal distinguishable particles. Only for small $N$ can we think of writing out the $N!$ permutations and determining the partition function via the explicit sum in eqn (4.47). It is better to adapt the recursion formula of Subsection 4.2.2 to the ideal-boson partition functions. Now, the cycle weights are given by the single-particle density matrices at temperature $k\beta$. Taking into account that the partition functions carry a factor $1/N!$ (see eqn (4.48)), we find

$$Z_N = \frac{1}{N} \sum_{k=1}^N z_k Z_{N-k} \quad \text{(with } Z_0 = 1\text{)}. \tag{4.52}$$

This recursion relation (Borrmann and Franke 1993) determines the partition function $Z_N$ of an ideal boson system with $N$ particles via the single-particle partition functions $z_k$ at temperatures $\{1/\beta, \ldots, 1/(N\beta)\}$ and the partition functions $\{Z_0, \ldots, Z_{N-1}\}$ of systems with fewer particles. For illustration, we shall compute the partition function of the five-boson bounded trap model at $T = 1/2$ $(\beta = 2)$ from the single-particle partition functions $z_k$ obtained from Table 4.1:

| $E$ | $\mathcal{N}(E)$ |
|-----|------------------|
| 0   | 1                |
| 1   | 3                |
| 2   | 6                |
| 3   | 10               |
| 4   | 15               |

$$\implies \quad z_k(\beta) = z(k\beta) = 1 + 3e^{-k\beta} + 6e^{-2k\beta} \\ + 10e^{-3k\beta} + 15e^{-4k\beta}. \quad (4.53)$$

**Table 4.7** Single-particle partition functions $z_k(\beta) = z(k\beta)$ at temperature $T = 1/\beta = \frac{1}{2}$, in the five-boson bounded trap model (from eqn (4.53))

| $k$ | $z_k$    |
|-----|----------|
| 1   | 1.545719 |
| 2   | 1.057023 |
| 3   | 1.007473 |
| 4   | 1.001007 |
| 5   | 1.000136 |

The single-particle partition functions $\{z_1(\beta), \ldots, z_5(\beta)\}$ (see Table 4.7, for $\beta = 2$), entered into the recursion relation, give

$$
\begin{aligned}
Z_0 &= \ldots & &= 1, \\
Z_1 &= (z_1 Z_0)/1 & &= \ldots & &= 1.5457, \\
Z_2 &= (z_1 Z_1 + z_2 Z_0)/2 = (1.5457 \times 1.5457 + \cdots)/2 &= 1.7231, \\
Z_3 &= (z_1 Z_2 + \cdots)/3 \; = (1.5457 \times 1.7231 + \cdots)/3 &= 1.7683, \\
Z_4 &= (z_1 Z_3 + \cdots)/4 \; = (1.5457 \times 1.7683 + \cdots)/4 &= 1.7782, \\
Z_5 &= (z_1 Z_4 + \cdots)/5 \; = (1.5457 \times 1.7782 + \cdots)/5 &= \underline{1.7802}. \quad (4.54)
\end{aligned}
$$

In a few arithmetic operations, we thus obtain the partition function of the five-boson bounded trap model at $\beta = 2$ $(Z_5 = Z_{\text{btm}} = 1.7802)$, that is, the same value that was earlier obtained through a laborious sum over 575 757 five-particle states (see Table 4.3 and Alg. 4.3 (`naive-bosons`)).

We can carry on with the recursion to obtain $\{Z_6, Z_7, \ldots\}$, and can also take to infinity the cutoff in the energy. There is no more combinatorial explosion. The partition function $z_k(\beta)$ in the harmonic trap, without any cutoff $(E_{\max} = \infty)$, is

$$
z_k(\beta) = \left( \sum_{E_x=0}^{\infty} e^{-k\beta E_x} \right) \left( \sum_{E_y=0}^{\infty} e^{-k\beta E_y} \right) \left( \sum_{E_z=0}^{\infty} e^{-k\beta E_z} \right)
$$

$$
= \left( \frac{1}{1 - e^{-k\beta}} \right)^3 \quad (4.55)
$$

(we note that the ground-state energy is now $E_0 = 0$). Naturally, the expansion of the final expression in eqn (4.55) starts with the terms in eqn (4.53). This formula goes way beyond the naive summation represented in Table 4.7 (compare with Alg. 4.3 (`naive-bosons`)). Together with the recursion formula (4.52), it gives a general method for computing the partition function of canonical ideal bosons (see Alg. 4.6 (`canonic-recursion`)).

**procedure** `canonic-recursion`
**input** $\{z_1, \ldots, z_N\}$ ($z_k \equiv z_k(\beta)$, from eqn (4.55))
$Z_0 \leftarrow 1$
**for** $M = 1, \ldots, N$ **do**
$\quad \{ \ Z_M \leftarrow (z_M Z_0 + z_{M-1} Z_1 + \cdots + z_1 Z_{M-1})/M$
**output** $Z_N$

---

**Algorithm 4.6** `canonic-recursion`. Obtaining the partition function for $N$ ideal bosons through the recursion in eqn (4.52) (see also Alg. 4.7).

A recursion formula analogous to eqn (4.18) allows us to compute the internal energy through its own recursion formula, and avoid numerical differentiation. For an $N$-particle system, we start from the definition of the internal energy and differentiate the recursion relation:

$$\langle E \rangle = -\frac{1}{Z_N} \frac{\partial Z_N}{\partial \beta} = -\frac{1}{N Z_N} \sum_{k=1}^{N} \frac{\partial}{\partial \beta} (z_k Z_{N-k})$$

$$= -\frac{1}{N Z_N} \sum_{k=1}^{N} \left( \frac{\partial z_k}{\partial \beta} Z_{N-k} + z_k \frac{\partial Z_{N-k}}{\partial \beta} \right).$$

This equation contains an explicit formula for the internal energy, but it also constitutes a recursion relation for the derivative of the partition function. To determine $\partial Z_N/\partial \beta$, one only needs to know the partition functions $\{Z_0, \ldots, Z_N\}$ and the derivatives for smaller particle numbers. (The recursion starts with $\partial Z_0/\partial \beta = 0$, because $Z_0$ is independent of the temperature.) We need to know only the single-particle density matrices $z_k$ and their derivatives $z_k'$. For the harmonic trap, with a single-particle partition function given by eqn (4.55), we obtain

$$\frac{\partial}{\partial \beta} z_k = \frac{\partial}{\partial \beta} \left( \frac{1}{1 - e^{-k\beta}} \right)^3 = -3 z_k \frac{k e^{-k\beta}}{1 - e^{-k\beta}}.$$

We pause for a moment to gain a better understanding of the recursion relation, and remember that each of its components relates to last-element cycles:

$$Z_N \propto \underbrace{z_N Z_0}_{\substack{\text{particle } N \\ \text{in cycle of} \\ \text{length } N}} + \cdots + \underbrace{z_k Z_{N-k}}_{\substack{\text{particle } N \\ \text{in cycle of} \\ \text{length } k}} + \cdots + \underbrace{z_1 Z_{N-1}}_{\substack{\text{particle } N \\ \text{in cycle of} \\ \text{length } 1}}. \tag{4.56}$$

It follows that the cycle probabilities satisfy

$$\left\{ \begin{array}{c} \text{probability of having particle} \\ N \text{ in cycle of length } k \end{array} \right\} : \pi_k = \frac{1}{N} \frac{z_k Z_{N-k}}{Z_N}. \tag{4.57}$$

The $N$th particle is in no way special, and the above expression gives the probability for any particle to be in a cycle of length $k$. (As a consequence of eqn (4.57), the mean number of particles in cycles of length $k$ is $z_k Z_{N-k}/Z_N$, and the mean number of cycles of length $k$ is $z_k Z_{N-k}/(k Z_N)$.)

We can compute the cycle probabilities $\{\pi_1, \ldots, \pi_N\}$ as a by-product of running Alg. 4.6 (`canonic-recursion`). For concreteness, we consider 40 particles in a harmonic trap (see Fig. 4.18). At high temperatures, only short cycles appear with reasonable probability, whereas at small temperatures also long cycles are probable. In the zero-temperature limit, the probability of a particle to be in a cycle of length $k$ becomes independent of $k$. We shall arrive at a better understanding of these cycle probabilities in Subsection 4.2.4.

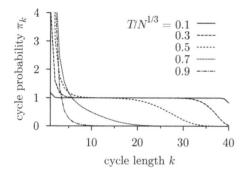

**Fig. 4.18** Cycle probabilities $\{\pi_1, \ldots, \pi_{40}\}$ for 40 ideal bosons in the harmonic trap (from modified Alg. 4.6 (`canonic-recursion`)).

## 4.2.4 Cycle-length distribution, condensate fraction

Using path integrals, we have so far computed partition functions and internal energies, finding a much more powerful algorithm than those studied earlier. It remains to be seen how Bose–Einstein condensation enters the path-integral picture. This is what we are concerned with in the present subsection. We shall see that the appearance of long cycles in the distribution of cycle lengths signals condensation into the ground state. Moreover, we shall find an explicit formula linking the distribution of cycle lengths to the distribution of condensed particles.

To derive the formula, we consider the restricted $N$-particle partition function $Y_{k,0}$, where at least $k$ particles are in the ground state. From Fig. 4.19, this partition function is

$$\left\{\begin{array}{c}\text{partition function with}\\ \geq k \text{ bosons in ground state}\end{array}\right\} = Y_{k,0} = e^{-\beta k E_0} Z_{N-k}.$$

Analogously, we may write, for $k+1$ instead of $k$,

$$\left\{\begin{array}{c}\text{partition function with}\\ \geq k+1 \text{ bosons in ground state}\end{array}\right\} = Y_{k+1,0} = e^{-\beta(k+1)E_0} Z_{N-k-1}.$$

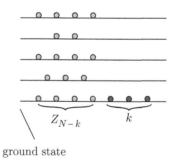

**Fig. 4.19** Restricted partition function $Y_{k,0}$ with at least $k = 3$ particles in the ground state.

Taking the difference between these two expressions, and paying attention to the special case $k = N$, we find

$$\left\{ \begin{array}{l} \text{partition function with} \\ k \text{ bosons in ground state} \end{array} \right\} = \begin{cases} Y_{k,0} - Y_{k+1,0} & \text{if } k < N \\ Y_{k,0} & \text{if } k = N \end{cases}. \quad (4.58)$$

Our choice of ground-state energy ($E_0 = 0$) implies $Y_{k,0} = Z_{N-k}$, and we may write the probability of having $N_0$ bosons in the ground state as

$$\pi(N_0) = \frac{1}{Z_N} \begin{cases} Z_{N-N_0} - Z_{N-(N_0+1)} & \text{if } N_0 < N \\ 1 & \text{if } N_0 = N \end{cases}. \quad (4.59)$$

This probability was earlier evaluated with more effort (see Fig. 4.14). The condensate fraction, the mean value of $N_0$, is given by

$$\langle N_0 \rangle = \sum_0^N N_0 \pi(N_0) = \frac{1}{Z_N} \left\{ \sum_{N_0=1}^{N-1} N_0 \cdot \left[ Z_{N-N_0} - Z_{N-(N_0+1)} \right] + N Z_0 \right\}.$$

This is a telescopic sum, where similar terms are added and subtracted. It can be written more simply as

$$\langle N_0 \rangle = \frac{Z_{N-1} + Z_{N-2} + \cdots + Z_0}{Z_N} \quad \text{(with } E_0 = 0\text{)}. \quad (4.60)$$

The calculations of the condensate fraction and the internal energy are incorporated into Alg. 4.7 (`canonic-recursion(patch)`), which provides us with the same quantities as the program which used the integral representation of the Kronecker $\delta$-function. It is obviously more powerful, and basically allows us to deal with as many particles as we like.

> **procedure** `canonic-recursion(patch)`
> **input** $\{z_1, \ldots, z_N\}, \{z_1', \ldots, z_N'\}$ (from eqn (4.51))
> $Z_0 \leftarrow 1$
> $Z_0' \leftarrow 0$
> **for** $M = 1, \ldots, N$ **do**
> $\quad \left\{ \begin{array}{l} Z_M \leftarrow (z_M Z_0 + z_{M-1} Z_1 + \cdots + z_1 Z_{M-1})/M \\ Z_M' \leftarrow \left[ (z_M' Z_0 + z_M Z_0') + \cdots + (z_1' Z_{M-1} + z_1 Z_{M-1}') \right]/M \end{array} \right.$
> $\langle E \rangle \leftarrow -Z_N'/Z_N$
> $\langle N_0 \rangle \leftarrow (Z_0 + \cdots + Z_{N-1})/Z_N$ (with $E_0 = 0$)
> **output** $\{Z_0, \ldots, Z_N\}, \langle E \rangle, \langle N_0 \rangle$
> ———

**Algorithm 4.7** `canonic-recursion(patch)`. Calculation of the partition function, the energy, and the condensate fraction for $N$ ideal bosons.

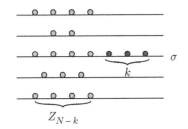

**Fig. 4.20** Restricted partition function $Y_{k,\sigma}$ with at least $k = 3$ particles in state $\sigma$ (for $N = 20$ particles).

We continue the analysis of restricted partition functions, by simply generalizing the concept of the restricted partition functions to a state $\sigma$, rather than only the ground state (see Fig. 4.20). From eqn (4.58), we arrive at

$$\left\{ \begin{array}{l} \text{partition function with} \\ \geq k \text{ bosons in state } \sigma \end{array} \right\} = Y_{k,\sigma} = e^{-\beta k E_\sigma} Z_{N-k}.$$

This equation can be summed over all states, to arrive at a crucial expression,

$$\sum_\sigma \left\{ \begin{array}{c} \text{partition function} \\ \text{with} \geq k \text{ bosons} \\ \text{in state } \sigma \end{array} \right\} = \underbrace{\sum_\sigma e^{-\beta k E_\sigma} Z_{N-k}}_{z_k, \text{ see eqn (4.51)}} \propto \left\{ \begin{array}{c} \text{cycle} \\ \text{weight} \\ \pi_k \end{array} \right\},$$

because it relates the energy-level description (on the left) with the description in terms of density matrices and cycle-length distributions (on the right). Indeed, the sum over the exponential factors gives the partition function of the single-particle system at temperature $1/(k\beta)$, $z_k = z(k\beta)$, and the term $z_k Z_{N-k}$ is proportional to the cycle weight $\pi_k$. This leads to a relation between occupation probabilities of states and cycle weights:

$$\sum_\sigma \left\{ \begin{array}{c} \text{partition function with} \\ k \text{ bosons in state } \sigma \end{array} \right\}$$

$$\propto \left\{ \begin{array}{c} \text{cycle weight} \\ \pi_k \end{array} \right\} - \left\{ \begin{array}{c} \text{cycle weight} \\ \pi_{k+1} \end{array} \right\}. \quad (4.61)$$

To interpret this equation, we note that the probability of having $k \gg 1$ in any state other than the ground state is essentially zero. This was discussed in the context of the saturation densities, in Subsection 4.1.3. It follows that the sum in eqn (4.61) is dominated by the partition function with $k$ particles in the ground state, and this relates the probability of having $k$ particles in the ground state (the distribution whose mean gives the condensate fraction) to the integer derivative of the cycle-length distribution. (The difference in eqn (4.61) constitutes a negative integer derivative: $-\Delta f(k)/\Delta k = f(k) - f(k+1)$.) We arrive at the conclusion that the condensate distribution is proportional to the integer derivative of the cycle length distribution (Holzmann and Krauth 1999).

For concreteness, we continue with a system of 1000 trapped bosons at $T/N^{1/3} = 0.5$ (see Fig. 4.21). We can compute the cycle length distribution, from eqn (4.56), and the distribution function $\pi(N_0)$ for the condensate (see eqn (4.59)). We are thus able to compute the condensate fraction from exact or sampled distributions of cycle lengths, as in Fig. 4.18.

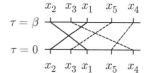

**Fig. 4.22** Boson configuration with positions $\{x_1, \ldots, x_5\}$ and permutation $P = (1, 2)(3, 5, 4)$.

### 4.2.5   Direct-sampling algorithm for ideal bosons

In the previous subsections, we have computed partition functions for ideal bosons by appropriately summing over all permutations and integrating over all particle positions. We now consider sampling, the twin brother of integration, in the case of the ideal Bose gas. Specifically, we discuss a direct-sampling algorithm for ideal bosons, which lies at the heart of some path-integral Monte Carlo algorithms for interacting bosons in the same way as the children's algorithm performed in a square on the beach underlies the Markov-chain Monte Carlo algorithm

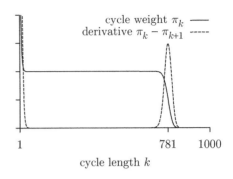

**Fig. 4.21** Cycle weights $\pi_k$, and derivative $\pi_k - \pi_{k+1}$, for 1000 trapped bosons at $T/N^{1/3} = 0.5$ (from Alg. 4.6 (`canonic-recursion`)).

for hard disks. A boson configuration consists of a permutation and a set of positions (see Fig. 4.22), which we sample in a two-step procedure.

We saw in Subsection 4.2.3 that the partition function of the canonical ideal Bose gas can be written as a sum over permutation-dependent partition functions. To pick a permutation, we might think of tower sampling in the $N!$ sectors. For large $N$, this strategy is not an option. However, we know the following from eqn (4.52):

$$Z_N = \frac{1}{N}(\underbrace{z_1 Z_{N-1}}_{\substack{\text{particle } N \\ \text{in cycle of} \\ \text{length } 1}} + \underbrace{z_2 Z_{N-2}}_{\substack{\text{particle } N \\ \text{in cycle of} \\ \text{length } 2}} + \cdots + \underbrace{z_{N-1} Z_1}_{\substack{\text{particle } N \\ \text{in cycle of} \\ \text{length } N-1}} + \underbrace{z_N Z_0}_{\substack{\text{particle } N \\ \text{in cycle of} \\ \text{length } N}}).$$

This equation allows us to sample the length $k$ of the last-element cycle from only $N$ choices, without knowing anything about the permutations of the $N - k$ particles in other cycles. These probabilities are already generated in the final step of computing $Z_N$ (see, for the five-boson bounded trap model, the last line of eqn (4.54)).

After sampling the length of the last-element cycle, we eliminate this cycle from the permutation $P$, and then continue to sample another cycle in the remaining permutation with $N - k$ particles, etc. This is iterated until we run out of particles (see Alg. 4.8 (`direct-cycles`)). Because particles are indistinguishable, we need only remember the lengths of cycles generated, that is, the histogram of cycle lengths $\{m_1, \ldots, m_N\}$ in one permutation of $N$ particles ($m_k$ gives the number of cycles of length $k$ (see Table 4.8)).

After sampling the permutation, we must determine the coordinates $\{\mathbf{x}_1, \ldots, \mathbf{x}_N\}$. Particles $\{l+1, l+2, \ldots, l+k\}$ on each permutation cycle of length $l$ form a closed path and their coordinates $\{\mathbf{x}_{l+1}(0), \ldots, \mathbf{x}_{l+k}(0)\}$ can be sampled using the Lévy construction of Subsection 3.3.2, at inverse temperature $k\beta$ and with a discretization step $\Delta_\tau = \beta$ (see Fig. 4.24 and Alg. 4.9 (`direct-harmonic-bosons`)).

The complete program for simulating ideal Bose–Einstein condensates

**Table 4.8** Typical output of Alg. 4.8 (`direct-cycles`) for $N = 1000$ and $T/N^{1/3} = 0.5$

| $k$ | $m_k$ |
| --- | --- |
| 1 | 147 |
| 2 | 18 |
| 3 | 4 |
| 4 | 2 |
| 5 | 1 |
| 8 | 1 |
| 25 | 1 |
| 73 | 1 |
| 228 | 1 |
| 458 | 1 |

**procedure** `direct-cycles`
**input** $\{z_1, \ldots, z_N\}, \{Z_0, \ldots, Z_{N-1}\}$ (from Alg. 4.6 (`canonic-recursion`))
$\{m_1, \ldots, m_N\} \leftarrow \{0, \ldots, 0\}$
$M \leftarrow N$
**while** $(M > 0)$ **do**
$\left\{ \begin{array}{l} k \leftarrow \texttt{tower-sample}\left(\{z_1 Z_{M-1}, \ldots, z_k Z_{M-k}, \ldots, z_M Z_0\}\right) \\ M \leftarrow M - k \\ m_k \leftarrow m_k + 1 \end{array} \right.$
**output** $\{m_1, \ldots, m_N\}$ ($m_k$: number of cycles of length $k$.)

———

**Algorithm 4.8** `direct-cycles`. Sampling a cycle-length distribution for $N$ ideal bosons, using Alg. 1.14 (`tower-sample`).

**procedure** `direct-harmonic-bosons`
**input** $\{z_1, \ldots, z_N\}, \{Z_0, \ldots, Z_N\}$ (for harmonic trap)
$\{m_1, \ldots, m_N\} \leftarrow \texttt{direct-cycles}\left(\{z_1, \ldots, z_N\}, \{Z_0, \ldots, Z_{N-1}\}\right)$
$l \leftarrow 0$
**for all** $m_k \neq 0$ **do**
$\left\{ \begin{array}{l} \textbf{for } i = 1, \ldots, m_k \textbf{ do} \\ \quad \left\{ \begin{array}{l} \Upsilon \leftarrow \texttt{gauss}\,(\ldots) \\ \{x_{l+1}, \ldots, x_{l+k}\} \leftarrow \texttt{levy-harmonic-path}(\Upsilon, \Upsilon, k\beta, k) \\ l \leftarrow l + k \end{array} \right. \end{array} \right.$
**output** $\{x_1, \ldots, x_N\}$

———

**Algorithm 4.9** `direct-harmonic-bosons`. Direct-sampling algorithm for ideal bosons in the harmonic trap. Only $x$-coordinates are shown.

with tens of thousands of particles in the harmonic trap takes no more than a few dozen lines of computer code. It allows us to represent the spatial distribution of particles (see Fig. 4.23 for a projection in two dimensions). The very wide thermal cloud at temperatures $T > T_{\rm c}$ suddenly shrinks below $T_{\rm c}$ because most particles populate the single-particle ground state or, in our path-integral language, because most particles are on a few long cycles. The power of the path-integral approach resides in the fact that the inclusion of interactions into the rudimentary Alg. 4.9 (`direct-harmonic-bosons`) is straightforward, posing only a few practical problems (most of them treated in Subsection 3.4.2). Conceptual problem are not met.

## 4.2.6   Homogeneous Bose gas, winding numbers

The single-particle partition function $z_k(\beta)$ in a three-dimensional cube is the product of the one-dimensional partition functions of a particle on a line with periodic boundary conditions. It can be computed from the free density matrix or as a sum over energy levels (see eqn (3.19)).

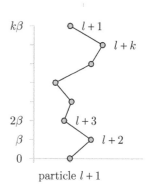

**Fig. 4.24** Cycle path in Alg. 4.9 (`direct-harmonic-bosons`).

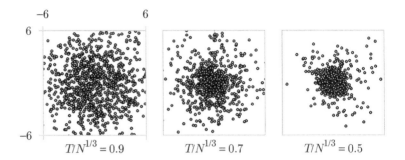

**Fig. 4.23** Two-dimensional snapshots of 1000 ideal bosons in a three-dimensional harmonic trap (from Alg. 4.9 (`direct-harmonic-bosons`)).

The latter approach gives

$$z^{\text{cube}}(k\beta) = \left[ \sum_{n=-\infty}^{\infty} \exp(-k\beta E_n) \right]^3. \tag{4.62}$$

We can differentiate with respect to $\beta$ to get the expressions which allow us to determine the internal energy:

$$\frac{\partial}{\partial \beta} z^{\text{cube}}(k\beta) = -3k \left[ \sum_n \exp(-k\beta E_n) \right]^2 \sum_n E_n \exp(-k\beta E_n). \tag{4.63}$$

(In the above two formulas, we use $E_n = 2n^2\pi^2/L^2$, see Table 4.9.)

**Table 4.9** Single-particle partition functions $z_k(\beta)$, and their derivative, in a cubic box with $L = 2$, at temperature $T = 2$ (from eqns (4.62) and (4.63))

| $k$ | $z_k(\beta)$ | $\frac{\partial}{\partial\beta} z_k(\beta)$ |
|---|---|---|
| 1 | 1.6004 | -3.4440 |
| 2 | 1.0438 | -0.4382 |
| 3 | 1.0037 | -0.0543 |
| 4 | 1.0003 | -0.0061 |
| 5 | 1.0000 | -0.0006 |
| 6 | 1.0000 | -0.0001 |
| 7 | 1.0000 | 0.0000 |
| 8 | 1.0000 | 0.0000 |

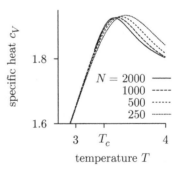

**Fig. 4.25** Specific heat computed for the density $= 1$, in the canonic ensemble (from Alg. 4.6 (`canonic-recursion`), adapted for ideal bosons)

The single-particle partition functions $z^{\text{cube}}(k\beta)$ and their derivatives can be entered into Alg. 4.6 (`canonic-recursion`). The permutations are sampled as for the harmonic trap. Positions are sampled in a similar fashion (see Alg. 4.10 (`direct-period-bosons`)). It uses Alg. 3.7

(`levy-periodic-path`) which itself contains as a crucial ingredient the sampling of winding numbers. As a consequence, the paths generated can thus wind around the box, in any of the three spatial dimensions (see Fig. 4.26 for a two-dimensional representation).

**procedure** `direct-period-bosons`
**input** $\{z_1, \ldots, z_N\}, \{Z_0, \ldots, Z_N\}$ (cube, periodic boundary conditions)
$\{m_1, \ldots, m_N\} \leftarrow$ `direct-cycles` $(\{z_1, \ldots, z_N\}, \{Z_0, \ldots, Z_{N-1}\})$
$l \leftarrow 0$
**for all** $m_k \neq 0$ **do**
$\left\{\begin{array}{l} \quad \textbf{for } i = 1, \ldots, m_k \textbf{ do} \\ \quad \left\{\begin{array}{l} \quad \Upsilon \leftarrow \texttt{ran}\,(0, L) \\ \quad \{w_x, \{x_{l+1}, \ldots, x_{l+k}\}\} \leftarrow \texttt{levy-period-path}\,(\Upsilon, \Upsilon, k\beta, k) \\ \quad l \leftarrow l + k \end{array}\right. \end{array}\right.$
**output** $\{x_1, \ldots, x_N\}$
————

**Algorithm 4.10** `direct-period-bosons`. Sampling ideal bosons in a periodic cube. Only $x$-coordinates are shown.

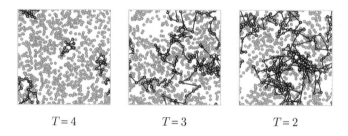

$T = 4$  $\qquad\qquad$  $T = 3$  $\qquad\qquad$  $T = 2$

**Fig. 4.26** Projected snapshots of 1000 ideal bosons in a cubic box with periodic boundary conditions (from Alg. 4.10 (`direct-period-bosons`)).

At high temperature, the mean squared winding number is zero because the lengths of cycles are very small. At lower temperatures, long cycles appear. Paths in long cycles can exit the periodic simulation box on one side and reenter through another side. These paths contribute to the intricate response of a quantum system to an outside motion. For each configuration, the winding number $\mathbf{w} = \{w_x, w_y, w_z\}$ is the sum of the winding numbers of the individual cycles. Using the results of Subsection 3.1.4, we can use the mean squared winding number to determine the fraction of the system which remains stationary under a small rotation, in other words the superfluid fraction:

$$\rho_{\mathrm{s}}/\rho = \frac{\langle w^2 \rangle L^2}{3\beta N} = \frac{\langle w_x^2 \rangle L^2}{\beta N}.$$

### 4.2.7  Interacting bosons

The structure of general path-integral Monte Carlo algorithms for inter-acting bosons is virtually unchanged with respect to the two sampling algorithms for ideal bosons, Algs 4.9 (`direct-harmonic-bosons`) and 4.10 (`direct-period-bosons`). Their detailed presentation and the dis-cussion of results that can be obtained with them are beyond the scope of this book. An interacting system has the same path configurations as the noninteracting system, but the configurations must be reweighted. This can be done using the perfect pair actions discussed in Chapter 3 (see eqn (3.59)). Markov-chain sampling methods must be employed for changing both the permutation state and the spatial configurations. The resulting approach has been useful in precision simulations for liquid he-lium (see Pollock and Ceperley (1987), and Ceperley (1998)), and many other systems in condensed matter physics. It can also be used for sim-ulations of weakly interacting bosons in atomic physics (Krauth, 1996). Cluster algorithms, which are discussed in several other chapter of this book, can also be adapted to path-integral simulations of bosons on a lattice (see Prokof'ev, Svistunov, and Tupitsyn (1998)). This is also be-yond the present scope of this book, but a discussion of these methods will certainly be included in subsequent editions.

# Exercises

## (Section 4.1)

(4.1) Generalize eqn (4.1), the single-particle density of states $\mathcal{N}(E)$ in the homogeneous trap, to the case of an inhomogeneous trap with frequencies $\{\omega_x, \omega_y, \omega_z\} = \{5, 1, 1\}$. Test your result with a modified version of Alg. 4.1 (**naive-degeneracy**). Use the Kronecker $\delta$-function to generalize the integral representation of $\mathcal{N}(E)$ to arbitrary integer values of $\{\omega_x, \omega_y, \omega_z\}$. For frequencies of your choice, evaluate this integral as a discrete Riemann sum (generalizing eqn (4.4)). Also determine the discrete Riemann sum in the complex plane, generalizing eqn (4.5), then, determine an analytic formula for the density of states, valid in the limit of large energies.
NB: Address the isotropic trap first and recover Table 4.1. The complex Riemann sum might incorporate the following fragment, with complex variables $\{z, z_{\text{old}}, \Delta_z\}$,

$$
\begin{aligned}
&\cdots \\
&z_{\text{old}} \leftarrow \{1, 0\} \\
&\textbf{for } \phi = \Delta_\phi, 2\Delta_\phi, \ldots, 2\pi \textbf{ do} \\
&\quad \left\{
\begin{array}{l}
z \leftarrow e^{i\phi} \\
\Delta_z \leftarrow z - z_{\text{old}} \\
\cdots \\
z_{\text{old}} \leftarrow z
\end{array}
\right. \\
&\cdots
\end{aligned}
$$

(4.2) Implement Alg. 4.3 (**naive-bosons**). Compare its output with the data in Table 4.3. Test the results for $Z(\beta)$ against a numerical integration of eqns (4.17) and (4.16). Are the two calculations of the partition function equivalent (in the limit of integration step size $\Delta_x \to 0$) or should you expect small differences? Include the calculation of the distribution function $\pi(N_0)$ in the five-boson bounded trap model. Implement Alg. 4.4 (**canonic-bosons**). Test it for the case of the five-boson bounded trap model. Finally, choose larger values for $E_{\text{max}}$ and increase the number of particles.

(4.3) Calculate the ideal-boson chemical potential $\mu$ vs. mean particle number in the grand canonical ensemble (Alg. 4.5 (**grandcan-bosons**)), for the harmonic trap. Plot the condensate fraction against the rescaled temperature, as in Fig. 4.7. Discuss why the macroscopic occupation of the ground state in the $T \to 0$ limit does not constitute Bose–Einstein condensation and that a more subtle limit is involved. Compute saturation numbers for the excited states.

(4.4) Familiarize yourself with the calculations of the critical temperature for Bose–Einstein condensation, and of the saturation densities, for the isotropic harmonic trap. Then, compute the critical temperature for an anisotropic harmonic trap with frequencies $\{\omega_x, \omega_y, \omega_z\}$. In that trap, the energy $E_x$ can take on values $\{0, \omega_x, 2\omega_x, 3\omega_x \ldots\}$, etc.

(4.5) Compute the specific heat capacity in the grand canonical ensemble for $N$ ideal bosons at density $\rho = 1$ in a box with periodic boundary conditions. First determine the chemical potential $\mu$ for a given mean number $\langle N \rangle$ of bosons (adapt Alg. 4.5 (**grandcan-bosons**) to the case of a cubic box—you may obtain the density of states from Alg. 4.2 (**naive-degeneracy-cube**)). Then compute the energy and the specific heat capacity. Compare your data, at finite $N$, with the results valid in the thermodynamic limit.

(4.6) Implement the first thirty wave functions $\{\psi_0^{\text{h.o.}}, \ldots, \psi_{29}^{\text{h.o.}}\}$ of the harmonic oscillator using Alg. 3.1 (**harmonic-wavefunction**) (see Subsection 3.1.1). Use these wave functions to naively calculate the density of $N$ bosons in the harmonic trap (see eqn (4.32)). (Either implement Alg. 4.5 (**grandcan-bosons**) for determining the chemical potential as a function of mean particle number, or use data from Table 4.5.) Is it possible to speak of a saturation density $\eta_{\text{sat}}(x, y, z)$ of all the particles in excited states (analogous to the saturation numbers discussed in Subsection 4.1.3)?

(4.7) Redo the calculation of the critical temperatures $T_c$ for Bose–Einstein condensation of the ideal Bose gas, in the harmonic trap and in the cube with periodic boundary conditions, but use physical units (containing particle masses, the Boltzmann and Planck constants, and the harmonic oscillator strengths $\omega$), rather than natural units (where $\hbar = m = \omega = 1$).

What is the Bose–Einstein condensation temperature, in kelvin, of a gas of $N = 1 \times 10^6$ Sodium atoms (atomic weight 23) in a harmonic trap of $\omega \simeq 1991$ Hz? Likewise, compute the critical tem-

perature of a gas of ideal bosons of the same mass as
$^4$He (atomic weight 4), and the same molar volume
(59.3 cm$^3$/mole at atmospheric pressure). Is this
transition temperature close to the critical temperature for the normal–superfluid transition in liquid
$^4$He (2.17 kelvin)?

**(Section 4.2)**

(4.8) (See also Exerc. 1.9). Implement Alg. 1.11
(`ran-perm`), as discussed in Subsection 1.2.2. Use
it to generate $1 \times 10^6$ random permutations with
$N = 1000$ elements. Write any of them in cycle-
representation, using a simple subroutine. Generate
histograms for the cycle length distribution, and
show that the probability of finding a cycle is inversely proportional to its length. Likewise, generate a histogram of the length of the cycle containing
an arbitrary element say $k = 1000$. Convince yourself that it is flat (see Subsection 4.2.2). Next, use
the recursion relation of eqn (4.46) to determine
the number of permutations of 1000 elements containing only cycles of length $l \leq 5$ (construct an
algorithm from eqn (4.46), test it with smaller permutations). Write a program to sample these permutations (again test this program for smaller $N$).
In addition, compute the number of permutations
of 1000 elements with cycles only of length 5. Determine the number of permutations of $N$ elements
with a cycle of length $l = 1000$.

(4.9) Enumerate the 120 permutations of five elements.
Write each of them in cycle representation, using
a simple subroutine, and generate the cycle length
distribution $\{m_1, \ldots, m_5\}$. Use this information to
compute nonrecursively the partition function of
the five-boson bounded trap model as a sum over
120 permutation partition functions $Z_P$, each of
them written as a product of single-particle partition functions (see eqns (4.47) and (4.50)).
NB: Repeatedly use the fragment

```
   ...
   input {P_1, ..., P_N}
   for k = N, N - 1, ..., 0 do
      { output {P_1, ..., P_k, N + 1, P_{k+1}, ..., P_N}
   ...
```

to generate all permutations of $N+1$ elements from
a list of permutations of $N$ elements. It generates
the following permutations of four elements from
one permutation of three elements, $P = \left(\begin{smallmatrix} 3 & 1 & 2 \\ 1 & 2 & 3 \end{smallmatrix}\right)$:

$$
312 \rightarrow \begin{cases} 312\underline{4} \\ 31\underline{4}2 \\ 3\underline{4}12 \\ \underline{4}312 \end{cases}
$$

(4.10) Compute the single-particle partition function in a
cubic box with periodic boundary conditions, and
enter it in the recursion relations for the internal
energy and the specific heat capacity of the ideal
Bose gas in the canonical ensemble. Implement re-
cursions for $Z(\beta)$ and for the derivative $\partial Z(\beta)/\partial \beta$,
and compute the mean energy as a function of tem-
perature. Determine the specific heat capacity of
the ideal Bose gas, as in Fig. 4.25, from the numerical derivative of the energy with respect to temperature. Determine the condensate fraction from
eqn (4.60) for various temperatures and temperatures. Compare your results with those obtained in
the grand canonical ensemble.

(4.11) Perform a quantum Monte Carlo simulation of
$N$ ideal bosons in the harmonic trap. First implement Alg. 4.6 (`canonic-recursion`) and then
sample permutations for the $N$-particle system,
using Alg. 4.8 (`direct-cycles`). Finally, sample positions of particles on each cycle with
the Lévy algorithm in a harmonic potential (see
Alg. 3.6 (`levy-harmonic-path`)). At various temperatures, reproduce snapshots of configurations as
in Fig. 4.23. Which is the largest particle number
that you can handle?

(4.12) (Compare with Exerc. 4.11.) Set up a quantum
Monte Carlo simulation of $N$ ideal bosons in a
three-dimensional cubic box with periodic boundary conditions. The algorithm is similar to that
in Exerc. 4.11, however, positions must be sampled using Alg. 3.7 (`levy-periodic-path`). (Use
a naive subroutine for sampling the winding numbers.) Show that below the transition temperature
nonzero winding numbers become relevant. Plot
paths of particles on a cycle, as in Fig. 4.26, for
several system sizes and temperatures. Next, use
this program to study quantum effects at high temperatures. Sample configurations at temperature
$T/T_c = 3$; compute the pair correlation function
between particles $k$ and $l$, using the periodically
corrected distance, $r$, between them (see Alg. 2.6
(`diff-vec`)). Generate a histogram of this pair correlation as a function of distance, to obtain $\pi(r)$.
The function $\pi(r)/(4\pi r^2)$ shows a characteristic increase at small distance, in the region of close encounter. Interpret this effect in terms of cycle length
distributions. What is its length scale? Discuss this
manifestation of the quantum nature of the Bose
gas at high temperature.

# References

Borrmann P., Franke G. (1993) Recursion formulas for quantum statistical partition functions, *Journal of Chemical Physics* **98**, 2484–2485

Ceperley D. M. (1995) Path-integrals in the theory of condensed helium, *Reviews of Modern Physics* **67**, 279–355

Feynman R. P. (1972) *Statistical Mechanics: A Set of Lectures*, Benjamin/Cummings, Reading, Massachusetts

Holzmann M., Krauth W. (1999) Transition temperature of the homogeneous, weakly interacting Bose gas, *Physical Review Letters* **83**, 2687–2690

Krauth W. (1996) Quantum Monte Carlo calculations for a large number of bosons in a harmonic trap, *Physical Review Letters* **77**, 3695–3699

Pollock E. L., Ceperley D. M. (1987) Path-integral computation of superfluid densities, *Physical Review B* **36**, 8343–8352

Prokof'ev N. V., Svistunov B. V., Tupitsyn I. S. (1998) "Worm" algorithm in quantum Monte Carlo simulations, *Physics Letters A* **238**, 253–257

# Order and disorder in spin systems

**5**

In this chapter, we approach the statistical mechanics and computational physics of the Ising model, which has inspired generations of physicists. This archetypal physical system undergoes an order–disorder phase transition. The Ising model shares this transition and many other properties with more complicated models which cannot be analyzed so well.

The first focus of this chapter is on enumeration, which applies to the Ising model because of its finite number of configurations, even though this number grows exponentially with the lattice size. We shall enumerate the spin configurations, and also the loop configurations of the Ising model's high-temperature expansion, which can be summed for very large and even infinite lattices, leading to Onsager's analytic solution in two dimensions. This chapter's second focus is on Monte Carlo algorithms. We shall start with a simple local implementation of the Metropolis sampling algorithm and move on to nontrivial realizations of the perfect-sampling approach described in Subsection 1.1.7 and to modern cluster algorithms. Cluster algorithms originated from the Ising model. They have revolutionized computation in many fields of classical and quantum statistical mechanics.

Theoretical and computational approaches to the Ising model have met with outstanding success. However, it suffices to modify a few parameters in the model, for example to let the sign of the interaction be sometimes positive and sometimes negative, to cause all combined approaches to get into trouble. The two-dimensional spin glass (where the interactions are randomly positive and negative) illustrates the difficulties faced by Monte Carlo algorithms. Remarkably, the above-mentioned enumeration of loop configurations still works. Onsager's analytic solution of the Ising model thus turns into a powerful algorithm for two-dimensional spin glasses.

In this chapter, we witness many close connections between theory and computation across widely different fields. This unity of physics is illustrated further in the final section through a relation between spin systems and classical liquids: we shall see that a liquid with pair interactions is in some sense equivalent to an Ising spin glass and can be simulated with exactly the same methods. This far-reaching equivalence makes it difficult to tell whether we are simulating a liquid (with particles moving around in a continuous space) or spins that are sometimes up and sometimes down.

In a ferromagnet, up spins want to be next to up spins and down spins want to be next to down spins. Likewise, colloidal particles on a liquid surface want to be surrounded by other particles (see Fig. 5.1). At high temperature, up and down spins are equally likely across the system and, likewise, the colloidal particles are spread out all over the surface. At low temperature, the spin system is magnetized, either mostly up or mostly down; likewise, most of the colloidal particles are in one big lump. This, in a nutshell, is the statistical physics of the Ising model, which describes magnets and lattice gases, and which we shall study in the present chapter.

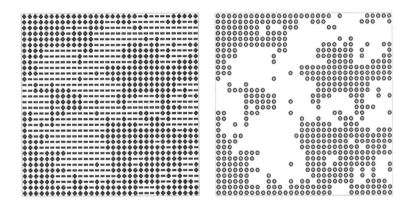

**Fig. 5.1** Configurations of the Ising model on a two-dimensional square lattice considered as a magnet (*left*) and as a lattice gas (*right*).

# 5.1   The Ising model—exact computations

The Ising model describes spins $\sigma_k \pm 1$, $k = 1, \ldots, N$, on a lattice, for example the two-dimensional square lattice shown in Fig. 5.1. In the simplest case, the ferromagnetic Ising model, neighboring spins prefer to align. This means that pairs $\{+, +\}$ and $\{-, -\}$ of neighboring spins direction have a lower energy than antiparallel spins (pairs $\{+, -\}$ and $\{-, +\}$), as expressed by the energy

$$E = -J \sum_{\langle k, l \rangle} \sigma_k \sigma_l. \tag{5.1}$$

The sum is over all pairs of neighbors. The parameter $J$ is positive, and we shall take it equal to one. In a two-dimensional square lattice, the sites $k$ and $l$ then differ by either a lattice spacing in $x$ or a lattice spacing in $y$. In a sum over pairs of neighbors, as in eqn (5.1), we consider each pair only once, that is, we pick either $\langle k, l \rangle$ or $\langle l, k \rangle$. Algorithm 5.1 (energy-ising) implements eqn (5.1) with the help of a neighbor scheme that we have encountered already in Chapter 1. The sum $n$ runs over half the neighbors, so that each pair $\langle l, k \rangle$ is indeed counted only once. We shall soon adopt better approaches for calculating the energy, but shall always keep Alg. 5.1 (energy-ising) for checking purposes. We also note that the lattice may either have peri-

procedure energy-ising
input $\{\sigma_1, \ldots, \sigma_N\}$
$E \leftarrow 0$
for $k = 1, \ldots, N$ do
$\left\{\begin{array}{l} \textbf{for } n = 1, \ldots, d \textbf{ do} \text{ (}d\text{: space dimension)} \\ \quad \left\{\begin{array}{l} j \leftarrow \mathrm{Nbr}(n, k) \\ \textbf{if } (j \neq 0) \textbf{ then} \\ \quad \{ E \leftarrow E - \sigma_k \sigma_j \end{array}\right. \end{array}\right.$
output $E$
———

Algorithm 5.1 energy-ising. Computing the energy of an Ising-model configuration. Nbr(., .) encodes the neighbor scheme of Fig. 5.2.

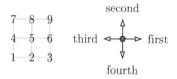

Fig. 5.2 Neighbor scheme in the two-dimensional Ising model. The first neighbor of 2 is $\mathrm{Nbr}(1, 2) = 3$, $\mathrm{Nbr}(4, 2) = 0$, etc.

odic boundary conditions or be planar.

The Ising model's prime use is for magnets. Figure 5.1, however, illustrates that it can also serve to describe particles on a lattice. Now, a variable $\tilde{\sigma}_k = 1, 0$ signals the presence or absence of a particle on site $k$. Let us suppose that particles prefer to aggregate: two particles next to each other have a lower energy than two isolated particles. The simplest configurational energy is

$$E = -4\tilde{J} \sum_{\langle k, l \rangle} \tilde{\sigma}_k \tilde{\sigma}_l.$$

However, the transformation $\tilde{\sigma}_k = \frac{1}{2}(\sigma_k + 1)$ brings us back to the original Ising model.

The main difference between the Ising model considered as a magnet and as a lattice gas is in the space of configurations: for a magnet, the spins can be up or down, more or less independently of the others, so that all of the $2^N$ configurations $\{\sigma_1, \ldots, \sigma_N\} = \{\pm 1, \ldots, \pm 1\}$ contribute to the partition function. For the lattice gas, the number of particles, equivalent to the proportions of up and down spins, must be kept constant, and the partition function is made up of all configurations with a fixed $M = \sum_k \sigma_k$. For large $N$, the two versions of the Ising model become more or less equivalent: it is sufficient to include a constant external magnetic field, which plays the same role here as the chemical potential in Section 4.1.3.

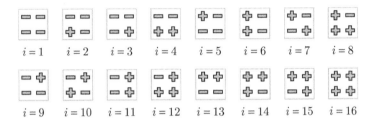

**Fig. 5.3** List of configurations of the Ising model on a $2 \times 2$ square lattice.

In Fig. 5.3, we list all configurations for a (magnetic) Ising model on a $2 \times 2$ lattice. Without periodic boundary conditions, configurations $i = 1$ and $i = 16$ have an energy $E = -4$, and configurations $i = 7$ and $i = 10$ have an energy $E = +4$. All others are zero-energy configurations.

### 5.1.1  Listing spin configurations

In the present subsection, we enumerate all the spin configurations of the Ising model; in fact, we list them one after another. Most simply, each configuration $i = 1, \ldots, 2^N$ of $N$ Ising spins is related to the binary representation of the number $i - 1$: in Fig. 5.3, zeros in the binary representation of $i - 1$ correspond to down spins, and ones to up spins. As an example, the binary representation of the decimal number 10 (configuration $i = 11$ in Fig. 5.3) is 1010, which yields a spin configuration $\{+, -, +, -\}$ to be translated to the lattice with our standard numbering scheme. It is a simple matter to count numbers from 0 to $2^N - 1$ if $N$ is not too large, to represent each number in binary form, and to compute the energy and statistical weight $e^{-\beta E}$ of each configuration with Alg. 5.1 (energy-ising).

It is often faster to compute the change of energy resulting from a spin-flip rather than the energy itself. In Fig. 5.4, for example, we can find out that $E_b = E_a - 4$, simply because the "molecular field" acting on the central site is equal to 2 (it is generated by three up spins and one down spin). The change in energy is equal to twice the value of the spin at the site times the molecular field.

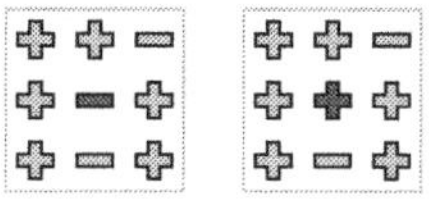

**Fig. 5.4** Two configurations of the Ising model connected by the flip of a single spin.

**procedure** gray-flip
**input** $\{\tau_0, \ldots, \tau_N\}$
$k \leftarrow \tau_0$
**if** $(k > N)$ **exit**
$\tau_{k-1} \leftarrow \tau_k$
$\tau_k \leftarrow k+1$
**if** $(k \neq 1)\ \tau_0 \leftarrow 1$
**output** $k, \{\tau_0, \ldots, \tau_N\}$
———

**Algorithm 5.2** gray-flip. Gray code for spins $\{1, \ldots, N\}$. $k$ is the next spin to flip. Initially, $\{\tau_0, \ldots, \tau_N\} = \{1, \ldots, N+1\}$.

On lattices of any size, the change in energy can be computed in a constant number of operations, whereas the effort for calculating the energy grows with the number of edges. Therefore it is interesting that all $2^N$ spin configurations can be enumerated through a sequence of $2^N$ spin-flips, one at a time. (Equivalently, one may enumerate all numbers $\{0, \ldots, 2^N - 1\}$ by changing a single digit at a time during the enumeration.) Algorithms that perform such enumerations are called Gray codes, and an application of a Gray code for four spins is shown in Table 5.1. How it works can be understood by (mentally) folding Table 5.1 along the horizontal line between configurations $i = 8$ and $i = 9$: the configurations of the first three spins $\{\sigma_1, \sigma_2, \sigma_3\}$ are folded onto each other (the first three spins are the same for $i = 8$ and $i = 9$, and also for $i = 7$ and $i = 10$, etc.). The spins $\{\sigma_1, \sigma_2, \sigma_3\}$ remain unchanged between $i = 8$ and $i = 9$, and this is the only moment at which $\sigma_4$ flips, namely from $-$ to $+$. To write down the Gray code for $N = 5$, we would fold Table 5.1 along the line following configuration $i = 16$, and insert $\{\sigma_5(i = 1), \ldots, \sigma_5(i = 16)\} = \{-, \ldots, -\}$, and $\{\sigma_5(i = 17), \ldots, \sigma_5(i = 32)\} = \{+, \ldots, +\}$. Algorithm 5.2 (gray-flip) provides a practical implementation. We may couple the Gray code enumeration to an update of the energy (see Alg. 5.3 (enumerate-ising)). Of course, the Gray code still has exponential running time, but the enumeration as in Fig. 5.5 gains a factor $\propto N$ with respect to naive binary enumeration.

Algorithm 5.3 (enumerate-ising) does not directly compute the partition function at inverse temperature $\beta$, but rather the number of configurations with energy $E$, in other words, the density of states $\mathcal{N}(E)$

**Table 5.1** Gray-code enumeration of spins $\{\sigma_1, \ldots, \sigma_4\}$. Each configuration differs from its predecessor by one spin only.

| $i$ | $\{\sigma_1, \ldots, \sigma_4\}$ | | | |
|----|----|----|----|----|
| 1  | $-$ | $-$ | $-$ | $-$ |
| 2  | $+$ | $-$ | $-$ | $-$ |
| 3  | $+$ | $+$ | $-$ | $-$ |
| 4  | $-$ | $+$ | $-$ | $-$ |
| 5  | $-$ | $+$ | $+$ | $-$ |
| 6  | $+$ | $+$ | $+$ | $-$ |
| 7  | $+$ | $-$ | $+$ | $-$ |
| 8  | $-$ | $-$ | $+$ | $-$ |
| 9  | $-$ | $-$ | $+$ | $+$ |
| 10 | $+$ | $-$ | $+$ | $+$ |
| 11 | $+$ | $+$ | $+$ | $+$ |
| 12 | $-$ | $+$ | $+$ | $+$ |
| 13 | $-$ | $+$ | $-$ | $+$ |
| 14 | $+$ | $+$ | $-$ | $+$ |
| 15 | $+$ | $-$ | $-$ | $+$ |
| 16 | $-$ | $-$ | $-$ | $+$ |

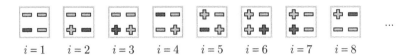

$i = 1$ $\quad$ $i = 2$ $\quad$ $i = 3$ $\quad$ $i = 4$ $\quad$ $i = 5$ $\quad$ $i = 6$ $\quad$ $i = 7$ $\quad$ $i = 8$

**Fig. 5.5** List of Ising-model configurations on a $2 \times 2$ square lattice, generated by the Gray code (only the dark spins flip, see Table 5.1).

**procedure** enumerate-ising
$\{\mathcal{N}(-2N),\ldots,\mathcal{N}(2N)\} \leftarrow \{0,\ldots,0\}$
$\{\sigma_1,\ldots,\sigma_N\} \leftarrow \{-1,\ldots,-1\}$
$\{\tau_0,\ldots,\tau_N\} \leftarrow \{1,\ldots,N+1\}$
$E \leftarrow -2N$
$\mathcal{N}(E) \leftarrow 2$
**for** $i = 1,\ldots,2^{N-1} - 1$ **do**
$\left\{\begin{array}{l} k \leftarrow \texttt{gray-flip}(\{\tau_0,\ldots,\tau_N\}) \\ h \leftarrow \sum_{\langle j,k \rangle} \sigma_j \text{ (field on site } k) \\ E \leftarrow E + 2 \cdot \sigma_k h \\ \mathcal{N}(E) \leftarrow \mathcal{N}(E) + 2 \\ \sigma_k \leftarrow -\sigma_k \end{array}\right.$
**output** $\{\mathcal{N}(E) > 0\}$

**Algorithm 5.3** enumerate-ising. Single spin-flip (Gray code) enumeration for the Ising model, using Alg. 5.2 (gray-flip).

**Table 5.2** Density of states $\mathcal{N}(E)$ for small square lattices with periodic boundary conditions (from Alg. 5.3 (enumerate-ising))

| | $\mathcal{N}(E) = \mathcal{N}(-E)$ | | |
|---|---|---|---|
| $E$ | $2 \times 2$ | $4 \times 4$ | $6 \times 6$ |
| 0 | 12 | 20 524 | 13 172 279 424 |
| 4 | 0 | 13 568 | 11 674 988 208 |
| 8 | 2 | 6 688 | 8 196 905 106 |
| 12 | . | 1 728 | 4 616 013 408 |
| 16 | . | 424 | 2 122 173 684 |
| 20 | . | 64 | 808 871 328 |
| 24 | . | 32 | 260 434 986 |
| 28 | . | 0 | 71 789 328 |
| 32 | . | 2 | 17 569 080 |
| 36 | . | . | 3 846 576 |
| 40 | . | . | 804 078 |
| 44 | . | . | 159 840 |
| 48 | . | . | 35 148 |
| 52 | . | . | 6 048 |
| 56 | . | . | 1 620 |
| 60 | . | . | 144 |
| 64 | . | . | 72 |
| 68 | . | . | 0 |
| 72 | . | . | 2 |

(see Table 5.2). We must take care in implementing this program because $\mathcal{N}(E)$ can easily exceed $2^{31}$, the largest integer that fits into a standard four-byte computer word. We note, in our case, that it suffices to generate only half of the configurations, because $E(\sigma_1,\ldots,\sigma_N) = E(-\sigma_1,\ldots,-\sigma_N)$.

## 5.1.2   Thermodynamics, specific heat capacity, and magnetization

The Ising-model partition function $Z(\beta)$ can be obtained by summing appropriate Boltzmann factors for all configurations, but it is better to start from the density of states, the number of configurations with energy $E$, as just calculated:

$$Z(\beta) = \overbrace{\sum_{\sigma_1=\pm 1,\ldots,\sigma_N=\pm 1} e^{-\beta E(\sigma_1,\ldots,\sigma_N)}}^{\propto\, 2^N \text{ terms}} = \overbrace{\sum_E \mathcal{N}(E)\, e^{-\beta E}}^{\propto\, N \text{ terms}}.$$

Similarly, the mean energy $\langle E \rangle$ can be computed from $Z(\beta)$ by numerical differentiation, that is,

$$\langle E \rangle = -\frac{\partial}{\partial \beta} \log Z, \tag{5.2}$$

but we are again better off using an average over the density of states:

$$\langle E \rangle = \frac{\sum_\sigma E_\sigma e^{-\beta E_\sigma}}{\sum_\sigma e^{-\beta E_\sigma}} = \frac{1}{Z} \sum_E E \mathcal{N}(E) e^{-\beta E}, \qquad (5.3)$$

where we have used $\sigma$ as a shorthand for $\{\sigma_1, \ldots, \sigma_N\}$. Higher moments of the energy can also be expressed via $\mathcal{N}(E)$:

$$\langle E^2 \rangle = \frac{\sum_\sigma E_\sigma^2 e^{-\beta E_\sigma}}{\sum_\sigma e^{-\beta E_\sigma}} = \frac{1}{Z} \sum_E E^2 \mathcal{N}(E) e^{-\beta E}. \qquad (5.4)$$

The specific heat capacity $C_V$, the increase in internal energy caused by an infinitesimal increase in temperature,

$$C_V = \frac{\partial \langle E \rangle}{\partial T} = \frac{\partial \beta}{\partial T} \frac{\partial \langle E \rangle}{\partial \beta} = -\beta^2 \frac{\partial \langle E \rangle}{\partial \beta}, \qquad (5.5)$$

can be expressed via eqn (5.2) as a second-order derivative of the partition function:

$$C_V = \beta^2 \frac{\partial^2}{\partial \beta^2} \log Z.$$

Again, there is a more convenient expression, which we write for the specific heat capacity per particle $c_V$,

$$\frac{c_V}{N} = -\frac{\beta^2}{N} \frac{\partial \langle E \rangle}{\partial \beta} = -\frac{\beta^2}{N} \frac{\partial}{\partial \beta} \left( \frac{\sum_\sigma E_\sigma e^{-\beta E_\sigma}}{\sum_\sigma e^{-\beta E_\sigma}} \right)$$

$$= \frac{\beta^2}{N} \frac{\sum_\sigma E^2 e^{-\beta E_\sigma} \sum_\sigma e^{-\beta E_\sigma} - \left( \sum_\sigma E_\sigma e^{-\beta E_\sigma} \right)^2}{\left( \sum_\sigma e^{-\beta E_\sigma} \right)^2} = \frac{\beta^2}{N} \left( \langle E^2 \rangle - \langle E \rangle^2 \right),$$

which can be evaluated with the second formulas in eqns (5.3) and (5.4) and is implemented in Alg. 5.4 (`thermo-ising`). We can recognize that the specific heat capacity, an experimentally measurable quantity, is proportional to the variance of the energy, a statistical measure of the distribution of energies. Specific-heat-capacity data for small two-dimensional lattices with periodic boundary conditions are shown in Fig. 5.6.

The density of states $\mathcal{N}(E)$ does not carry complete information about the Ising model. We must modify Alg. 5.3 (`enumerate-ising`) in a straightforward way to track the magnetization $M = \left\langle \sum_{k=1}^N \sigma_k \right\rangle$ of the system and to find the probability $\pi_M$. This probability is obtained, at any temperature, from the density of states as a function of energy and magnetization, $\mathcal{N}(E, M)$ (see Fig. 5.7). The probability distribution of the magnetization per spin is always symmetric around $M/N = 0$, featuring a single peak at $M = 0$ at high temperature, where the system is paramagnetic, and two peaks at magnetizations $\pm \tilde{M}/N$ at low temperature, where the system is in the ferromagnetic state. The critical temperature,

$$T_c = \frac{2}{\log \left(1 + \sqrt{2}\right)} = 2.269 \quad (\beta_c = 0.4407), \qquad (5.6)$$

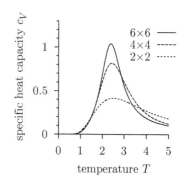

Fig. 5.6 Specific heat capacity of the Ising model on small square lattices with periodic boundary conditions (from Alg. 5.4 (`thermo-ising`)).

**procedure** `thermo-ising`
**input** $\{\mathcal{N}(E_{\min}),\ldots,\mathcal{N}(E_{\max})\}$ (from Alg. 5.3 (`enumerate-ising`))
$Z \leftarrow 0$
$\langle E' \rangle \leftarrow 0$
$\langle E'^2 \rangle \leftarrow 0$
**for** $E = E_{\min},\ldots,E_{\max}$ **do**
$\left\{ \begin{array}{l} E' \leftarrow E - E_{\min} \\ Z \leftarrow Z + \mathcal{N}(E)\,\mathrm{e}^{-\beta E'} \\ \langle E' \rangle \leftarrow \langle E' \rangle + E'\mathcal{N}(E)\,\mathrm{e}^{-\beta E'} \\ \langle E'^2 \rangle \leftarrow \langle E'^2 \rangle + E'^2\mathcal{N}(E)\,\mathrm{e}^{-\beta E'} \end{array} \right.$
$\langle E' \rangle \leftarrow \langle E' \rangle / Z$
$\langle E'^2 \rangle \leftarrow \langle E'^2 \rangle / Z$
$Z \leftarrow Z\mathrm{e}^{-\beta E_{\min}}$
$c_V \leftarrow \beta^2(\langle E'^2 \rangle - \langle E' \rangle^2)/N$
$\langle e \rangle \leftarrow (\langle E' \rangle + E_{\min})/N$
**output** $\{Z, \langle e \rangle, c_V\}$

---

**Algorithm 5.4** `thermo-ising`. Thermodynamic quantities for the Ising model at temperature $T = 1/\beta$ from enumeration data.

**Table 5.3** Thermodynamic quantities for the Ising model on a $6 \times 6$ lattice with periodic boundary conditions (from Alg. 5.4 (`thermo-ising`))

| $T$ | $\langle e \rangle$ | $c_V$ |
|-----|------|------|
| 0.5 | −1.999 | 0.00003 |
| 1.0 | −1.997 | 0.02338 |
| 1.5 | −1.951 | 0.19758 |
| 2.0 | −1.747 | 0.68592 |
| 2.5 | −1.280 | 1.00623 |
| 3.0 | −0.887 | 0.55665 |
| 3.5 | −0.683 | 0.29617 |
| 4.0 | −0.566 | 0.18704 |

separates the two regimes. It is at this temperature that the specific heat capacity diverges. Our statement about the distribution of the magnetization is equivalent to saying that below $T_c$ the Ising model acquires a spontaneous magnetization (per spin), equal to one of the peak values of the distribution $\pi(M/N)$.

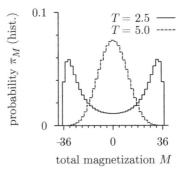

**Fig. 5.7** Probability $\pi_M$ on a $6 \times 6$ square lattice with periodic boundary conditions (from modified Alg. 5.3 (`enumerate-ising`)).

### 5.1.3   Listing loop configurations

The word "enumeration" has two meanings: it refers to listing items (configurations), but it also applies to simply counting them. The difference between the two is of more than semantic interest: in the list generated by Alg. 5.3 (**enumerate-ising**), we were able to pick out any information we wanted, for example the number of configurations of energy $E$ and magnetization $M$, that is, the density of states $\mathcal{N}(E, M)$. In this subsection we discuss an alternative enumeration for the Ising model. It does not list the spin configurations, but rather all the loop configurations which appear in the high-temperature expansion of the Ising model. This program will then turn, in Subsection 5.1.4, into an enumeration of the second kind (Kac and Ward, 1954). It counts configurations and obtains $Z(\beta)$ for a two-dimensional Ising system of any size (Kaufman, 1949), and even for the infinite system (Onsager, 1944). However, it then counts without listing. For example, it finds the number $\mathcal{N}(E)$ of configurations with energy $E$ but does not tell us how many of them have a magnetization $M$.

Van der Waerden, in 1941, noticed that the Ising-model partition function,

$$
\begin{aligned}
Z &= \sum_{\sigma} \exp\left(J\beta \sum_{\langle k,l\rangle} \sigma_k\sigma_l\right) \\
&= \sum_{\sigma} \prod_{\langle k,l\rangle} e^{J\beta\sigma_k\sigma_l},
\end{aligned}
\tag{5.7}
$$

allows each term $e^{J\beta\sigma_k\sigma_l}$ to be expanded and rearranged into just two terms, one independent of the spins and the other proportional to $\sigma_k\sigma_l$:

$$
\begin{aligned}
e^{\beta\sigma_k\sigma_l} &= 1 + \beta\sigma_k\sigma_l + \frac{\beta^2}{2!}\underbrace{(\sigma_k\sigma_l)^2}_{=1} + \frac{\beta^3}{3!}\underbrace{(\sigma_k\sigma_l)^3}_{=\sigma_k\sigma_l} + \cdots - \cdots \\
&= \underbrace{\left(1 + \frac{\beta^2}{2!} + \frac{\beta^4}{4!} + \cdots\right)}_{\cosh\beta} - \sigma_k\sigma_l\underbrace{\left(\beta + \frac{\beta^3}{3!} + \frac{\beta^5}{5!} + \cdots\right)}_{\sinh\beta} \\
&= (\cosh\ \beta)\,(1 + \sigma_k\sigma_l\tanh\ \beta)\,.
\end{aligned}
$$

Inserted into eqn (5.7), with $J = +1$, this yields

$$
Z(\beta) = \sum_{\sigma}\prod_{\langle k,l\rangle}\left((\cosh\ \beta)\,(1 + \sigma_k\sigma_l\tanh\ \beta)\right). \tag{5.8}
$$

For concreteness, we continue with a $4\times 4$ square lattice without periodic boundary conditions (with $J = 1$). This lattice has 24 edges and 16 sites, so that, by virtue of eqn (5.8), its partition function $Z_{4\times4}(\beta)$ is the product of 24 parentheses, one for each edge:

$$
Z_{4\times4}(\beta) = \sum_{\{\sigma_1,\ldots,\sigma_{16}\}} \cosh^{24}\beta \overbrace{(1 + \sigma_1\sigma_2\tanh\ \beta)}^{\text{edge 1}}\overbrace{(1 + \sigma_1\sigma_5\tanh\ \beta)}^{\text{edge 2}}
$$

$$
\times\ldots(1 + \sigma_{14}\sigma_{15}\tanh\ \beta)\underbrace{(1 + \sigma_{15}\sigma_{16}\tanh\ \beta)}_{\text{edge 24}}. \tag{5.9}
$$

We multiply out this product: for each edge (parenthesis) $k$, we have a choice between a "one" and a "tanh" term. This is much like the option of a spin-up or a spin-down in the original Ising-model enumeration, and can likewise be expressed through a binary variable $n_k$:

$$n_k = \begin{cases} 0 & (\equiv \text{ edge } k \text{ in eqn (5.9) contributes 1}) \\ 1 & (\equiv \text{ edge } k \text{ contributes } (\sigma_{s_k}\sigma_{s'_k}\tanh\beta)) \end{cases},$$

where $s_k$ and $s'_k$ indicate the sites at the two ends of edge $k$. Edge $k = 1$ has $\{s_1, s'_1\} = \{1, 2\}$, and edge $k = 24$ has, from eqn (5.9), $\{s_{24}, s'_{24}\} = \{15, 16\}$. Each factored term can be identified by variables

$$\{n_1, \ldots, n_{24}\} = \{\{0, 1\}, \ldots, \{0, 1\}\}.$$

For $\{n_1, \ldots, n_{24}\} = \{0, \ldots, 0\}$, each parenthesis picks a "one". Summed over all spin configurations, this gives $2^{16}$. Most choices of $\{n_1, \ldots, n_{24}\}$ average to zero when summed over spin configurations because the same term is generated with $\sigma_k = +1$ and $\sigma_k = -1$. Only choices leading to spin products $\sigma_s^0, \sigma_s^2, \sigma_s^4$ at each lattice site $s$ remain finite after summing over all spin configurations. The edges of these terms form loop configurations, such as those shown for the $4 \times 4$ lattice in Fig. 5.8. The list of all loop configurations may be generated by Alg. 5.5 (edge-ising), a recycled version of the Gray code for 24 digits, coupled to an incremental calculation of the number of spins on each site. The $\{o_1, \ldots, o_{16}\}$ count the number of times the sites $\{1, \ldots, 16\}$ are present. The numbers in this vector must all be even for a loop configuration, and for a nonzero contribution to the sum in eqn (5.9).

**Table 5.4** Numbers of loop configurations in Fig. 5.8 with given numbers of edges (the figure contains one configuration with 0 edges, 9 with 4 edges, etc). (From Alg. 5.5 (edge-ising)).

| # Edges | # Configs |
|---------|-----------|
| 0 | 1 |
| 4 | 9 |
| 6 | 12 |
| 8 | 50 |
| 10 | 92 |
| 12 | 158 |
| 14 | 116 |
| 16 | 69 |
| 18 | 4 |
| 20 | 1 |

```
procedure edge-ising
input {(s₁, s'₁), ..., (s₂₄, s'₂₄)}
{n₁, ..., n₂₄} ← {0, ..., 0}
{τ₀, ..., τ₂₄} ← {1, ..., 25}
{o₁, ..., o₁₆} ← {0, ..., 0}
output {n₁, ..., n₂₄}
for i = 1, 2²⁴ − 1 do
  ⎧ k ← gray-flip({τ₀, ..., τ₂₄})
  ⎪ nₖ ← mod(nₖ + 1, 2)
  ⎨ oₛₖ ← oₛₖ + 2·nₖ − 1
  ⎪ oₛ'ₖ ← oₛ'ₖ + 2·nₖ − 1
  ⎪ if ({o₁, ..., o₁₆} all even) then
  ⎩   { output {n₁, ..., n₂₄}
```

**Algorithm 5.5** edge-ising. Gray-code enumeration of the loop configurations in Fig. 5.8. The edge $k$ connects neighboring sites $\sigma_k$ and $\sigma'_k$.

For the thermodynamics of the $4 \times 4$ Ising model, we only need to keep track of the number of edges in each configuration, not the configurations themselves. Table 5.4, which shows the number of loop configurations

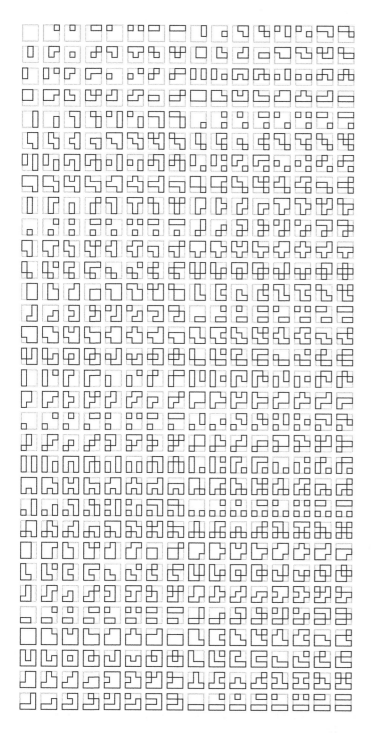

**Fig. 5.8** The list of all 512 loop configurations for the $4 \times 4$ Ising model without periodic boundary conditions (from Alg. 5.5 (`edge-ising`)).

for any given number of edges, thus yields the exact partition function for the $4 \times 4$ lattice without periodic boundary conditions:

$$Z_{4 \times 4}(\beta) = \left(2^{16} \cosh^{24} \beta\right) \left(1 + 9 \tanh^4 \beta + 12 \tanh^6 \beta \right.$$
$$\left. + \cdots + 4 \tanh^{18} \beta + 1 \tanh^{20} \beta\right). \quad (5.10)$$

Partition functions obtained from this expression are easily checked against the Gray-code enumeration.

### 5.1.4  Counting (not listing) loops in two dimensions

Following Kac and Ward (1952), we now construct a matrix whose determinant counts the number of loop configurations in Fig. 5.8. This is possible because the determinant of a matrix $U = (u_{kl})$ is defined by a sum of permutations $P$ (with signs and weights). Each permutation can be written as a collection of cycles, a "cycle configuration". Our task will consist in choosing the elements $u_{kl}$ of the matrix $U$ in such a way that the signs and weights of each cycle configurations correspond to the loop configurations in the two-dimensional Ising model. We shall finally arrive at a computer program which implements the correspondence, and effectively solves the enumeration problem for large two-dimensional lattices. For simplicity, we restrict ourselves to square lattices without periodic boundary conditions, and consider the definition of the determinant of a matrix $U$,

$$\det U = \sum_{\text{permutations}} (\text{sign } P) u_{1P_1} u_{2P_2} \ldots u_{NP_N}.$$

We now represent $P$ in terms of cycles. The sign of a permutation $P$ of $N$ elements with $n$ cycles is $\text{sign } P = (-1)^{N+n}$ (see our detailed discussion in Section 1.2.2). In the following, we shall consider only matrices with even $N$, for which $\text{sign } P = (-1)^{\# \text{ of cycles}}$. The determinant is thus

$$\det U = \sum_{\substack{\text{cycle} \\ \text{configs}}} (-1)^{\# \text{ of cycles}} \underbrace{u_{P_1 P_2} u_{P_2 P_3} \ldots u_{P_M P_1}}_{\text{weight of first cycle}} \underbrace{u_{P_1' P_2'} \cdots}_{\text{other cycles}}$$

$$= \sum_{\substack{\text{cycle} \\ \text{configs}}} \left(\left\{\begin{matrix} (-1) \cdot \text{ weight of} \\ \text{first cycle} \end{matrix}\right\}\right) \times \cdots \times \left(\left\{\begin{matrix} (-1) \cdot \text{ weight of} \\ \text{last cycle} \end{matrix}\right\}\right).$$

It follows from this representation of a determinant in terms of cycle configurations that we should choose the matrix elements $u_{kl}$ such that each cycle corresponding to a loop on the lattice (for example $(P_1, \ldots, P_M)$) gets a negative sign (this means that the sign of $u_{P_1 P_2} u_{P_2 P_3} \ldots u_{P_M P_1}$ should be negative). All cycles not corresponding to loops should get zero weight.

We must also address the problem that cycles in the representation of the determinant are directed. The cycle $(P_1, P_2, \ldots, P_{M-1}, P_M)$ is different from the cycle $(P_M, P_{M-1}, \ldots, P_2, P_1)$, whereas the loop configurations in Fig. 5.8 have no sense of direction.

For concreteness, we start with a $2 \times 2$ lattice without periodic boundary conditions, for which the partition function is

$$Z_{2\times 2} = \left(2^4 \cosh^4 \beta\right) \left(1 + \tanh^4 \beta\right). \tag{5.11}$$

The prefactor in this expression ($2^N$ multiplied by one factor of $\cosh \beta$ per edge) was already encountered in eqn (5.10). We can find naively a $4 \times 4$ matrix $\hat{U}_{2\times 2}$ whose determinant generates cycle configurations which agree with the loop configurations. Although this matrix cannot be generalized to larger lattices, it illustrates the problems which must be overcome. This matrix is given by

$$\hat{U}_{2\times 2} = \begin{bmatrix} 1 & \gamma \tanh \beta & \cdot & & \cdot \\ \cdot & 1 & \cdot & & \gamma \tanh \beta \\ \gamma \tanh \beta & \cdot & 1 & & \cdot \\ \cdot & \cdot & \gamma \tanh \beta & & 1 \end{bmatrix}.$$

(In the following, zero entries in matrices are represented by dots.) The matrix must satisfy

$$Z_{2\times 2} = \left(2^4 \cosh^4 \beta\right) \det \hat{U}_{2\times 2},$$

and because of

$$\det \hat{U}_{2\times 2} = 1 - \gamma^4 \tanh^4 \beta,$$

we have to choose $\gamma = e^{i\pi/4} = \sqrt[4]{-1}$. The value of the determinant is easily verified by expanding with respect to the first row, or by naively going through all the 24 permutations of 4 elements (see Fig. 4.16 for a list of them). Only two permutations have nonzero contributions: the unit permutation ($\binom{1234}{1234}$), which has weight 1 and sign 1 (it has four cycles), and the permutation, ($\binom{2431}{1234}$) $= (1, 2, 4, 3)$, which has weight $\gamma^4 \tanh^4 \beta = -\tanh^4 \beta$. The sign of this permutation is $-1$, because it consists of a single cycle.

The matrix $\hat{U}_{2\times 2}$ cannot be generalized directly to larger lattices. This is because it sets $u_{21}$ equal to zero because $u_{12} \neq 0$, and sets $u_{13} = 0$ because $u_{31} \neq 0$; in short it sets $u_{kl} = 0$ if $u_{lk}$ is nonzero (for $k \neq l$). In this way, no cycles with hairpin turns are retained (which go from site $k$ to site $l$ and immediately back to site $k$). It is also guaranteed that between a permutation and its inverse (in our case, between the permutation ($\binom{1234}{1234}$) and ($\binom{2431}{1234}$)), at most one has nonzero weight. For larger lattices, this strategy is too restrictive. We cannot generate all loop configurations from directed cycle configurations if the direction in which the edges are gone through is fixed. We would thus have to allow both weights $u_{kl}$ and $u_{lk}$ different from zero, but this would reintroduce the hairpin problem. For larger $N$, there is no $N \times N$ matrix whose determinant yields all the loop configurations.

Kac and Ward's (1951) solution to this problem associates a matrix index, not with each lattice site, but with each of the four directions

**Table 5.5** Correspondence between lattice sites and directions, and the indices of the Kac–Ward matrix $U$

| Site | Direction | Index |
|------|-----------|-------|
| 1 | $\rightarrow$ | 1 |
| | $\uparrow$ | 2 |
| | $\leftarrow$ | 3 |
| | $\downarrow$ | 4 |
| 2 | $\rightarrow$ | 5 |
| | $\uparrow$ | 6 |
| | $\leftarrow$ | 7 |
| | $\downarrow$ | 8 |
| $\vdots$ | $\vdots$ | $\vdots$ |
| | $\rightarrow$ | $4k - 3$ |
| $k$ | $\uparrow$ | $4k - 2$ |
| | $\leftarrow$ | $4k - 1$ |
| | $\downarrow$ | $4k$ |

on each lattice site (see Table 5.5), and a matrix element with each pair of directions and lattice sites. Matrix elements are nonzero only for neighboring sites, and only for special pairs of directions (see Fig. 5.9), and hairpin turns can be suppressed.

For concreteness, we continue with the $2 \times 2$ lattice, and its $16 \times 16$ matrix $U_{2\times2}$. We retain from the preliminary matrix $\hat{U}_{2\times2}$ that the nonzero matrix element must essentially correspond to terms $\tanh \beta$, but that there are phase factors. This phase factor is 1 for a straight move (case $a$ in Fig. 5.9); it is $e^{i\pi/4}$ for a left turn, and $e^{-i\pi/4}$ for a right turn.

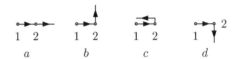

**Fig. 5.9** Graphical representation of the matrix elements in the first row of the Kac–Ward matrix $U_{2\times2}$ (see Table 5.6).

**Table 5.6** The matrix elements of Fig. 5.9 that make up the first row of the Kac–Ward matrix $U_{2\times2}$ (see eqn (5.12)).

| Case | Matrix element | value | type |
|------|----------------|-------|------|
| $a$ | $u_{1,5}$ | $\nu = \tanh \beta$ | (straight move) |
| $b$ | $u_{1,6}$ | $\alpha = e^{i\pi/4} \tanh \beta$ | (left turn) |
| $c$ | $u_{1,7}$ | $0$ | (hairpin turn) |
| $d$ | $u_{1,8}$ | $\overline{\alpha} = e^{-i\pi/4} \tanh \beta$ | (right turn) |

The nonzero elements in the first row of $U_{2\times2}$ are shown in Fig. 5.9, and taken up in Table 5.6. We arrive at the matrix

$$
U_{2\times2} = \begin{bmatrix}
1 & \cdot & \cdot & \cdot & \cdot & \nu & \alpha & \cdot & \overline{\alpha} & \cdot & \cdot & \cdot & \cdot & \cdot & \cdot & \cdot \\
\cdot & 1 & \cdot & \cdot & \cdot & \cdot & \cdot & \cdot & \overline{\alpha} & \nu & \alpha & \cdot & \cdot & \cdot & \cdot & \cdot \\
\cdot & \cdot & 1 & \cdot & \cdot & \cdot & \cdot & \cdot & \cdot & \cdot & \cdot & \cdot & \cdot & \cdot & \cdot & \cdot \\
\cdot & \cdot & \cdot & 1 & \cdot & \cdot & \cdot & \cdot & \cdot & \cdot & \cdot & \cdot & \cdot & \cdot & \cdot & \cdot \\
\cdot & \cdot & \cdot & \cdot & 1 & \cdot & \cdot & \cdot & \cdot & \cdot & \cdot & \cdot & \overline{\alpha} & \nu & \alpha & \cdot \\
\cdot & \overline{\alpha} & \nu & \alpha & \cdot & 1 & \cdot & \cdot & \cdot & \cdot & \cdot & \cdot & \cdot & \cdot & \cdot & \cdot \\
\cdot & \cdot & \cdot & \cdot & \cdot & \cdot & 1 & \cdot & \cdot & \cdot & \nu & \alpha & \cdot & \overline{\alpha} & \cdot & \cdot \\
\cdot & \cdot & \cdot & \cdot & \cdot & \cdot & \cdot & 1 & \cdot & \cdot & \cdot & \cdot & \cdot & \cdot & \cdot & \cdot \\
\alpha & \cdot & \overline{\alpha} & \nu & \cdot & \cdot & \cdot & \cdot & 1 & \cdot & \cdot & \cdot & \cdot & \cdot & \cdot & \cdot \\
\cdot & \cdot & \cdot & \cdot & \cdot & \cdot & \cdot & \cdot & \cdot & 1 & \cdot & \cdot & \cdot & \cdot & \cdot & \cdot \\
\cdot & \cdot & \cdot & \cdot & \cdot & \cdot & \cdot & \cdot & \cdot & \cdot & 1 & \cdot & \cdot & \cdot & \cdot & \cdot \\
\cdot & \cdot & \cdot & \cdot & \cdot & \cdot & \cdot & \overline{\alpha} & \nu & \alpha & \cdot & \cdot & 1 & \cdot & \cdot & \cdot \\
\cdot & \cdot & \cdot & \cdot & \cdot & \cdot & \cdot & \cdot & \cdot & \cdot & \cdot & \cdot & \cdot & 1 & \cdot & \cdot \\
\cdot & \cdot & \cdot & \cdot & \cdot & \alpha & \cdot & \overline{\alpha} & \nu & \cdot & \cdot & \cdot & \cdot & \cdot & \cdot & 1
\end{bmatrix}.
\tag{5.12}
$$

The matrix $U_{2\times2}$ contains four nonzero permutations, which we can generate with a naive program (in each row of the matrix, we pick one term out of $\{1, \nu, \alpha, \overline{\alpha}\}$, and then check that each column index appears exactly once). We concentrate in the following on the nontrivial cycles in each permutation (that are not part of the identity). The identity permutation, $P^1 = \left( \begin{smallmatrix} 1 & \cdots & 16 \\ 1 & \cdots & 16 \end{smallmatrix} \right)$, one of the four nonzero permutations, has only trivial cycles. It is characterized by an empty nontrivial cycle configuration $c_1$. Other permutations with nonzero weights are

$$
c_2 \equiv \begin{pmatrix} \text{site} & 1 & 2 & 4 & 3 \\ \text{dir.} & \rightarrow & \uparrow & \leftarrow & \downarrow \\ \text{index} & 1 & 6 & 15 & 12 \end{pmatrix}
$$

and

$$
c_3 \equiv \begin{pmatrix} \text{site} & 1 & 3 & 4 & 2 \\ \text{dir.} & \uparrow & \rightarrow & \downarrow & \leftarrow \\ \text{index} & 2 & 9 & 16 & 7 \end{pmatrix}.
$$

Finally, the permutation $c_4$ is put together from the permutations $c_2$ and $c_3$, so that we obtain

$$
\begin{aligned}
c_1 &\equiv 1, \\
c_2 &\equiv u_{1,6} u_{6,15} u_{15,12} u_{12,1} = \alpha^4 = -\tanh^4 \beta, \\
c_3 &\equiv u_{2,9} u_{9,16} u_{16,7} u_{7,2} = \overline{\alpha}^4 = -\tanh^4 \beta, \\
c_4 &\equiv c_2 c_3 = \alpha^4 \overline{\alpha}^4 = \tanh^8 \beta.
\end{aligned}
$$

We thus arrive at

$$
\det U_{2\times 2} = 1 + 2\tanh^4 \beta + \tanh^8 \beta = \underbrace{\left(1 + \tanh^4 \beta\right)^2}_{\text{see eqn (5.11)}}, \qquad (5.13)
$$

and this is proportional to the square of the partition function in the $2 \times 2$ lattice (rather than the partition function itself).

The cycles in the expansion of the determinant are oriented: $c_2$ runs anticlockwise around the pad, and $c_3$ clockwise. However, both types of cycles may appear simultaneously, in the cycle $c_4$. This is handled by drawing two lattices, one for the clockwise, and one for the anticlockwise cycles (see Fig. 5.10). The cycles $\{c_1, \ldots, c_4\}$ correspond to all the loop configurations that can be drawn simultaneously in both lattices. It is thus natural that the determinant in eqn (5.13) is related to the partition function in two independent lattices, the square of the partition function of the individual systems.

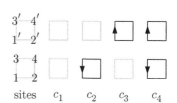

Before moving to larger lattices, we note that the matrix $U_{2\times 2}$ can be written in more compact form, as a matrix of matrices:

**Fig. 5.10** Neighbor scheme and cycle configurations in two independent $2 \times 2$ Ising models.

$$
U_{2\times 2} = \begin{bmatrix} 1 & u_{\rightarrow} & u_{\uparrow} & \cdot \\ u_{\leftarrow} & 1 & \cdot & u_{\uparrow} \\ u_{\downarrow} & \cdot & 1 & u_{\rightarrow} \\ \cdot & u_{\downarrow} & u_{\leftarrow} & 1 \end{bmatrix} \quad \begin{array}{l} \text{(a } 16 \times 16 \text{ matrix,} \\ \text{see eqn (5.15))} \end{array}, \qquad (5.14)
$$

where $1$ is the $4 \times 4$ unit matrix, and furthermore, the $4 \times 4$ matrices

$u_\rightarrow$, $u_\uparrow$, $u_\leftarrow$, and $u_\downarrow$ are given by

$$
u_\rightarrow = \begin{bmatrix} \nu & \alpha & \cdot & \overline{\alpha} \\ \cdot & \cdot & \cdot & \cdot \\ \cdot & \cdot & \cdot & \cdot \\ \cdot & \cdot & \cdot & \cdot \end{bmatrix}, \quad
u_\uparrow = \begin{bmatrix} \cdot & \cdot & \cdot & \cdot \\ \overline{\alpha} & \nu & \alpha & \cdot \\ \cdot & \cdot & \cdot & \cdot \\ \cdot & \cdot & \cdot & \cdot \end{bmatrix},
$$

$$
u_\leftarrow = \begin{bmatrix} \cdot & \cdot & \cdot & \cdot \\ \cdot & \cdot & \cdot & \cdot \\ \cdot & \overline{\alpha} & \nu & \alpha \\ \cdot & \cdot & \cdot & \cdot \end{bmatrix}, \quad
u_\downarrow = \begin{bmatrix} \cdot & \cdot & \cdot & \cdot \\ \cdot & \cdot & \cdot & \cdot \\ \cdot & \cdot & \cdot & \cdot \\ \alpha & \cdot & \overline{\alpha} & \nu \end{bmatrix}.
$$

(5.15)

The difference between eqns (5.12) and (5.14) is purely notational.

The $2 \times 2$ lattice is less complex than larger lattices. For example, one cannot draw loops in this lattice which sometimes turn left, and sometimes right. (On the level of the $2 \times 2$ lattice it is unclear why left turns come with a factor $\alpha$ and right turns with a factor $\overline{\alpha}$.) This is what we shall study now, in a larger matrix. Cycle configurations will come up that do not correspond to loop configurations. We shall see that they sum up to zero.

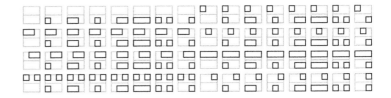

**Fig. 5.11** All 64 loop configurations for two uncoupled $4 \times 2$ Ising models without periodic boundary conditions (a subset of Fig. 5.8).

For concreteness, we consider the $4 \times 2$ lattice (without periodic boundary conditions), for which the Kac–Ward matrix can still be written down conveniently. We understand by now that the matrix and the determinant describe pairs of lattices, one for each sense of orientation, so that the pair of $4 \times 2$ lattices corresponds to a single $4 \times 4$ lattice with a central row of links eliminated. The 64 loop configurations for this case are shown in Fig. 5.11. We obtain

$$
U_{4\times 2} = \begin{bmatrix}
\mathbb{1} & u_\rightarrow & \cdot & \cdot & u_\uparrow & \cdot & \cdot & \cdot \\
u_\leftarrow & \mathbb{1} & u_\rightarrow & \cdot & \cdot & u_\uparrow & \cdot & \cdot \\
\cdot & u_\leftarrow & \mathbb{1} & u_\rightarrow & \cdot & \cdot & u_\uparrow & \cdot \\
\cdot & \cdot & u_\leftarrow & \mathbb{1} & \cdot & \cdot & \cdot & u_\uparrow \\
u_\downarrow & \cdot & \cdot & \cdot & \mathbb{1} & u_\rightarrow & \cdot & \cdot \\
\cdot & u_\downarrow & \cdot & \cdot & u_\leftarrow & \mathbb{1} & u_\rightarrow & \cdot \\
\cdot & \cdot & u_\downarrow & \cdot & \cdot & u_\leftarrow & \mathbb{1} & u_\rightarrow \\
\cdot & \cdot & \cdot & u_\downarrow & \cdot & \cdot & u_\leftarrow & \mathbb{1}
\end{bmatrix}.
$$

(5.16)

Written out explicitly, this gives a $32 \times 32$ complex matrix $U_{4\times 2} = (u_{k,l})$

with elements

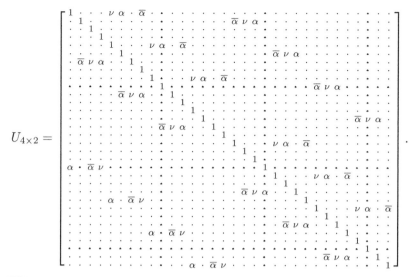

$$U_{4\times2} =$$

This matrix is constructed according to the same rules as $U_{2\times2}$, earlier.

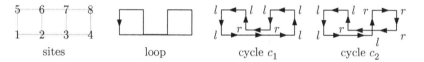

**Fig. 5.12** A loop in the $4 \times 2$ system, not present in Fig. 5.11. Weights of $c_1$ and $c_2$ cancel.

The cycle $c_2$ in Fig. 5.12 can be described by the following trajectory:

$$\text{cycle } c_2 \equiv \begin{pmatrix} \text{site} & 1 & 2 & 3 & 7 & 8 & 4 & 3 & 2 & 6 & 5 \\ \text{dir.} & \rightarrow & \rightarrow & \uparrow & \rightarrow & \downarrow & \leftarrow & \leftarrow & \uparrow & \leftarrow & \downarrow \\ \text{index} & 1 & 5 & 10 & 25 & 32 & 15 & 11 & 6 & 23 & 20 \end{pmatrix}.$$

This cycle thus corresponds to the following product of matrix elements:

$$\{\text{weight of } c_2\} : u_{1,5}u_{5,10}\ldots u_{23,20}u_{20,1}.$$

The cycle $c_2$ makes four left and four right turns (so that the weight is proportional to $\bar{\alpha}^4\alpha^4 \propto +1$) whereas the cycle $c_1$ turns six times to the left and twice to the right, with weight $\bar{\alpha}^6\alpha^2 \propto -1$, canceling $c_2$.

A naive program easily generates all of the nontrivial cycles in $U_{4\times2}$ (in each row of the matrix, we pick one term out of $\{1, \nu, \alpha, \bar{\alpha}\}$, and then check that each column index appears exactly once). This reproduces the loop list, with 64 contributions, shown in Fig. 5.11. There are in addition 80 more cycle configurations, which are either not present in the figure, or are equivalent to cycle configurations already taken into account. Some examples are the cycles $c_1$ and $c_2$ in Fig. 5.12. It was the good fortune of Kac and Ward that they all add up to zero.

**procedure** combinatorial-ising
**input** $\{u_\rightarrow, u_\uparrow, u_\leftarrow, u_\downarrow\}$ (see eqn (5.15))
$\{U(j, j')\} \leftarrow \{0, \dots, 0\}$
**for** $k = 1, \dots, N$ **do**
$\left\{\begin{array}{l} \textbf{for } n = 1, \dots, 4 \textbf{ do} \\ \left\{\begin{array}{l} j \leftarrow 4 \cdot (k - 1) + n \\ U(j, j) \leftarrow 1 \\ \textbf{for } n' = 1, \dots, 4 \textbf{ do} \\ \left\{\begin{array}{l} k' \leftarrow \text{Nbr}(1, k) \\ \textbf{if } (k' \neq 0) \textbf{ then} \\ \quad \left\{\begin{array}{l} j' \leftarrow 4 \cdot (k' - 1) + n' \\ U(j, j') \leftarrow u_\rightarrow(n, n') \end{array}\right. \\ k' \leftarrow \text{Nbr}(2, k) \\ \textbf{if } (k' \neq 0) \textbf{ then} \\ \quad \left\{\begin{array}{l} j' \leftarrow 4 \cdot (k' - 1) + n' \\ U(j, j') \leftarrow u_\uparrow(n, n') \end{array}\right. \\ k' \leftarrow \text{Nbr}(3, k) \\ \textbf{if } (k' \neq 0) \textbf{ then} \\ \quad \left\{\begin{array}{l} j' \leftarrow 4 \cdot (k' - 1) + n' \\ U(j, j') \leftarrow u_\leftarrow(n, n') \end{array}\right. \\ k' \leftarrow \text{Nbr}(4, k) \\ \textbf{if } (k' \neq 0) \textbf{ then} \\ \quad \left\{\begin{array}{l} j' \leftarrow 4 \cdot (k' - 1) + n' \\ U(j, j') \leftarrow u_\downarrow(n, n') \end{array}\right. \end{array}\right. \end{array}\right. \end{array}\right.$
**output** $\{U(j, j')\}$

——

**Algorithm 5.6** combinatorial-ising. The $4N \times 4N$ matrix $U$, for which $\sqrt{\det U} \propto Z(\beta)$ (Ising model without periodic boundary conditions).

On larger than $4 \times 2$ lattices, there are more elaborate loops. They can, for example, have crossings (see, for example, the loop in Fig. 5.13). There, the cycle configurations $c_1$ and $c_2$ correspond to loops in the generalization of Fig. 5.11 to larger lattices, whereas the cycles $c_3$ and $c_4$ are superfluous. However, $c_3$ makes six left turns and two right turns, so that the overall weight is $\alpha^4 = -1$, whereas the cycle $c_4$ makes three left turns and three right turns, so that the weight is $+1$, the opposite of that of $c_3$. The weights of $c_3$ and $c_4$ thus cancel.

For larger lattices, it becomes difficult to establish that the sum of cycle configurations in the determinant indeed agrees with the sum of loop configurations of the high-temperature expansion, although rigorous proofs exist to that effect. However, at our introductory level, it is more rewarding to proceed heuristically. We can, for example, write down the $144 \times 144$ matrix $U_{6\times6}$ of the $6 \times 6$ lattice for various temperatures (using Alg. 5.6 (combinatorial-ising)), and evaluate the determinant $\det U_{6\times6}$ with a standard linear-algebra routine. Partition functions thus obtained are equivalent to those resulting from Gray-code enumeration, even though the determinant is evaluated in on the order

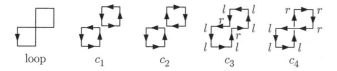

**Fig. 5.13** Loop and cycle configurations. The weights of $c_3$ and $c_4$ cancel.

of $144^3 \simeq 3{\times}10^6$ operations, while the Gray code goes over $2^{35} \simeq 3{\times}10^{10}$ configurations. The point is that the determinant can be evaluated for lattices that are much too large to go through the list of all configurations.

The matrix $U_{L \times L}$ for the $L \times L$ lattice contains the key to the analytic solution of the two-dimensional Ising model first obtained, in the thermodynamic limit, by Onsager (1944). To recover Onsager's solution, we would have to compute the determinant of $U$, not numerically as we did, but analytically, as a product over all the eigenvalues. Analytic expressions for the partition functions for Ising models can also be obtained for finite lattices with periodic boundary conditions. To adapt for the changed boundary conditions, one needs four matrices, generalizing the matrix $U$ (compare with the analogous situation for dimers in Chapter 6). Remarkably, evaluating $Z(\beta)$ on a finite lattice reduces to evaluating an explicit function (see Kaufman (1949) and Fisher and Ferdinand (1969); see also Exerc. 5.9).

The analytic solutions of the Ising model have not been generalized to higher dimensions, where only Monte Carlo simulations, high-temperature expansions, and renormalization-group calculations allow to compute to high precision the properties of the phase transition. These properties, as mentioned, are universal, that is, they are the same for a wide class of systems, called the Ising universality class.

## 5.1.5   Density of states from thermodynamics

The direct and indirect enumeration algorithms in this chapter differ in the role played by the density of states. In Alg. 5.3 (`enumerate-ising`), it was appropriate to first compute $\mathcal{N}(E)$, and later determine partition functions, internal energies, and specific heat capacities at any temperature, in $\propto N$ operations. In contrast, the indirect enumerations in Section 5.1.4 determine the partition function $Z(\beta)$, not the density of states. Computing $Z(\beta)$ from $\mathcal{N}(E)$ is straightforward, but how to recover $\mathcal{N}(E)$ from $Z(\beta)$ requires some thought:

$$\mathcal{N}(E) \underset{\text{this subsection}}{\overset{\text{Subsection 5.1.2}}{\rightleftarrows}} Z(\beta).$$

The mathematical problem of the present section is common to many basic problems in statistical and solid state physics, and appears also in the interpretation of experimental or Monte Carlo data. In the presence of statistical uncertainties, it is very difficult to solve, and may often be ill-defined. This means, in our case, that algorithms exist for computing $\mathcal{N}(E)$ if the partition functions were computed exactly. If, however, $Z(\beta)$ is known only to limited precision, the output generated by the slightly perturbed input can be drastically different from the exact output.

For a two-dimensional Ising model on a finite lattice, the exact partition function $Z(\beta)$ can be obtained from the matrix $U$ of Alg. 5.6 (`combinatorial-ising`), or better from Kaufman's explicit formula (see Exerc. 5.9). For concreteness, we consider the case of periodic boundary conditions, where the $\Delta_E = 4$, and where there are $N$ levels of excited states. In this case, the Boltzmann weight of the $k$th excited state is $x^k e^{-\beta E_0}$, where $x = e^{-4\beta}$. The partition function can be expressed as a polynomial in $x$, where the prefactors are the densities of state,

$$\tilde{Z}(x) = \frac{Z(\beta)}{e^{-\beta E_0}} = \mathcal{N}(0) + \mathcal{N}(1)\,x + \mathcal{N}(2)\,x^2 + \cdots + \mathcal{N}(N)\,x^N.$$

It now suffices to compute the partition functions of the Ising model at $N+1$ different temperatures, $x_0 = e^{-4\beta_0}, \ldots, x_N = e^{-4\beta_N}$, to arrive at a matrix equation relating the partition functions to the densities of state:

$$\underbrace{\begin{bmatrix} 1 & x_0 & x_0^2 & \cdots & x_0^N \\ 1 & x_1 & x_1^2 & \cdots & x_1^N \\ \vdots & \vdots & \vdots & & \vdots \\ 1 & x_N & x_N^2 & \cdots & x_N^N \end{bmatrix}}_{A \text{ (Vandermonde matrix)}} \begin{bmatrix} \mathcal{N}(0) \\ \mathcal{N}(1) \\ \vdots \\ \mathcal{N}(N) \end{bmatrix} = \begin{bmatrix} \tilde{Z}_0 \\ \tilde{Z}_1 \\ \vdots \\ \tilde{Z}_N \end{bmatrix}. \tag{5.17}$$

In principle, we can multiply both sides of eqn (5.17) (from the left) by $A^{-1}$ to obtain the densities of states. ( The special type of matrix in eqn (5.17) is called a Vandermonde matrix.)

As an alternative to matrix inversion, we could also obtain the densities of state $\{\mathcal{N}(0), \ldots, \mathcal{N}(N)\}$ from repeated interpolations of the partition function to zero temperature, where only the ground state contributes to the partition function:

$$Z(x) \simeq \mathcal{N}(E_0)\,e^{-\beta E_0} \quad \text{for } x \to 0.$$

Extrapolating $Z(\beta)$ to zero temperature thus gives the energy and the degeneracy of the ground state. We can now go further and, so to speak, peel off the result of this interpolation from our original problem, that is, interpolate $(\tilde{Z}_k - \mathcal{N}(0))/x_k$ through a polynomial of order $N-1$, in order to determine $\mathcal{N}(1)$, etc.

For lattices beyond $4 \times 4$ or $6 \times 6$, the matrix inversion in eqn (5.17) and the naive interpolation scheme both run into numerical instabilities, unless we increase the precision of our calculation much beyond the standard 32- or 64-bit floating-point formats discussed in Subsection 2.1.2. Overall, we may be better off using symbolic rather than numerical computations, and relying on commercial software packages, which are beyond the scope of this book. A compact procedure by Beale (1996) allows one to obtain the exact density of states of the Ising model for systems of size up to $100 \times 100$.

# 5.2 The Ising model—Monte Carlo algorithms

Gray-code enumerations, as discussed in Subsection 5.1.1, succeed only for relatively small systems, and high-temperature expansions, the subject of Subsection 5.1.4, must usually be stopped after a limited number of terms, before they get too complicated. Only under exceptional circumstances, as in the two-dimensional Ising model, can these methods be pushed much further. Often, Monte Carlo methods alone are able to obtain exact results for large sizes. The price to be paid is that configurations are sampled rather than enumerated, so that statistical errors are inevitable. The Ising model has been a major test bed for Monte Carlo simulations and algorithms, methods and recipes.

Roughly speaking, there have been two crucial periods in Ising-model computations, since it was realized that local Monte Carlo calculations are rather imprecise in the neighborhood of the critical point. It took a long time to appreciate that the lack of precision has two origins: First, spins are spatially correlated on a certain length scale, called the correlation length $\xi$. On this length scale, spins mostly point in the same direction. It is difficult to turn a spin around, or to turn a small patch of correlated spins, embedded in a larger region, where the spins are themselves correlated. This large-scale correlation of spins is directly responsible for the "critical slowing down" of simulations close to $T_c$. Large correlation times are the first origin of the lack of precision. In the infinite system, $\xi$ diverges at the critical temperature, but in a finite lattice, this correlation length cannot exceed the system size, so that the simulation has a typical length scale which depends on system size. This is the second origin of the lack of precision. In summary, we find that, on the one hand, observables are difficult to evaluate close to $T_c$ because of critical slowing down. On the other hand, the observables of the infinite lattice are difficult to evaluate, because they differ from those computed for a finite system if we are close to the critical point. Extrapolation to an infinite system size is thus nontrivial. The critical slowing down of Monte Carlo algorithms is a natural consequence of a diverging correlation length, and it can also be seen in many experiments, in liquids, magnets, etc.

The second crucial period of simulations of the Ising model started in the late 1980s, when it was finally understood that critical slowing down is not an inescapable curse of simulation, as it is for experiments. It was overcome through the construction of cluster algorithms which enforce the detailed-balance condition with nonintuitive rules which are unrelated to the flipping of single spins. These methods have spread from the Ising model to many other fields of statistical physics.

## 5.2.1 Local sampling methods

A basic task in statistical physics is to write a local Metropolis algorithm for the Ising model. This program is even simpler than a basic

Markov-chain simulation for hard disks (in Chapter 2). The Ising model has a less immediate connection with classical mechanics (there is no molecular dynamics algorithm for the model). Its phase transition is better understood than that of hard disks, and the results can be compared with exact solutions in two dimensions, even for finite systems, so that very fine tests of the algorithm on lattices of any size are possible. Analogously to Alg. 2.9 (`markov-disks`), we randomly pick a site and attempt to flip the spin at that site (see Fig. 5.14). The proposed

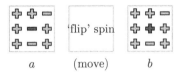

$$a \qquad \text{(move)} \qquad b$$

**Fig. 5.14** Local Monte Carlo move $a \rightarrow b$ in the Ising model, to be accepted with probability $\min\left[1, \mathrm{e}^{-\beta(E_b - E_a)}\right]$.

move between configuration $a$, with energy $E_a$, and configuration $b$, with energy $E_b$, must be accepted with probability $\min\left\{1, \mathrm{e}^{-\beta(E_b - E_a)}\right\}$, as straightforwardly implemented in Alg. 5.7 (`markov-ising`). We must beware of factors of two in evaluating the energy and thoroughly check the results on small lattices against exact enumeration, before moving on to larger-scale simulations (see Table 5.7).

**procedure markov-ising**
**input** $\{\sigma_1, \ldots, \sigma_N\}, E$
$k \leftarrow \mathbf{nran}\,(1, N)$
$h \leftarrow \sum_l \sigma_{\mathrm{Nbr}(l,k)}$
$\Delta_E \leftarrow 2h\sigma_k$
$\Upsilon \leftarrow \mathrm{e}^{-\beta\Delta_E}$
**if** $(\mathbf{ran}\,(0, 1) < \Upsilon)$ **then**
$\left\{ \begin{array}{l} \sigma_k \leftarrow -\sigma_k \\ E \leftarrow E + \Delta_E \end{array} \right.$
**output** $\{\sigma_1, \ldots, \sigma_N\}, E$

**Algorithm 5.7** `markov-ising`. Local Metropolis algorithm for the Ising model in $d$ dimensions.

**Table 5.7** Results of five runs of Alg. 5.7 (`markov-ising`) on a $6 \times 6$ lattice with periodic boundary conditions at temperature $T = 2.0$ with, each time, $1\times10^6$ samples (see also Table 5.3)

| Run | $\langle E/N \rangle$ | $c_V$ |
|-----|------|------|
| 1 | -1.74772 | 0.68241 |
| 2 | -1.74303 | 0.70879 |
| 3 | -1.75058 | 0.66216 |
| 4 | -1.74958 | 0.68106 |
| 5 | -1.75075 | 0.66770 |

This program easily recovers the phase transition between a paramagnet and a ferromagnet, which takes place in two dimensions at an inverse temperature $\beta_\mathrm{c} = \log(1 + \sqrt{2})/2 \simeq 0.4407$ (see Fig. 5.17). A naive approach for detecting the phase transition consists in plotting the mean absolute magnetization as a function of temperature (see Fig. 5.16). However, there are more expert approaches for locating $T_\mathrm{c}$ (see Exerc. 5.6).

Around the critical temperature, the local Monte Carlo algorithm is

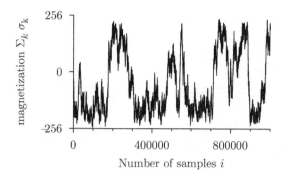

**Fig. 5.15** Total magnetization during a run for a $16 \times 16$ Ising model with periodic boundary conditions at $\beta = 0.42$ (from Alg. 5.7 (`markov-ising`)).

increasingly slow. This is because the distribution of the total magnetization becomes wide: between the high-temperature regime where it is sharply peaked at zero magnetization, and the low-temperature regime with its double peak structure, the system passes through a regime where the probability distribution of the magnetization is essentially flat for almost all values of $M$. Such a distribution is extremely difficult to sample with a single spin-flip algorithm.

To illustrate this point, we consider, in Fig. 5.15, the trajectory of $1 \times 10^6$ iterations (number of samples) of a $16 \times 16$ Ising model with periodic boundary conditions. Visual inspection of the trajectory reveals that magnetizations in the range from $-256$ to $+256$ appear with roughly equal probabilities. This means that a Markov chain for $M$ is more or less equivalent to a random walk on an interval of of length 512, with a step width $\Delta_M = 2$. For a random walk, the distance covered ($\Delta_M$) grows as the square root of the number of iterations $\Delta_i$ ($\Delta_M \propto \sqrt{\Delta_i}$, see the analogous discussion in Subsection 3.5.2). One independent sample is generated when the distance covered is of the order of the length of the interval, measured in units of the step width. We thus find that an independent sample is generated every $\Delta_i \simeq 256^2$ steps, so that we can expect the following:

$$\left\{ \begin{array}{c} \text{number of independent} \\ \text{samples in Fig. 5.15} \end{array} \right\} \simeq 1 \times 10^6 / 256^2 \simeq 15.$$

This again agrees with what we would conclude from a visual inspection of the figure.

The slowing-down of the local Monte Carlo algorithm close to $T_c$ can be interpreted as the effect of the divergence of the correlation length as we approach the critical point. In other words, the local algorithm slows down because it changes the magnetization by a small amount $\Delta_M$ only, in a situation where the distribution of magnetizations is wide, if measured in units of $\Delta_M$.

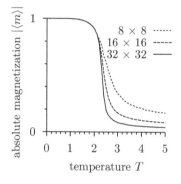

**Fig. 5.16** Mean absolute magnetization per spin $\langle |m| \rangle$ as a function of temperature, from Alg. 5.7 (`markov-ising`).

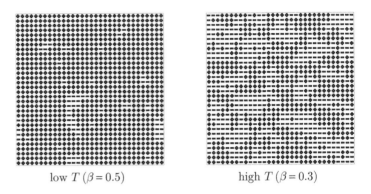

low $T$ ($\beta = 0.5$)        high $T$ ($\beta = 0.3$)

**Fig. 5.17** Ising-model configurations in a $32 \times 32$ square lattice with periodic boundary conditions (from Alg. 5.7 (`markov-ising`)).

## 5.2.2   Heat bath and perfect sampling

In this subsection, we discuss an alternative to the Metropolis Monte Carlo method, the heat bath algorithm. Rather than flipping a spin at a random site, we now thermalize this spin with its local environment (see Fig. 5.18). In the presence of a molecular field $h$ at site $k$, the spin points up and down with probabilities $\pi_h^+$ and $\pi_h^-$, respectively, where

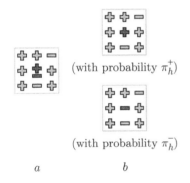

(with probability $\pi_h^+$)

(with probability $\pi_h^-$)

$a$          $b$

**Fig. 5.18** Heat bath algorithm for the Ising model. The spin on the central site has a molecular field $h = 2$ (see Alg. 5.8 (`heatbath-ising`)).

$$\pi_h^+ = \frac{e^{-\beta E^+}}{e^{-\beta E^+} + e^{-\beta E^-}} = \frac{1}{1 + e^{-2\beta h}},$$
$$\pi_h^- = \frac{e^{-\beta E^-}}{e^{-\beta E^+} + e^{-\beta E^-}} = \frac{1}{1 + e^{+2\beta h}}. \tag{5.18}$$

These probabilities are normalized ($\pi_h^+ + \pi_h^- = 1$). To sample $\{\pi_h^+, \pi_h^-\}$, we can pick a random number $\Upsilon = \mathtt{ran}\,(0,1)$ and make the spin point up if $\Upsilon < \pi_h^+$ and down otherwise. The action taken is independent of the spin's orientation before the move (see Alg. 5.8 (`heatbath-ising`)). The heat bath algorithm implements a priori probabilities for the smallest possible subsystem, a single site:

$$\mathcal{A}(\pm \to +) = \pi_h^+,$$
$$\mathcal{A}(\pm \to -) = \pi_h^-.$$

The heat bath algorithm is local, just like the Metropolis algorithm, and its performance is essentially the same. The algorithm is conveniently represented in a diagram of the molecular field $h$ against the random number $\Upsilon$ (see Fig. 5.20).

We now discuss an interesting feature of the heat bath algorithm, which allows it to function in the context of the perfect-sampling approach of Subsection 1.1.7. We first discuss the concept of half-order for configurations in the Ising model (see Fig. 5.19). We may say that, for a site, an up spin is larger than a down spin, and a whole configuration

```
procedure heatbath-ising
input {σ₁,...,σ_N}, E
k ← nran (1, N)
h ← Σ_n σ_Nbr(n,k)
σ' ← σ_k
Υ ← ran (0, 1)
if (Υ < π⁺_h) then (see eqn (5.18))
    { σ_k ← 1
else
    { σ_k ← −1
if (σ' ≠ σ_k) E ← E − 2hσ_k
output {σ₁,...,σ_N}
```

**Algorithm 5.8** heatbath-ising. Heat bath algorithm for the Ising model.

of spins $\{\sigma_1,\ldots,\sigma_N\}$ is larger than another configuration $\{\sigma'_1,\ldots,\sigma'_N\}$ if the spins on each site $k$ satisfy $\sigma_k \geq \sigma'_k$.

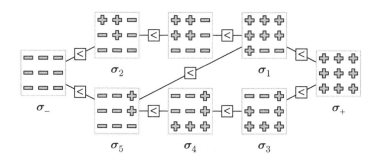

**Fig. 5.19** Half-order in the Ising model: configuration $\boldsymbol{\sigma}_-$ is smaller and $\boldsymbol{\sigma}_+$ is larger than all other configurations. $\boldsymbol{\sigma}_4$ and $\boldsymbol{\sigma}_1$ are unrelated.

In Fig. 5.19, $\boldsymbol{\sigma}_+$ and $\boldsymbol{\sigma}_-$ are the two ground states of the Ising model, but $\boldsymbol{\sigma}_+$ is larger and $\boldsymbol{\sigma}_-$ is smaller than all other configurations. Let us now apply the heat bath algorithm, with the same values of $k$ and $\Upsilon$ (see Alg. 5.8 (heatbath-ising)), to two configurations of spins $\boldsymbol{\sigma} = \{\sigma_1,\ldots,\sigma_N\} \geq \boldsymbol{\sigma}' = \{\sigma'_1,\ldots,\sigma'_N\}$. Because of the half-ordering, the molecular field $h_k$ of configuration $\boldsymbol{\sigma}$ is equal to or larger than the field $h'_k$, and on site $k$, the new spin $\sigma_k$ picked will be larger than or equal to the spin $\sigma'_k$ (see Fig. 5.20). The heat bath algorithm thus preserves the half-order of Fig. 5.19. In short, this is because the ordering of spin configurations induces an ordering of molecular fields, and vice versa.

We can apply the heat bath algorithm to all possible configurations of Ising spins and be sure that they will remain "herded in" by what results from applying the algorithm to the configuration $\boldsymbol{\sigma}_+ = \{+,\ldots,+\}$ and the configuration $\boldsymbol{\sigma}_- = \{-,\ldots,-\}$. This property allows us to use the

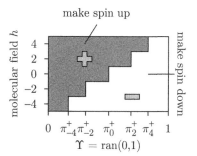

**Fig. 5.20** Action to be taken in the heat bath algorithm as a function of the molecular field $h$ and the random number $\Upsilon$.

heat bath algorithm as a time-dependent random map in the coupling-from-the-past framework of Subsection 1.1.7. At each iteration $i$, we simply sample values $\{k, \Upsilon\}$. This allows us to apply the heat bath algorithm, as a map, to any configuration (see Fig. 5.21).

$i = -\infty$

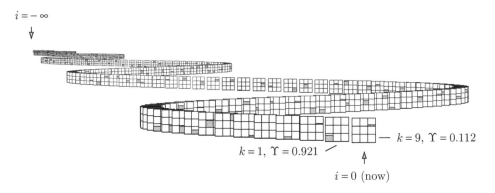

$k = 1, \Upsilon = 0.921$ — $k = 9, \Upsilon = 0.112$

$i = 0$ (now)

**Fig. 5.21** A random-map Markov chain for a $3 \times 3$ Ising model that has been running since iteration $i = -\infty$ (compare with Fig. 1.27).

We recall the basic idea of coupling from the past: in a Markov chain, a configuration (at the present time, $i = 0$) is perfectly decorrelated only with respect to configurations going back an infinite number of iterations (see Fig. 5.21). However, it is possible to infer the configuration at $i = 0$ by backtracking a finite number of steps (see Propp and Wilson (1996)). Half-order makes this practical for the Ising model: during the backtracking, we need not check all configurations (as we did for the pebble game in Subsection 1.1.7). It suffices to apply the heat bath algorithm for the all-up and the all-down configurations. The coming together of these two extremal configurations at $i = 0$ indicates a general merging of all configurations, and signals that the configuration at $i = 0$ is again a perfect sample.

For the ferromagnetic Ising model, perfect sampling is of fundamental interest, but of limited practical importance because of the rapid convergence of cluster algorithms (see Subsection 5.2.3). Nagging doubts about convergence come up in closely related models (see Section 5.3), and direct-sampling algorithms would be extremely valuable. In two dimensions, indirect counting methods using the Kac–Ward matrix $U$ also lead to direct sampling methods (similar to the algorithm for dimer configurations in Subsection 6.2.3).

### 5.2.3   Cluster algorithms

Algorithm 5.7 (`markov-ising`) and its variants, the classic simulation methods for spin models, have gradually given way to cluster algorithms, which converge much faster. These algorithms feature large-scale moves. In the imagery of the heliport game, they propose and accept displacements on the scale of the system, rather than walk about the landing pad in millimeter-size steps. In this subsection, we discuss cluster methods in a language stressing the practical aspect of a priori probabilities.

We recall that single-spin-flip Monte Carlo algorithms are slow close to $T_c$, because the histogram of essential values of the magnetization is wide and the step width of the magnetization is small. To sample faster, we must foster moves which change the magnetization by more than $\pm 2$. However, using the single-spin-flip algorithm in parallel, on several sites at a time, only hikes up the rejection rate. Neither can we, so to speak, solidly connect all neighboring spins of the same orientation and flip them all at once. Doing so would quickly lead to a perfectly aligned state, from which there would be no escape.

Let us analyze a more sophisticated rule for flipping spins. We suppose that, starting from a random initial spin, a cluster is constructed by adding, with probability $p$, neighboring sites with spins of the same orientation. For the moment, this probability is an arbitrary parameter. The above solid connection between neighboring spins corresponds to $p = 1$. During the cluster construction, we keep a list of cluster sites, but also one containing pocket sites, that is, new members of the cluster that can still make the cluster grow. The cluster construction algorithm picks one pocket site and removes it from the pocket. It then checks all of this site's neighbors outside the cluster with spins of like sign and adds these neighbors, with probability $p$, to the pocket and the cluster (see Fig. 5.22). After completion of the construction of the cluster, when the pocket is empty, all spins in the cluster are flipped. This brings us from the initial configuration $a$ to the final configuration $b$ (see Fig. 5.23). From our experience with a priori probabilities, we know beforehand that a suitable acceptance rule will ensure detailed balance between $a$ and $b$, for any $0 < p < 1$. In going from $a$ to $b$, the a priori construction probabilities $\mathcal{A}(a)$ and $\mathcal{A}(b)$, the acceptance probabilities $P(a \to b)$ and $P(b \to a)$, and the Boltzmann weights $\pi(a)$ and $\pi(b)$, must respect the generalized detailed-balance condition of Subsection 1.1.6:

$$\pi(a)\mathcal{A}(a \to b)P(a \to b) = \pi(b)\mathcal{A}(b \to a)P(b \to a). \qquad (5.19)$$

We must now compute the a priori probability $\mathcal{A}(a \to b)$, the probability of stopping the cluster construction process at a given stage rather than continuing and including more sites (see the cluster of gray sites in configuration $a$ in Fig. 5.23). $\mathcal{A}(a \to b)$ is given by an interior part (the two neighbors inside the cluster) and the stopping probability at the boundary: each sites on the boundary of the cluster was once a pocket site and the construction came to a halt because none of the possible new edges was included. Precisely, the boundary $\partial \mathcal{C}$ of the cluster (with one spin inside and its neighbor outside) involves two types of edge:

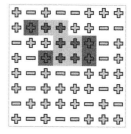

**Fig. 5.22** Ising configuration with 10 cluster sites (the dark *and* the light gray sites). The dark sites are pocket sites.

$$\begin{Bmatrix} \text{cluster in } a \\ \text{in Fig. 5.23} \end{Bmatrix} : \quad \overbrace{\begin{bmatrix} \text{inside} & \text{outside} & \# \\ + & - & n_1 \\ + & + & n_2 \end{bmatrix}}^{\text{edges across } \partial \mathcal{C}} \quad E|_{\partial \mathcal{C}} = n_1 - n_2 \quad (5.20)$$

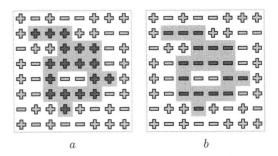

**Fig. 5.23** Ising-model configurations connected through a cluster flip. In $a$, 16 edges $\{+,-\}$ and 14 edges $\{+,+\}$ cross the boundary.

(in the example of Fig. 5.23, $n_1 = 16$ and $n_2 = 14$). The a priori probability is $\mathcal{A}(a \to b) = \mathcal{A}_{\text{in}} \cdot (1-p)^{n_2}$ because there were $n_2$ opportunities to let the cluster grow and none was taken. To evaluate the Boltzmann weight, we concentrate on the energy across the boundary $\partial \mathcal{C}$, given in eqn (5.20). It follows that $\pi(a) = \pi_{\text{in}} \pi_{\text{out}} e^{-\beta(n_1 - n_2)}$.

We consider the return move from configuration $b$ back to $a$ (see Fig. 5.23 again), and evaluate the return probability $\mathcal{A}(b \to a)$ and the Boltzmann weight $\pi(b)$. In the cluster for configuration $b$, the edges across the boundary $\partial \mathcal{C}$ are now

$$
\left\{ \begin{array}{l} \text{cluster in } b \\ \text{in Fig. 5.23} \end{array} \right\} : \quad \overbrace{\begin{bmatrix} \text{inside} & \text{outside} & \# \\ - & - & n_1 \\ - & + & n_2 \end{bmatrix}}^{\text{edges across } \partial \mathcal{C}} \quad E|_{\partial \mathcal{C}} = -n_1 + n_2.
$$

The cluster construction probability $\mathcal{A}(b \to a)$ contains the same interior part as before, but a new boundary part $\mathcal{A}(b \to a) = \mathcal{A}_{\text{in}} \cdot (1-p)^{n_1}$, because there were $n_1$ opportunities to let the cluster grow and again none was accepted. By an argument similar to that above, the statistical weight of configuration $b$ is $\pi(b) = \pi_{\text{in}} \pi_{\text{out}} e^{-\beta(n_2 - n_1)}$. The interior and exterior contributions to the Boltzmann weight are the same as for configuration $a$. All the ingredients of the detailed-balance condition in eqn (5.19) are now known:

$$
e^{-\beta(n_1 - n_2)}(1-p)^{n_2} \mathcal{P}(a \to b) = e^{-\beta(n_2 - n_1)}(1-p)^{n_1} \mathcal{P}(b \to a). \quad (5.21)
$$

The acceptance probability is

$$
\mathcal{P}(a \to b) = \min \left[ 1, \frac{e^{-\beta(n_2 - n_1)}(1-p)^{n_1}}{e^{-\beta(n_1 - n_2)}(1-p)^{n_2}} \right] = \ldots. \quad (5.22)
$$

This equation will soon be simplified, but its main property is that it can be evaluated explicitly for arbitrary $p$, like any other acceptance probability stemming from the generalized detailed-balance condition: with the $p$ of our choice, we run the cluster construction, which terminates with certain numbers $n_1$ and $n_2$ of satisfied and unsatisfied edges.

Equation (5.22) then yields the probability with which the constructed move is accepted. Otherwise, we have to stay with the original configuration, as we have done (with piles of pebbles on the heliport) since the first pages of this book. On closer inspection of eqn (5.22), we write the acceptance probability as

$$P(a \to b) = \min \left[ 1, \left( \frac{e^{-2\beta}}{1-p} \right)^{n_2} \left( \frac{1-p}{e^{-2\beta}} \right)^{n_1} \right].$$

The algorithm is even simpler (and at its peak efficiency) at the magic value $p = 1 - e^{-2\beta}$, when the acceptance probability is equal to one: we simply construct a cluster, then flip it, build another one, turns it over .... This is the algorithm of Wolff (1989), which generalizes the original cluster method of Swendsen and Wang (1987). The algorithm is easily implemented with the help of a pocket $\mathcal{P}$ containing the active sites (see Alg. 5.9 (cluster-ising)).

We need to work out one technical detail: how to check whether a site $k$ is already in $\mathcal{C}$. As always, there is a simple solution: we may go through the cluster and search it for $k$. It is better to set up a "flag" on each site: on running Alg. 5.9 (cluster-ising) for the $n$th time, the flag of a site entering the cluster is set to $n$, signaling that it is already inside.

**procedure** cluster-ising
**input** $\{\sigma_1, \ldots, \sigma_N\}$
$j \leftarrow \texttt{nran}(1, N)$
$\mathcal{C} \leftarrow \{j\}$
$\mathcal{P} \leftarrow \{j\}$
**while** $(\mathcal{P} \neq \emptyset)$ **do**
$\quad \left\{ \begin{array}{l} k \leftarrow \text{any element of } \mathcal{P} \\ \textbf{for } (\forall \, l \notin \mathcal{C} \text{ with } l \text{ neighbor of } k, \ \sigma_l = \sigma_k) \textbf{ do} \\ \quad \left\{ \begin{array}{l} \textbf{if } (\texttt{ran}(0,1) < p) \textbf{ then} \\ \quad \left\{ \begin{array}{l} \mathcal{P} \leftarrow \mathcal{P} \cup \{l\} \\ \mathcal{C} \leftarrow \mathcal{C} \cup \{l\} \end{array} \right. \\ \mathcal{P} \leftarrow \mathcal{P} \setminus \{k\} \end{array} \right. \end{array} \right.$
**for** $\forall k \in \mathcal{C}$ **do**
$\quad \{ \ \sigma_k \leftarrow -\sigma_k$
**output** $\{\sigma_1, \ldots, \sigma_N\}$

————

**Algorithm 5.9** cluster-ising. Cluster algorithm for the Ising model at the magic value $p = 1 - e^{-2\beta}$.

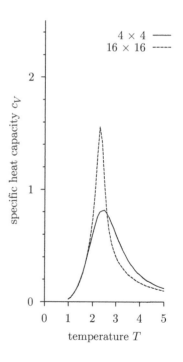

Fig. 5.24 Specific heat capacity of the two-dimensional Ising model with periodic boundary conditions (from Alg. 5.9 (cluster-ising); compare with Fig. 5.6).

The cluster algorithm moves through configuration space with breathtaking speed (for results, see Fig. 5.24). It far outpaces the local Markovchain algorithm, Alg. 5.7 (markov-ising), which suffers from critical slowing down owing to its small step size (see Fig. 5.15). A typical cluster flip easily involves $\sim 10^3$ spins in a $64 \times 64$ Ising model (see Fig. 5.25) and has the system make a giant leap. Running such a insightful code

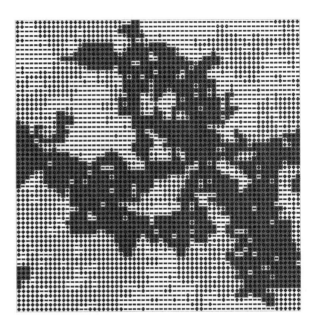

**Fig. 5.25** Large cluster with 1548 up spins in a $64 \times 64$ Ising model with periodic boundary conditions (from Alg. 5.9 (`cluster-ising`), $\beta = 0.43$).

makes us understand the great potential payoff from investments in algorithm design. The implementation of cluster algorithms such as Alg. 5.9 (`cluster-ising`) is straightforward, and the writing of the code takes no more than a few hours. It is the understanding, especially the operational handling of probabilities, which is difficult to obtain. It is on this point that we have been focusing.

In this context, it is essential to realize that powerful Monte Carlo methods which allow one to reach huge system sizes, and obtain millions of essentially independent samples, are the exception rather than the rule. As stressed throughout this book, one often has to face severe restrictions on the number of statistically independent samples which can be produced even during long runs. Moreover, even in cases where Monte Carlo methods work well (as in the Ising model), there is tight competition with other methods, such as transfer matrix approaches, exact enumerations, and high-temperature expansions (which are usually, however, less versatile). These methods only work for small lattices, but they make up much ground with respect to the Monte Carlo approach because they produce numerically exact results for small systems, and can be extrapolated much better because they have no statistical uncertainties. It takes dedication and programming skills, good understanding, and fair judgment to find one's way through this maze of models and methods.

# 5.3   Generalized Ising models

Interactions in nature vary in strength with distance but usually do not change sign. This applies to the four fundamental forces in nature, and also to many effective (induced) forces. The exchange interaction, due to the overlap of $d$-electron orbitals on different lattice sites, which is responsible for ferromagnetism, is of this type. It strives to align spins. In addition, it falls off very quickly with distance. This explains why, in the Ising model, the effective model for ferromagnetism, only nearest-neighbor interactions are retained and longer-range interactions are neglected.

   Some other interactions are more complicated. One example will be the depletion interaction of colloids considered in Chapter 6. We shall see that at some distances, particles are attracted to each other, and at other nearby distances they are repelled from each other. The dominant interaction between ferromagnetic impurities in many materials is also of this type. It couples spins over intermediate distances, but sometimes favors them to be aligned, sometimes to be of opposite sign. Materials for which this interaction is dominant are called spin glasses.

   The theory of spin glasses, and more generally of disordered systems, is an active field of research whose basic model is the Ising spin glass, where the interaction parameter $J$ is replaced by a term $J_{kl}$ which is different for any two neighbors $k$ and $l$. More precisely, each piece of material (each experimental "sample") is modeled by a set of interactions $\{J_{kl}\}$, which are random, because their values depend on the precise distances between spins, which differ from sample to sample. Each experimental sample has its own set of random parameters which do not change during the life of the sample (the $\{J_{kl}\}$ are "quenched" random variables). Most commonly, the interaction $J_{kl}$ between neighboring sites is taken as randomly positive and negative, $\pm 1$. One of the long-lasting controversies in the field of spin glasses concerns the nature of the low-temperature spin-glass phase in three spatial dimensions.

## 5.3.1   The two-dimensional spin glass

In this subsection, we pay a lightning visit to the two-dimensional spin glass. Among the many interesting problems posed by this model, we restrict ourselves to running through the battery of computational methods, enumeration (listing and counting), local Monte Carlo sampling, and cluster sampling. For concreteness, we consider a single Ising spin glass sample on a $6 \times 6$ lattice without periodic boundary conditions (see Fig. 5.26) with an energy

$$E = -J_{kl} \sum_{\langle k,l \rangle} \sigma_k \sigma_l,$$

where the parameters $J_{kl}$ are defined in Fig. 5.26.

   Algorithm 5.3 (enumerate-ising) is easily modified to generate the density of states $\mathcal{N}(E)$ of this system (see Table 5.8). The main difference

**Table 5.8** Number of configurations with energy $E$ of the two-dimensional spin glass shown in Fig. 5.26 (from modified Alg. 5.3 (enumerate-ising))

| $E$ | $\mathcal{N}(E) = \mathcal{N}(-E)$ |
|---|---|
| 0 | 6 969 787 392 |
| −2 | 6 754 672 256 |
| ⋮ | ⋮ |
| −34 | 59 456 |
| −36 | 6 912 |
| −38 | 672 |

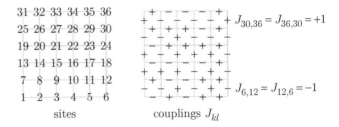

Fig. 5.26 Neighbor scheme and coupling strengths of a two-dimensional ±1 spin glass sample without periodic boundary conditions.

**Table 5.9** Logarithm of the partition function and mean energy per particle of the two-dimensional spin glass shown in Fig. 5.26 (from modified Alg. 5.3 (`enumerate-ising`))

| $T$ | $\log Z$ | $\langle E \rangle /N$ |
|---|---|---|
| 1. | 46.395 | −0.932 |
| 2. | 31.600 | −0.665 |
| 3. | 28.093 | −0.495 |
| 4. | 26.763 | −0.389 |
| 5. | 26.126 | −0.319 |

with the ferromagnetic Ising model resides in the existence of a large number of ground states. In our example, there are 672 ground states; some of them shown in Fig. 5.27. The thermodynamics of this model follows from $\mathcal{N}(E)$, using Alg. 5.4 (`thermo-ising`) (see Table 5.9). For a quantitative study of spin glasses, we would have to average the energy, the free energy, etc., over many realizations of the $\{J_{kl}\}$. However, this is beyond the scope of this book.

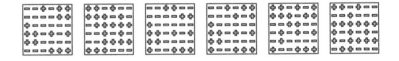

Fig. 5.27 Several of the 672 ground states (with $E = -38$) of the two-dimensional spin glass shown in Fig. 5.26.

Our aim is to check how the computational algorithms carry over from the Ising model to the Ising spin glass. We can easily modify the local Monte Carlo algorithm (see Alg. 5.10 (`markov-spin-glass`)), and reproduce the data in Table 5.9. For larger systems, the local Monte Carlo algorithm becomes very slow. This is due, roughly, to the existence of a large number of ground states, which lie at the bottoms of valleys in a very complicated energy landscape. At low temperature, Alg. 5.10 (`markov-spin-glass`) becomes trapped in these valleys, so that the local algorithm takes a long time to explore a representative part of the configuration space. In more than two dimensions, this time is so large that the algorithm, in the language of Subsection 1.4.1, is practically nonergodic for large system sizes.

The cluster algorithm of Subsection 5.2.3 can be generalized to the case of spin glasses (see Alg. 5.11 (`cluster-spin-glass`)) by changing a single line in Alg. 5.9 (`cluster-ising`) (instead of building a cluster with spins of same sign, we consider neighboring spins $\sigma_j$ and $\sigma_k$ that satisfy $\sigma_j J_{jk} \sigma_k > 0$). Algorithm 5.11 (`cluster-spin-glass`) allows to

**procedure** `markov-spin-glass`
**input** $\{\sigma_1, \ldots, \sigma_N\}, E$
$k \leftarrow \text{nran}(1, N)$
$\Delta_E \leftarrow \sigma_k \sum_l J_{kl} \sigma_{\text{Nbr}(l,k)}$   (matrix $\{J_{kl}\}$ from Fig. 5.26)
$\Upsilon \leftarrow \text{e}^{-\beta \Delta_E}$
**if** $\left(\text{ran}(0,1) < \Upsilon\right)$ **then**
$\left\{ \begin{array}{l} \sigma_k \leftarrow -\sigma_k \\ E \leftarrow E + \Delta_E \end{array} \right.$
**output** $\{\sigma_1, \ldots, \sigma_N\}, E$

---

**Algorithm 5.10** `markov-spin-glass`. Local Metropolis algorithm for the Ising spin glass.

recover the data in Table 5.9, but it does not lead to the spectacular performance gains that we witnessed in the ferromagnetic Ising model.

**procedure** `cluster-spin-glass`
**input** $\{J_{kl}\}$
$\vdots$
**while** $(\mathcal{P} \neq \emptyset)$ **do**
$\left\{ \begin{array}{l} k \leftarrow \text{any element of } \mathcal{P} \\ \textbf{for } (\forall \, l \notin \mathcal{C} \text{ with } l \text{ neighbor of } k, \sigma_l J_{lk} \sigma_k > 0) \textbf{ do} \\ \vdots \end{array} \right.$

---

**Algorithm 5.11** `cluster-spin-glass`. Lines that must be changed in order in Alg. 5.9 (`cluster-ising`) to allow it to be used for spin glasses.

The reason for this lack of efficiency is the following. The cluster algorithm for the Ising model was constructed with the aim of making large strides in magnetization at each step. This enables quick moves between the two ground states of the Ising model, but does not facilitate moves between the large number of valleys in the spin glass.

Finally, Alg. 5.6 (`combinatorial-ising`) can also be generalized to the two-dimensional spin glass, and we again must modify only a few lines (see Alg. 5.12 (`combinatorial-spin-glass`)). This algorithm (Saul and Kardar 1993) can reproduce the data in Table 5.9 exactly. It works for large two-dimensional spin glasses, where Gray-code enumeration is not an option. It represents the best computational method for studying the thermodynamics of two-dimensional spin glasses, allowing one to reach very large system sizes and to average over many samples. However, the method cannot be generalized to three dimensions.

In conclusion, in this subsection we have briefly discussed the Ising spin glass in two dimensions, with the aim of testing our algorithms. Local Monte Carlo methods slow down so much that they are practically useless, and cluster algorithms do not improve the convergence. However, the purely theoretical combinatorial approach of Kac and Ward,

**procedure** `combinatorial-spin-glass`

$$\left\{ \left\{ \left\{ \begin{array}{l} \vdots \\ U(j, j') \leftarrow J_{kk'} u_{\rightarrow}(n, n') \text{ (etc.)} \\ \vdots \end{array} \right. \right. \right.$$

**Algorithm 5.12** `combinatorial-spin-glass`. One of the four changes that allows Alg. 5.6 (`combinatorial-ising`) to be used for spin glasses.

which is closely related to Onsager's solution of the two-dimensional Ising model, turns into a computational algorithm. Improved versions of Alg. 5.12 (`combinatorial-spin-glass`) nowadays run on supercomputers, attacking problems as yet unsolved. The metamorphosis of the combinatorial solution into a practical algorithm illustrates once more that there is no separation between computational and theoretical physics.

### 5.3.2   Liquids as Ising-spin-glass models

The present chapter's main focus—the Ising model—is doubly universal. First, the Ising model is universal in the sense of critical phenomena. Near the critical temperature, the correlation length is much larger than the lattice spacing, the range of interactions and all other length scales. All detailed properties of the lattice structure and the interaction then become unimportant, and whole classes of microscopic models become equivalent to the Ising model.

Second, the Ising model is universal because it appears as a fundamental model in many branches of physics and beyond: it is a magnet and a lattice gas, as already discussed, but also a prominent model of associative memory in the field of theoretical neuroscience (the Hopfield model). Furthermore, it appears in nuclear physics, in coding theory, in short, wherever pairs of coupled binary variables play a role. In this subsection, we describe a universality of this second kind. We show how interacting liquids of particles in continuous space, with complicated classical pair potentials, can be mapped onto the Ising spin glass. This mapping allows us to extend the pivot cluster algorithm of Chapter 2 to general liquids, as first proposed by Liu and Luijten (2004).

The pivot cluster algorithm of Subsection 2.5.2 moves particles via a random symmetry operation, a reflection or rotation about a random pivot (see Fig. 5.28). Let us apply it, not to hard spheres, but to a liquid of particles interacting with a pair potential $V(r_{kl})$, where $r_{kl}$ is the Cartesian distance between particles $k$ and $l$. For concreteness, we consider $N$ particles in a box with periodic boundary conditions. The symmetry operation is a reflection with respect to a vertical axis. Because of the boundary conditions, the axis can be scrolled so that it comes to lie in the center of the box (see Fig. 5.28, and the discussion in Subsection 2.5.2). Only pivot transformations (flips) are allowed, and each particle can have only two positions, so that there are $2^N$ config-

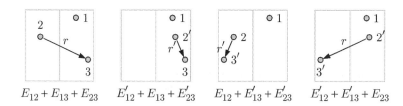

$$E_{12} + E_{13} + E_{23} \qquad E'_{12} + E_{13} + E'_{23} \qquad E_{12} + E'_{13} + E'_{23} \qquad E'_{12} + E'_{13} + E_{23}$$

**Fig. 5.28** A liquid as an Ising spin glass. The distance between particles 2 and 3 is either $r$ (with energy $E_{23}$) or $r'$ (with energy $E'_{23}$).

urations (the same as the number of configurations of $N$ Ising spins). Furthermore, the interaction energy between particle $k$ and $l$ can take only two values:

$E_{kl}$ : when both $k$ and $l$ are in original position or are flipped,

$E'_{kl}$ : when either $k$ or $l$ is flipped.

A spin variable $\sigma_k$ can be associated with the two possible positions of particle $k$:

$$\sigma_k = \begin{cases} 1 & \text{if } k \text{ is in its original position} \\ -1 & \text{if } k \text{ is flipped} \end{cases}.$$

For a fixed pivot position, the particle (spin) $k$ and the particle (spin) $l$ are coupled through an interaction parameter

$$J_{kl} = \frac{1}{2}\left(E_{kl} - E'_{kl}\right),$$

and the total energy of the system is

$$\left\{ \begin{matrix} \text{energy} \\ \text{of system} \end{matrix} \right\} = \underbrace{\sum_{\langle k,l \rangle} J_{kl}\sigma_k\sigma_l}_{\text{Ising spin glass}} + \underbrace{\frac{1}{2}\sum_{\langle k,l \rangle}\left(E_{kl} + E'_{kl}\right)}_{\text{const}}.$$

With this definition, the liquid of $N$ particles with arbitrary pair interaction is isomorphic to an Ising spin glass. It can be simulated with Alg. 5.11 (**cluster-spin-glass**) (see Liu and Luijten (2004)). Pivots are chosen randomly, and for each pivot, a single cluster is flipped. The pivot cluster algorithm for liquids is a powerful tool for studying interacting binary mixtures, and it has great potential for other problems in soft condensed matter physics. However, systems must be chosen carefully in order to avoid constructing clusters which comprise the whole system. As mentioned, the algorithm points to a far-reaching equivalence between spin models and classical liquids.

# Exercises

(5.1) Create the neighbor table $\text{Nbr}(n, k)$ of Fig. 5.2 for an $L \times L'$ square lattice with or without periodic boundary conditions. The subroutine implementing the algorithm may incorporate the following fragment

> **input** $\{L, L', \text{lattice}\}$
> **if** (lattice = "period") **then**
> $\quad \{\ \dots$
> **else**
> $\quad \{\ \dots$
> $\quad \dots$

This subroutine is the only lattice-specific component in many programs discussed in the following exercises.

## (Section 5.1)

(5.2) Generate the list of all configurations of $N$ Ising spins from the binary representation of the numbers $\{0, \dots, 2^N - 1\}$. Determine the density of states $\mathcal{N}(E)$ on small $L \times L$ square lattices with periodic boundary conditions (compare with Table 5.2). Compute the basic thermodynamic quantities. Can you enumerate all the configurations on the $6 \times 6$ lattice with this program?

(5.3) Implement the Gray code (Alg. 5.2 (**gray-flip**)) for $N$ spins and test it by printing all the configurations for small $N$ (as in Table 5.1). Use the Gray code together with Alg. 5.3 (**enumerate-ising**) to generate the density of states of the two-dimensional $2 \times 2$, $4 \times 4$, and $6 \times 6$ square lattices with and without periodic boundary conditions. NB: For the $6 \times 6$ lattice, make sure your data format allows you to handle the very large numbers appearing in $\mathcal{N}(E)$ for some of the energies.

(5.4) Implement Alg. 5.4 (**thermo-ising**) and compute the mean energy and specific heat capacity of the Ising model (use the density of states from Table 5.2 or from Alg. 5.3 (**enumerate-ising**), as in Exerc. 5.3). Test your implementation by alternatively computing $E$ and $c_V$ through discrete derivatives of $\log Z$ (see eqns (5.2) and (5.5)).

(5.5) A lattice is called bipartite if its sites can be partitioned into two sets, $\mathcal{S}_1$ and $\mathcal{S}_2$, such that the neighbors of a site are never in the same set as the site itself. The square lattice without periodic boundary conditions is always bipartite, but does this also hold for the $L \times L'$ square lattice with periodic boundary conditions? Show that the density of states of the Ising model on a bipartite lattice satisfies $\mathcal{N}(E) = \mathcal{N}(-E)$. Can this relation be satisfied on lattices which are not bipartite?

(5.6) Use single-spin-flip Gray code enumeration to generate the histogram of the number of configurations with an energy $E$ and a magnetization $M$, $\mathcal{N}(E, M)$ in the Ising model on $2 \times 2$, $4 \times 4$, and $6 \times 6$ square lattices with periodic boundary conditions. Recover the data of Table 5.2 by summing over all $M$. Generate from $\mathcal{N}(E, M)$ the temperature-dependent probability distribution $\pi_M$ of the total magnetization per spin $m = M/N$ (compare with Fig. 5.7). Discuss the qualitative change of $\pi_m$ between the single-peak and the double-peak regimes, which is well captured by the Binder cumulant $B(T) = \frac{1}{2} \left[ 3 - \left\langle m^4(T) \right\rangle / \left\langle m^2(T) \right\rangle^2 \right]$ (see Binder (1981)). Plot $B(T)$ for the three lattices. Determine the high- and low-temperature limits of $B(T)$. Using your numerical results, confirm that the Binder cumulants for different lattice sizes intersect almost exactly at $T_c$ (see eqn (5.6)).

(5.7) Implement Alg. 5.5 (**edge-ising**), generating all the loop configurations on small lattices with and without periodic boundary conditions. Use histograms (as in Table 5.2) to compute the partition function for small lattices (check against direct enumeration). Why do some of these lattices have the same number of configurations with $e$ edges and with $E - e$ edges, where $E$ is the total number of edges on the lattice? Finally, use the results obtained with Alg. 5.5 (**edge-ising**) on small lattices to determine the number of loop configuration with 4, 6, and 8 edges on very large $L \times L$ lattices with periodic boundary conditions. Determine the partition function at very high temperatures (low $\beta$) on these lattices.

(5.8) Consider the $32 \times 32$ matrix $U_{4 \times 2}$ in Subsection 5.1.4. Use a naive computer program to determine all cycle configurations with nonzero weights. Compute the weight for each cycle configuration and show that the 64 terms which do not cancel correspond to the loop configurations in Fig. 5.11 (You should find a total of 144 cycle configurations with nonzero weights.) Generate the matrix $U_{L\times L}$ on larger lattices using Alg. 5.6 (`combinatorial-ising`). Compute det $U_{L\times L}$ using a standard numerical linear-algebra routine. Compare with the results of Gray-code enumerations for small lattices.

NB: To simplify the enumeration of cycle configurations in the naive program, note that a row vector $\{u_{k,1},\ldots,u_{k,16}\}$ which differs from zero only on the diagonal contributes to trivial cycles only.

(5.9) Implement Kaufman's formula (Kaufman 1949) for the partition function of the Ising model on a $L \times L$ lattice with periodic boundary conditions from the following fragment:

$$\ldots$$
$$\gamma_0 \leftarrow \log\left(\mathrm{e}^{2\beta}\tanh\beta\right)$$
$$\textbf{for } k=1,\ldots,2L-1 \textbf{ do}$$
$$\left\{\begin{array}{l}\Upsilon \leftarrow \cosh^2(2\beta)/\sinh(2\beta)-\cos(k\pi/L)\\ \gamma_k \leftarrow \log\left(\Upsilon+\sqrt{\Upsilon^2-1}\right)\end{array}\right.$$
$$\Upsilon \leftarrow \sinh^{L^2/2}(2\beta)$$
$$\{Y_1,\ldots,Y_4\} \leftarrow \{\Upsilon,\ldots,\Upsilon\}$$
$$\textbf{for } k=0,\ldots,L-1 \textbf{ do}$$
$$\left\{\begin{array}{l}Y_1 \leftarrow 2Y_1\cosh(\gamma_{2k+1}L/2)\\ Y_2 \leftarrow 2Y_2\sinh(\gamma_{2k+1}L/2)\\ Y_3 \leftarrow 2Y_3\cosh(\gamma_{2k}L/2)\\ Y_4 \leftarrow 2Y_4\sinh(\gamma_{2k}L/2)\end{array}\right.$$
$$Z \leftarrow 2^{L^2/2-1}(Y_1+Y_2+Y_3+Y_4)$$
$$\ldots$$

Test output of this program against exact enumeration data, from your own Gray-code enumeration on the lattice with periodic boundary conditions, or from Table 5.3. Then consult Kaufman's original paper. For a practical application of these formulas, see, for example, Beale (1996).

**(Section 5.2)**

(5.10) Implement Alg. 5.7 (`markov-ising`) (local Metropolis algorithm for the Ising model) and test it against the specific heat capacity and the energy for small lattices from exact enumeration to at least four significant digits (see Table 5.3). Improve your program as follows. The exponential function evaluating the Boltzmann weight may take on only a few values: replace it by a table to avoid repeated function evaluations. Also, a variable $\Upsilon > 1$ will never be smaller than a random number $\mathrm{ran}(0,1)$: avoid the superfluous generation of a random number and comparison in this case. Again test the improved program against exact results. Generate plots of the average absolute magnetization against

temperature for lattices of different sizes (compare with Fig. 5.16).

(5.11) Implement Alg. 5.9 (`cluster-ising`) with sets $\mathcal{C}, \mathcal{F}_{\mathrm{old}}$, and $\mathcal{F}_{\mathrm{new}}$ simply programmed as vectors. Check for cluster membership through simple look-up. Test your program against specific-heat-capacity and mean-energy data obtained by exact enumeration on small lattices. Improvements of this program depend on the way your computer language treats vectors and lists. If possible, handle the initial conditions as follows (compare with Exerc. 1.3): at each temperature $T$, let the program search for an initial configuration generated at that same temperature (choose file names which encode $T$). If such a file does not exist, choose random initial spins. The final configuration of each run should be made into an initial configuration for the next run. Use this improved program to compute histograms of the magnetization and to plot the Binder cumulant as a function of $T$ (compare with Exerc. 5.6). Reconfirm that Binder cumulants, at different lattice sizes, intersect almost exactly at $T_c$.

**(Section 5.3)**

(5.12) Implement the local Monte Carlo algorithm for the two-dimensional $\pm 1$ spin glass. Thoroughly test it in the specific case of Fig. 5.26 (compare the mean energies per particle with Table 5.9, for the choice of $J_{kl}$ given). Compute the specific heat capacity, and average over many samples. Study the behavior of the ensemble-averaged specific heat capacity for square lattices of sizes between $2\times 2$ and $32\times 32$.

(5.13) Consider $N$ particles, constrained onto a unit circle, with positions $\{\mathbf{x}_1,\ldots,\mathbf{x}_N\}$, satisfying $|\mathbf{x}_k| = 1$. Particles interact with a Lennard-Jones potential

$$E_{kl} = |\Delta_{\mathbf{x}}|^{12} - |\Delta_{\mathbf{x}}|^6,$$

where $\Delta_{\mathbf{x}} = \mathbf{x}_k - \mathbf{x}_l$ is the two-dimensional distance vector. Implement the spin-glass cluster algorithm of Liu and Luijten (2004) for this problem. To test your program, compute the mean energy per particle, and compare it, for $N \leq 6$, with the exact value obtained by Riemann integration. (You may also compare with the results of a local Monte Carlo simulation.)

NB: If Alg. 5.11 (`cluster-spin-glass`) exists, adapt it for the particle simulation. Otherwise, write a naive version of the cluster algorithm for a few particles.

# References

Beale P. D. (1996) Exact distribution of energies in the two-dimensional Ising model, *Physical Review Letters* **76**, 78–81

Binder K. (1981) Finite size scaling analysis of Ising-model block distribution-functions, *Zeitschrift für Physik B–Condensed Matter* **43**, 119–140

Ferdinand A. E., Fisher M. E. (1969) Bounded and inhomogeneous Ising models. I. specific-heat anomaly of a finite lattice, *Physical Review* **185**, 832–846

Kac M., Ward J. C. (1952) A combinatorial solution of the two-dimensional Ising model, *Physical Review* **88**, 1332–1337

Kaufman B. (1949) Crystal Statistics. II. Partition function evaluated by spinor analysis, *Physical Review* **76**, 1232–1243

Liu J. W., Luijten E. (2004) Rejection-free geometric cluster algorithm for complex fluids, *Physical Review Letters* **92**, 035504

Onsager L. (1944) Crystal Statistics. I. A two-dimensional model with an order-disorder transition, *Physical Review* **65**, 117–149

Propp J. G., Wilson D. B. (1996) Exact sampling with coupled Markov chains and applications to statistical mechanics, *Random Structures & Algorithms* **9**, 223–252

Saul L., Kardar M. (1993) Exact integer algorithm for the two-dimensional $\pm J$ Ising spin glass, *Physical Review E* **48**, R3221–R3224

Swendsen R. H., Wang J. S. (1987) Nonuniversal critical-dynamics in Monte-Carlo simulations, *Physical Review Letters* **58**, 86–88

Wolff U. (1989) Collective Monte-Carlo updating for spin systems, *Physical Review Letters* **62**, 361–364

# Entropic forces

In the present chapter, we revisit classical entropic models, where all configurations have the same probability. This sight is familiar from the hard-core systems of Chapter 2, where we struggled with the foundations of statistical mechanics, and reached a basic understanding of molecular dynamics and Monte Carlo algorithms. At present, we are more interested in describing entropic forces, which dominate many effects in soft condensed matter (the physics of biological systems, colloids, and polymers) and are also of fundamental interest in the context of order–disorder transitions in quantum mechanical electronic systems.

The chapter begins with a complete mathematical and algorithmic solution of the problem of one-dimensional hard spheres (the "random clothes-pins" model) and leads us to a discussion of the Asakura–Oosawa depletion interaction, one of the fundamental forces in nature, at least in the realm of biological systems, colloids, and polymers. This multifaceted, mysterious interaction is at its strongest in binary systems of large and small particles, where interesting effects and phase transitions appear even at low density for no other reason than the presence of constituents of different sizes. Binary oriented squares in two dimensions provide the simplest model for colloids where the depletion interaction is sufficiently strong to induce a demixing (flocculation) transition. This extremely constrained model has not been solved analytically, and resists attacks by standard local Monte Carlo simulation. However, a simple cluster algorithm—set up in a few dozen lines—samples the binary-squares model without any problems.

In the second part of this chapter, we consider dimers on a lattice, the archetypal model of a discrete entropic system where orientation effects dominate. This lattice system has profound connections with the Ising model of a magnet (see Chapter 5). Ever since Pauling's theory of the benzene molecule in terms of resonating singlets (dimers), the theory of dimers has had important applications and extensions in molecular physics and condensed matter theory.

From a computational point of view, dimers lead us to revisit the issue of enumeration, as in Chapter 5, but here with the chance to assimilate systematic "breadth-first" and "depth-first" techniques, which are of general use. These tree-based methods are second only to the extremely powerful enumeration method based on Pfaffians, which we shall go through in detail. Again, we discuss Monte Carlo methods, and conclude with discussions of order and disorder in the monomer–dimer model.

Clothes-pins are randomly distributed on a line (see Fig. 6.1): any possible arrangements of pins is equally likely. What is the probability of a pin being at position $x$? Most of us would guess that this probability is independent of position, but this is not the case: pins are much more likely to be close to a boundary, as if attracted by it. They are also more likely to be close to each other. In this chapter, we study clothes-pin attractions and other entropic interactions, which exist even though there are no charges, currents, springs, etc. These interactions play a major role in soft condensed matter, the science of colloids, membranes, polymers, etc., but also in solid state physics.

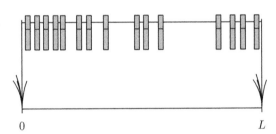

**Fig. 6.1** 15 randomly positioned pins on a segment of length $L$.

# 6.1 Entropic continuum models and mixtures

In this section, we treat two continuum models which bring out clearly the entropic interactions between hard particles. We concentrate first on a random-pin model (equivalent to hard spheres in one dimension), and then on a model of a binary mixture of hard particles, where the interaction between particles is strong enough to induce a phase transition. This model is easily simulated with the pivot cluster algorithm of Chapter 2.

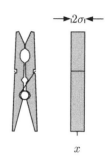

**Fig. 6.2** Clothes-pin in side view (*left*) and front view (*right*).

## 6.1.1 Random clothes-pins

We consider clothes-pins of width $2\sigma$ on a washing line (a line segment) between boundaries at $x = 0$ and $x = L$, as in Fig. 6.1. The pins are placed, one after another, at random positions, but if an overlap is generated, we take them all off the line and start anew (see Alg. 6.1 (`naive-pin`)). This process places all pins with a flat probability distribution

$$\pi(x_1, \ldots, x_N) = \begin{cases} 1 \text{ if legal} \\ 0 \text{ otherwise} \end{cases}. \qquad (6.1)$$

Two pins overlap if they are less than $2\sigma$ away from each other. Likewise, a pin overlaps with a boundary if it is less than $\sigma$ away from $x = 0$ or $x = L$ (see Fig. 6.2). Equation (6.1) corresponds to a trivial Boltzmann weight with zero energy for each nonoverlapping configuration and infinite energy for arrangements of pins which are illegal—our pins are one-dimensional hard spheres.

---

**procedure naive-pin**

1  **for** $k = 1, \ldots, N$ **do**
$\left\{ \begin{array}{l} x_k \leftarrow \mathbf{ran}\,(\sigma, L - \sigma) \\ \mathbf{for}\ l = 1, \ldots, k - 1\ \mathbf{do} \\ \quad \{\ \mathbf{if}\ (|x_k - x_l| < 2\sigma)\ \mathbf{goto}\ 1\ \text{(reject sample—tabula rasa)} \end{array} \right.$
**output** $\{x_1, \ldots, x_N\}$

---

**Algorithm 6.1 naive-pin.** Direct-sampling algorithm for $N$ pins of width $2\sigma$ on a segment of length $L$ (see Alg. 2.7 (`direct-disks`)).

The partition function of this system is

$$Z_{N,L} = \int_\sigma^{L-\sigma} dx_1 \ldots \int_\sigma^{L-\sigma} dx_N\ \pi(x_1, \ldots, x_N). \qquad (6.2)$$

(One often multiplies this partition function with a factor $1/N!$, in order to avoid a problem called the Gibbs paradox. However, for distinguishable pins, the definition in eqn (6.2) is preferable.) The $N$-particle probability $\pi(x_1, \ldots, x_N)$ is totally symmetric in its arguments and thus satisfies the following for any permutation $P$ of the indices $\{1, \ldots, N\}$:

$$\pi(x_1, \ldots, x_N) = \pi(x_{P_1}, \ldots, x_{P_N}).$$

The complete domain of integration separates into $N!$ sectors, one for each permutation $x_{P_1} < \cdots < x_{P_N}$. Each sector gives the same contribution to the integral, and we may select one ordering and multiply by the total number of sectors:

$$Z_{N,L} = N! \int_\sigma^{L-\sigma} dx_1 \ldots \int_\sigma^{L-\sigma} dx_N \pi(x_1, \ldots, x_N) \Theta(x_1, \ldots, x_N) \quad (6.3)$$

(the function $\Theta$ is equal to one if $x_1 < x_2 < \cdots < x_N$ and zero otherwise). We know that pin $k$ has $k-1$ pins to its left, so that we may shift all the arguments $\{x_1, \ldots, x_N\}$ by $k-1$ pin widths and by $\sigma$, because of the left boundary. Likewise, to the right of pin $k$, there are $N-k-1$ pins, so that, with the transformations

$$y_1 = x_1 - \sigma, \ \ldots, \ y_k = x_k - (2k-1)\sigma, \ \ldots, \ y_N = x_N - (2N-1)\sigma,$$

we obtain the integral

$$Z_{N,L} = N! \int_0^{L-2N\sigma} dy_1 \ \ldots \int_0^{L-2N\sigma} dy_N \ \Theta(y_1, \ldots, y_N), \quad (6.4)$$

from which the weight function $\pi$ has disappeared, and only the order $y_1 < \cdots < y_N$ remains enforced by the function $\Theta$. Undoing the trick which took us from eqn (6.2) to eqn (6.3), we find that $Z_{N,L}$ is equal to the $N$th power of the effective length of the interval:

$$Z_{N,L} = \begin{cases} (L - 2N\sigma)^N & \text{if } L > 2N\sigma \\ 0 & \text{otherwise} \end{cases}. \quad (6.5)$$

From eqns (6.2) and (6.5), it follows that the acceptance rate of Alg. 6.1 (naive-pin) is extremely small:

$$p_{\text{accept}} = (L - 2N\sigma)^N / (L - 2\sigma)^N, \quad (6.6)$$

which makes us look for more successful sampling approaches using the transformed variables $\{y_1, \ldots, y_N\}$ in eqn (6.4). However, sampling this integral literally means picking uniform random numbers $\{y_1, \ldots, y_N\}$ between 0 and $L - 2N\sigma$ but accepting them only if $y_1 < \cdots < y_N$. This does even a worse job than the naive random-pin program, with an acceptance rate of $1/N!$, an all-time low.

It may be intuitively clear that this worst of all algorithms is in fact the best of all ..., if we generate the random numbers $\{y_1, \ldots, y_N\}$ as before, and then sort them. To justify this trick, we decompose the sampling of $N$ random numbers $\{y_1, \ldots, y_N\} = \{\text{ran}(0,1), \ldots, \text{ran}(0,1)\}$ into two steps: we first find the set of the values of the $N$ numbers (without deciding which one is the first, the second, etc.), and then sample a permutation $P$ giving the order of the numbers. The integral in eqn (6.4) can then be written, in terms of the permutation $P$, as

$$\text{eqn (6.4)} = \sum_P \int_0^{L-2N\sigma} d(y_{P_1}) \cdots \int_0^{L-2N\sigma} d(y_{P_N}) \, \Theta(y_{P_1}, \ldots, y_{P_N}).$$

Among all the permutations, only $P$ (obtained by sorting the numbers) gives a nonzero contribution, in agreement with our intuition above. The generation of random numbers $y_k$, the sorting and the back-transformation from $y_k$ to $x_k$ are implemented in Alg. 6.2 (direct-pin), a rejection-free direct-sampling algorithm for one-dimensional hard spheres.

**procedure** direct-pin
  **for** $k = 1, \ldots, N$ **do**
    $\{\ \tilde{y}_k \leftarrow \text{ran}\,(0, L - 2N\sigma)$
  $\{y_1, \ldots, y_N\} \leftarrow \text{sort}[\{\tilde{y}_1, \ldots, \tilde{y}_N\}]$
  **for** $k = 1, \ldots, N$ **do**
    $\{\ x_k \leftarrow y_k + (2k - 1)\sigma$
  **output** $\{x_1, \ldots, x_N\}$

**Algorithm 6.2** direct-pin. Rejection-free direct-sampling algorithm for $N$ pins of width $2\sigma$ on a line segment of length $L$.

To check our reasoning about permutation sectors and our sorting trick, we compare the rejection rate in eqn (6.6) with the empirical rate of Alg. 6.1 (naive-pin) (see Table 6.1), and generate with both algorithms the histograms of the probability $\pi(x)$ for a pin to be at position $x$ (see Fig. 6.3).

**Table 6.1** Number of accepted configurations for $10^7$ attempts of Alg. 6.1 (naive-pin)

| $N$ | $2\sigma/L$ | Accepted | Eqn (6.6) |
|---|---|---|---|
| 3 | 0.1 | 4 705 436 | 4 705 075 |
| 10 | 0.03 | 383 056 | 383 056 |
| 30 | 0.01 | 311 | 305 |

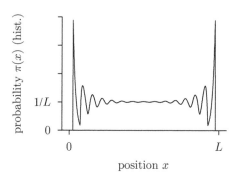

**Fig. 6.3** Probability $\pi(x)$ for a pin to be at position $x$, with $N = 15$, $2\sigma = 0.1$, and $L = 2$ (from Alg. 6.2 (direct-pin), or Alg. 6.1 (naive-pin)).

Remarkably, the probability $\pi(x)$ of finding a pin at position $x$ on the washing line depends strongly on position, even though we have put them there with uniform probability (see eqn (6.1)). In particular, we are more likely to find a pin close to the boundaries than in the middle of the line. There are intricate oscillations as a function of $x$. At one particular point, the probability of finding a pin falls almost to zero.

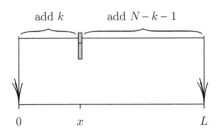

**Fig. 6.4** Adding $k$ pins to the left and $N - k - 1$ others to the right of the first pin, at $x$, in order to compute the probability $\pi(x)$.

All of this is quite mysterious, and we would clearly be more comfortable with an analytic solution for $\pi(x)$. This is what we shall obtain now. We first place a pin at position $x$, and then add $k$ other pins to its left and another $N - k - 1$ pins to its right. This special arrangement (see Fig. 6.4) has a probability $\pi_k(x)$, and the sum over all types of arrangement, putting $k$ pins to the left (for all $k$) and $N - 1 - k$ pins to the right of the particle already present at $x$, gives $\pi(k)$. We need to include a combinatorial factor $\binom{N-1}{k}$, reflecting the number of choices for picking $k$ out of the $N - 1$ remaining pins, and to account for the statistical weights $Z_{k,x-\sigma}$ and $Z_{N-1-k,L-x-\sigma}$:

$$\pi(x) = \sum_{k=0}^{N-1} \underbrace{\frac{1}{Z_{N,L}} \binom{N-1}{k} Z_{k,x-\sigma} Z_{N-1-k,L-x-\sigma}}_{\pi_k(x)}. \tag{6.7}$$

In eqn (6.8), we have used the fact that

$$\int_0^a dx\, x^k (a-x)^l = a^{k+l+1} \frac{k!\, l!}{(k+l+1)!}.$$

The probability is normalized to $\int dx\, \pi(x) = 1$ if $\int dx\, \pi_k(x) = 1/N$, as can be seen from

$$\int_0^L dx\, \pi_k(x) = \int_{\sigma(1+2k)}^{\sigma[1+2(k-N)]+L} dx\, \pi_k(x) = \frac{1}{N}. \tag{6.8}$$

We analyze $\pi(x)$ close to the left boundary where $x \gtrsim \sigma$, and where only the $k = 0$ term contributes, and obtain the following from eqns (6.5) and (6.7):

$$\pi(x \gtrsim \sigma) = \pi_0(x) = \frac{Z_{N-1,L-x-\sigma}}{Z_{N,L}} \simeq \frac{1}{L - 2N\sigma}\left[1 - \frac{N-1}{L - 2N\sigma}(x - \sigma)\right].$$

At the left boundary $(x = \sigma)$, the probability is much larger than the average value $1/L$ (in Fig. 6.3, $\pi(x = \sigma) = 4/L$). For $x \gtrsim \sigma$, $\pi(x)$ decreases simply because the remaining space to put other particles in diminishes. A little farther down the line, the peaks in $\pi(x)$ (see Fig. 6.3) arise from the setting-in of sectors $k = 1, 2, 3 \ldots$, that is, every $2\sigma$. They die away rapidly.

The time has come to compare our system of pins on a line segment with the analogous system on a circle, that is, with periodic boundary conditions. In the first case, the probability $\pi(x)$ depends explicitly on $x$. This is the quantity that we computed in eqn (6.7), and checked with the results of numerical simulations. Periodic boundary conditions (obtained by bending the line into a ring) make the system homogeneous, and guarantee that the density, the probability $\pi(x)$ of having a pin at $x$, is everywhere the same. However, the pair correlation function $\pi(x, x')$ on a ring (the probability of having one pin at $x$ and another at $x'$) will be inhomogeneous, for exactly the same reason that $\pi(x)$ was inhomogeneous on the line (see Fig. 6.5). We may cut one of the pins in two for it to take the place of the two boundaries. This transformation is exact: the boundary interacts with a pin in exactly the same way as two pins with each other.

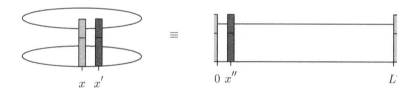

**Fig. 6.5** A pair of pins on a ring (*left*) and a single pin on a line (*right*).

More generally, the pair correlation function $\pi(x, x')$ on a ring of length $L$, for a system of $N$ pins, agrees with the boundary correlation function $\pi(x'' = |x' - x| - \sigma)$ on a line of length $L' = L - 2\sigma$, for $N - 1$ pins:

$$\underbrace{\pi(x_k, x_l, L) = \pi_2(|x_k - x_l|, L)}_{\substack{\text{pair correlation} \\ \text{on ring}}} = \underbrace{\pi(|x_k - x_l| - \sigma, L - 2\sigma)}_{\substack{\text{boundary correlation} \\ \text{on line}}}. \qquad (6.9)$$

## 6.1.2   The Asakura–Oosawa depletion interaction

Hard pins are the one-dimensional cousins of the hard-sphere systems of Chapter 2. Without periodic boundary conditions, the density in both systems is much higher at the boundaries than in the bulk. In contrast, with periodic boundary conditions, the density is necessarily constant in both systems. Nontrivial structure can be observed first in two-particle properties, for example the pair correlation functions. For pins, we have seen that the pair correlations were essentially identical to the correlations next to a wall (see Fig. 6.5). In more than one dimension, the pair correlations in a periodic system have no reason to agree with the single-particle density in a finite box, because the mapping sketched in Fig. 6.5 cannot be generalized.

In order to better follow the variations of density and pair correlation functions, we now turn to an effective description, which concentrates on a single particle or a pair of particles, and averages over all the remaining particles. In our random-clothes-pin model, we interpret the probability $\pi(x)$—the raw computational output shown in Fig. 6.3—as stemming from the Boltzmann distribution of a single particle $\pi(x) \propto \exp[-\beta V(x)]$, in an effective potential $V(x)$, giving rise to an (effective) force $F(x) = -(\partial/\partial x) V(x)$. So we can say that the increased probability $\pi(x)$ is due to an attractive force on a pin even though there are no forces in the underlying model. The entropic force $F(x)$ attracting the pin to the wall is counterbalanced by an effective temperature. The force is "effective": it is not generated by springs, fields or charges. Likewise, the temperature does not correspond to the kinetic energy of the particles. The forces and the temperature are created by integrating out other particles. (In $d = 1$, we must imagine the pin as able to hop over its neighbors, and not being limited in a fixed order.)

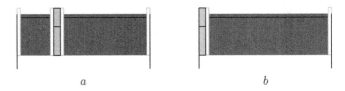

*a*                               *b*

**Fig. 6.6** Two configurations of a clothes-pin on a line, with "halos" drawn in white, and the accessible region in dark gray.

In drawing configurations of clothes-pins, it is useful to distinguish between excluded regions of two types. The first is the "core", the space occupied by the pin itself. In addition, there is a second type of excluded region, drawn in Fig. 6.6 as a white "halo". The center of another pin can penetrate into neither the core nor the halo. The total added width of all pins is fixed, but the total added length of the halos is not: in configuration $a$ in Fig. 6.6, the halos add up to $4\sigma$, whereas, in $b$, there is only $2\sigma$ of white space (we also draw halos at the boundaries). It follows that configuration $b$ has more accessible space for other particles than has $a$, and should have a higher probability: the pin is attracted to the boundary. Analogously, the halo picture explains why two pins prefer to be close to each other (their halos overlap and leave more space for other particles).

We begin to see that hard-core objects quite generally have a halo around them. This excluded region of the second type depends on the configuration, and this gives rise to the interesting behavior of spheres, disks, pins, etc. The simple halo picture nicely explains the density increase near the boundaries, but fails to capture the intricate oscillations in $\pi(x)$ (see the plot of $\pi(x)$ in Fig. 6.3). Halos are, however, a priceless

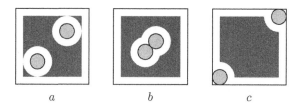

**Fig. 6.7** Configurations of two disks in a box, with "halos". The region accessible to other disks is shown in dark gray.

guide to our intuition in more complex situations, whenever we need qualitative arguments because there is no exact solution.

We now consider two-dimensional disks and spheres of same radius: the halo forms a ring of width $\sigma$ around a disk, which itself has a radius $\sigma$. In Fig. 6.7, we consider three different configurations of two disks. Evidently, the configuration $c$ has a larger accessible area than has $b$, but configuration $a$ has the smallest of them all. We thus expect the density to be higher in the corners than in the center of the box, as already shown by our earlier simulations of hard disks in a box (in Subsection 2.1.3).

Particles in the corners or pairs of particles in close contact gain space for other particles, but this space is quickly used up: unless the density is really high, and space very precious, the effects described so far only lead to local modulations of densities and correlation functions, not to phase transitions.

Mixtures of particles of different size and, more specifically, mixtures of small and large particles behave differently. To get a first rough idea, we consider two large disks, with radius $\sigma_1$, surrounded by a large number of small ones, of radius $\sigma_s$ (see Fig. 6.8).

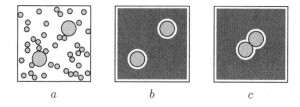

**Fig. 6.8** Large disks in a box, surrounded by many small disks ($a$), and effective description in terms of regions excluded for small disks ($b$, $c$).

**Fig. 6.9** Elimination of accessible areas (in dark gray) for a pair of disks (*left*) and for squares (*right*).

We notice that now the halo around a large disk forms a ring of width $\sigma_s$. The halo looks smaller than before, but is in fact much larger than before, if we count how many small particles can be placed in it.

We now compute (in the limit $\sigma_s/\sigma_1 \to 0$) how much halo is lost if two disks touch (the dark gray area in Fig. 6.9):

$$\cos \phi = \frac{\sigma_1}{\sigma_1 + \sigma_s} \simeq 1 - \frac{\sigma_s}{\sigma_1},$$

which implies that $\phi \propto \sqrt{\sigma_s/\sigma_1}$. We easily check that the area of the dark gray disk segments behaves like

$$\left\{ \begin{array}{c} \text{lost area} \\ \text{(disk)} \end{array} \right\} \propto \sigma_1^2 \phi^3 \propto \sigma_1^2 \left( \frac{\sigma_s}{\sigma_1} \right)^{3/2} = \sqrt{\frac{\sigma_1}{\sigma_s}} \sigma_s^2.$$

As the area per small disk is $\propto \sigma_s^2$, the lost halo area corresponds to the space occupied by $\propto \sqrt{\sigma_1/\sigma_s}$ small particles. This number (slowly) goes to infinity as $\sigma_s/\sigma_1 \to 0$. We can redo this calculation for squares, rather than disks (see Fig. 6.9 again), with the result:

$$\left\{ \begin{array}{c} \text{lost area} \\ \text{(squares)} \end{array} \right\} \propto \sigma_1 \sigma_s = \frac{\sigma_1}{\sigma_s} \sigma_s^2,$$

This lost halo area corresponds to many more small particles than in the case of binary disks, and we can expect the depletion effects to be much stronger in binary mixtures of squares than in disks.

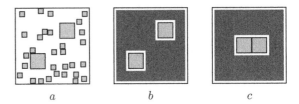

$a$            $b$            $c$

**Fig. 6.10** Large and small squares ($a$), and regions that are excluded for small squares ($b, c$). Depletion effects are stronger than for disks.

Asakura and Oosawa (1954) first described the effective interaction between hard-sphere particles, and pointed out its possible relevance to biological systems, macromolecules, and proteins. Their theoretical description was on the level of our halo analysis. In the clothes-pin model, we have already seen that the true interaction between two particles is oscillatory, and richer than what the simple halo model suggests. But there is complexity from another angle: even if we computed the exact effective interaction potential of the two large particles in the boxes in Figs 6.8 and 6.5, we would fail to be able to describe mixtures of large particles in a sea of small ones. The probability of having three or more particles at positions $\{\mathbf{x}_1, \mathbf{x}_2, \mathbf{x}_3, \dots\}$ is not given by products of pair-interaction terms. There is some benefit in keeping large and small particles all through our theoretical description, as modern computational tools allow us to do.

### 6.1.3 Binary mixtures

We have come to understand that hard particles—pins, disks, spheres, etc.—experience a curious depletion interaction because configurations of a few particles which leave a lot of space for the rest have a higher probability of appearing than others. This causes two hard particles to strongly attract each other at small distance, then repel each other more weakly as the distance between them becomes a little larger, then attract again, etc.

As the large particles in a mixture of large and small ones have a certain propensity for each other, it is legitimate to ask whether they do not simply *all* get closer together in part of the simulation box (together with a few small ones), leaving the system phase-separated—demixed into one phase rich in large particles and one rich in small particles. In nature, phase separation, driven by energy gains at the expense of entropy, is a very common phenomenon: gravitationally interacting particles underwent (incomplete) phase separation when forming stars and planets (and leaving empty space) in the early universe; likewise, water below the triple point phase-separates into a liquid phase (rich in water) and a vapor phase (rich in air). In the case of hard particles, the effect that we shall seek (and find) is special, and clearly paradoxical, as the particles, at the expense of entropy, attempt to gain interaction energy, which does not really exist, because it is only an effective quantity for particles which move freely in space.

For ease of visualization, we discuss phase separation in two-dimensional mixtures, where the effective interaction is stronger than in one dimension, but less pronounced than in the three-dimensional case. To directly witness the phase separation transition with a basic simulation, we shall eventually have to enhance the interaction: we consider large and small squares, rather than disks. As discussed, the depletion effects are stronger in squares than in disks.

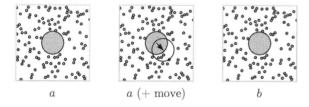

$a$       $a$ (+ move)       $b$

**Fig. 6.11** "Pope in the crowd" effect, which slows down the local Monte Carlo motion of a large particle surrounded by many small ones.

Exact solutions for hard-sphere systems do not exist in more than one dimension, and direct-sampling algorithms are unknown. We thus turn to Markov-chain Monte Carlo simulation for models of large and small particles at comparable, not necessarily high, densities. However, because of the difference in size, there are many more small particles than large ones. The local Monte Carlo approach then encounters a curious difficulty: virtually all attempted moves of large particles lead to overlaps with one of the many small particles and must be rejected (see Fig. 6.11). This "pope in the crowd" effect immobilizes the large particles in the midst of the small ones, even though the densities are quite low. This problem arises because the configurations are highly constrained, not because they are densely packed.

In this situation (many constraints, and rather low density), the rejection-free pivot cluster algorithm of Subsection 2.5.2 is the method of choice. This algorithm uses a transformation that maps the simulation box onto itself and, when applied twice to a particle, returns it to its original position. With periodic boundary conditions, reflections with respect to any line parallel to the coordinate axes and also any point-reflection can be used. In a square box, reflections with respect to a diagonal are also possible.

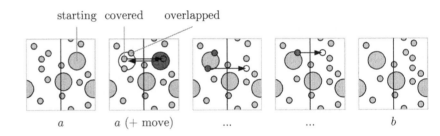

**Fig. 6.12** Pivot cluster algorithm for binary disks.

The transformation, when applied to a particle, provisionally breaks the hard-core rule (see Fig. 6.12). A move starts with the choice of the transformation and an arbitrary starting particle, in an initial configuration $a$. Any particle (large or small) that has generated an overlap must eventually be moved, possibly generating new overlaps, etc. The process is guaranteed to terminate with a legal configuration (as in the final configuration $b$ in Fig. 6.12). It satisfies detailed balance, as was discussed in Section 2.5.

In this algorithm, two disks overlap if the periodic distance between their centers is smaller than the sum of their radii (this condition is checked using Alg. 2.6 (diff-vec)). Algorithm 6.3 (pocket-binary) provides a simple implementation in a few dozen lines of code (the set $\mathcal{P}$, the pocket, contains the dark particles in Fig. 6.12, which still have to be moved). Grid/cell schemes (see Subsection 2.4.1) can speed up the check for overlap, but we do not need them for our modest system sizes. Another highly efficient speedup is implemented in Alg. 6.3 (pocket-binary): disks that are fully covered by another disk are immediately moved, without passing through the pocket (see Fig. 6.12). These fully covered disks drop out of the consideration because they cannot generate additional moves.

```
procedure pocket-binary
input {x₁,...,xₙ}
k ← nran (1, N)
𝒫 ← {k}
𝒜 ← {1,...,N} \ {k}
while (𝒫 ≠ {}) do
⎰ i ← any element of 𝒫
⎥ xᵢ ← T(xᵢ)
⎥ for ∀ j ∈ 𝒜 do
⎥   ⎰ if (i covers j) then
⎥   ⎥   ⎰ 𝒜 ← 𝒜 \ {j}
⎥   ⎥   ⎱ xⱼ ← T(xⱼ)
⎥   ⎥ else if (i overlaps j) then
⎥   ⎥   ⎰ 𝒜 ← 𝒜 \ {j}
⎥   ⎱   ⎱ 𝒫 ← 𝒫 ∪ {j}
⎱ 𝒫 ← 𝒫 \ {i}
output {x₁,...,xₙ}
```

**Algorithm 6.3** pocket-binary. Pocket–cluster algorithm for $N$ particles. $T$ is a random symmetry transformation of the simulation box.

The choice of initial configuration plays no role, as in any Markov-chain algorithm (see the discussion of the initial condition at the club-house, in Subsection 1.1.2): we start either with the last configuration of a previous simulation with identical parameters or with a configuration whose only merit is that it is legal. An initial configuration can have the small and large particles in separate halves of the simulation box. Configurations obtained after a few thousand pivots are shown in Fig. 6.13. For this system of disks, we notice no qualitative change in behavior as a function of density.

To directly witness the phase separation transition with a basic simulation, we now consider large and small squares, rather than disks. As discussed, the depletion effects are stronger in squares than in disks. A system of two-dimensional squares contains the essential ingredients of entropic mixtures. We consider a square box with periodic boundary conditions, as in Fig. 6.14, with $N_l$ large squares and $N_s$ small ones. It is for simplicity only that we suppose that the squares are oriented, that is, they cannot turn. Strikingly, we see a transition between the uniform configuration at low densities and the phase-separated configuration at high density.

Experiments usually involve a solvent, in addition to the small and large particles, and the phase transition manifests itself because the big clusters come out of solution (they flocculate) and can be found either at the bottom of the test tube or on its surface (Dinsmore, Yodh, and Pine 1995). Our naive two-dimensional simulation lets us predict a transition for hard squares between the densities $\eta_l = \eta_s = 0.18$ and $\eta_l = \eta_s = 0.26$ for a size ratio of 20, but this precise experiment is not likely to be

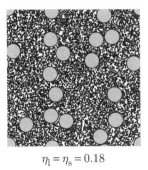
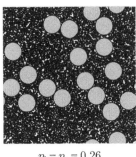

$\eta_l = \eta_s = 0.18$          $\eta_l = \eta_s = 0.26$

**Fig. 6.13** Large and small disks in a square box with periodic boundary conditions ($N_l = 18$, $N_s = 7200$, size ratio $\sigma_s/\sigma_l = 1/20$).

done anytime soon (Buhot and Krauth 1999). Direct comparison with experiment is easier in three dimensions, and with spheres rather than cubes.

Mixtures, in particular, binary mixtures, form an active field of experimental and theoretical research, with important technological and industrial applications. Most of this research is concentrated on three dimensions, where the depletion interaction is ubiquitous. We do not have to play tricks (such as study cubes rather than spheres) in order to increase the contact area and strengthen the effective interaction between large particles. For serious work (trying to go to the largest systems possible and scanning many densities) we would be well advised to speed up the calculation of overlaps with the grid/cell scheme of Section 2.4.1. All this has been done already, and theory and experiment agree beautifully.

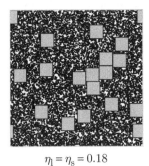

$\eta_l = \eta_s = 0.18$          $\eta_l = \eta_s = 0.26$

**Fig. 6.14** Large and small oriented squares in a square box with periodic boundary conditions ($N_l = 18$, $N_s = 7200$, size ratio $\sigma_s/\sigma_l = 1/20$).

# 6.2 Entropic lattice model: dimers

The dimer model that we discuss in the present section is the mother of all entropic systems in which orientation plays a role. It appears in solid state physics, in soft condensed matter (polymers and liquid crystals), and in pure mathematics. Its earliest incarnation was in the chemistry of the benzene molecule ($C_6H_6$), where six carbon atoms form a ring with electron orbitals sticking out. These orbitals tend to dimerize, that is, form orbitals which are common to two neighboring carbon atoms (see Fig. 6.15). Pauling first realized that there are two equivalent lowest-energy configurations, and proposed that the molecule was neither really in the one configuration nor in the other, that it in fact resonates between the two configurations. This system is entropic in the same sense as the hard spheres of Chapter 2, because the two dimer configurations have the same energy.

**Fig. 6.15** Equal-energy conformations of the benzene molecule (*left*), and a complete dimer configuration on a $4 \times 4$-checkerboard (*right*).

In the present section, we shall work on lattices where dimerized electronic orbitals cover the whole system. Again, all configurations appear with the same probability. In the closely related Ising model of Chapter 5, we were interested in studying a symmetry-breaking phase transition of spins: at high temperature, the typical configurations of the Ising model have the same proportions of up spins and down spins, whereas in the ferromagnetic phase, at low temperature, there is a majority of spins in one direction. The analogous symmetry-breaking transition for dimers concerns their orientation. One could imagine that dimers in a typical large configuration would be preferentially oriented in one direction (horizontal or vertical). However, we shall see in Subsection 6.2.5 that this transition never takes place, whatever the underlying lattice.

## 6.2.1 Basic enumeration

We first study dimers by enumeration. Our first dimer enumeration program works on bipartite lattices, which can be partitioned into two classes (such as the gray and white squares of the checkerboard in Fig. 6.15) such that each dimer touches both classes. We are interested both in complete coverings (with $M = N/2$ dimers on $N$ lattice sites) and in monomer–dimer coverings, where the dimers do not cover all the sites of the lattice and the number of configurations is usually much larger. We may occupy each dark square either with a monomer or with one extremity of a dimer, and check out all the possible directions of a dimer (or monomer) starting from that site (see Fig. 6.16).

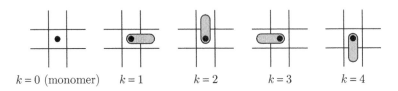

$k = 0$ (monomer)   $k = 1$   $k = 2$   $k = 3$   $k = 4$

**Fig. 6.16** The five possible directions of a monomer or dimer starting on a given site of a two-dimensional square lattice.

While going through all possible directions for all of the sites of one sublattice, one simply needs to reject all arrangements that have overlaps or violate the boundary conditions. An initial algorithm to enumerate dimers usually produces code such as the awkward Alg. 6.4 (naive-dimer): there as many nested loops as there are dimers to place. This program would have to nest loops deeper and deeper as the lattice gets larger, and we would have to type it anew for any other lattice size! (Taking this to extremes, we might generate the computer program with the appropriate depth of nested loops by another program ....) Nevertheless, Alg. 6.4 (naive-dimer) does indeed successfully generate the number of inequivalent ways to place $M$ dimers on a lattice of $N$ sites (see Table 6.2). Some of the legal configurations generated on a $4 \times 4$ lattice without periodic boundary conditions are shown in Fig. 6.17; the sites $\{1, \ldots, 8\}$ are numbered row-wise from lower left to upper right, as usual.

**procedure naive-dimer**
**for** $i_1 = 1, \ldots, 4$ **do**
$\left\{\begin{array}{l} \textbf{for } i_2 = 1, \ldots, 4 \textbf{ do} \\ \left\{\begin{array}{l} \textbf{for } i_3 = 1, \ldots, 4 \textbf{ do} \\ \left\{\begin{array}{l} \ldots \\ \left\{\begin{array}{l} \textbf{for } i_7 = 1, \ldots, 4 \textbf{ do} \\ \left\{\begin{array}{l} \textbf{for } i_8 = 1, \ldots, 4 \textbf{ do} \\ \left\{\begin{array}{l} \textbf{if } (\{i_1, \ldots, i_8\} \textbf{ legal}) \textbf{ then} \\ \quad \{ \textbf{ output } \{i_1, \ldots, i_8\} \end{array}\right. \end{array}\right. \end{array}\right. \end{array}\right. \end{array}\right.$

**Table 6.2** Number of dimer configurations for a $4 \times 4$ square lattice (from Alg. 6.4 (naive-dimer))

| # of dimers | # of configurations Periodic boundary | |
|---|---|---|
| | With | Without |
| 0 | 1 | 1 |
| 1 | 32 | 24 |
| 2 | 400 | 224 |
| 3 | 2496 | 1044 |
| 4 | 8256 | 2593 |
| 5 | 14208 | 3388 |
| 6 | 11648 | 2150 |
| 7 | 3712 | 552 |
| 8 | 272 | 36 |

**Algorithm 6.4 naive-dimer.** Enumeration of dimer configurations with an awkward loop structure (see Alg. 6.5 (naive-dimer(patch))).

Evidently, Alg. 6.4 (naive-dimer) merely enumerates all numbers from 11 111 111 to 44 444 444, with digits from 1 to 4 (for the problem of monomers and dimers, we need to count from 00 000 000 to 44 444 444, with digits from 0 to 4). Let us consider, for a moment, how we usually count in the decimal system. Counting from $i$ to $i + 1$, i.e. adding one

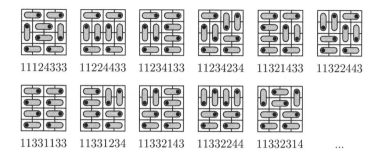

11124333    11224433    11234133    11234234    11321433    11322443

11331133    11331234    11332143    11332244    11332314    ...

**Fig. 6.17** Complete dimer coverings on a $4 \times 4$ square lattice without periodic boundary conditions (from Alg. 6.4 (**naive-dimer**); the values of $\{i_1, \ldots, i_8\}$ are shown).

to the number $i$, is done as we can explain for $i = 4999$:

$$
\begin{array}{rl}
4999 & \leftarrow i \\
+ \quad 1 & \\
\hline
5000 & \leftarrow i + 1.
\end{array}
$$

Here, we checked the least significant digits of $i$ (4999) (on the right) and set any 9 to 0 until a digit different from 9 could be incremented. This basic counting algorithm—the beginning of mathematics in first grade—can be adapted for running through $\{i_1, \ldots, i_8\}$ with digits from 1 to 4 (see Alg. 6.5 (**naive-dimer(patch)**)). It avoids an awkward loop structure and is easily adapted to other lattices.

Both the awkward and the patched enumeration program must test whether $\{i_1, \ldots, i_8\}$ is a legal dimer configuration. Violations of boundary conditions are trivially detected. To test for overlaps, we use an occupation vector $\{o_1, \ldots, o_N\}$, which, before testing, is set to $\{0, \ldots, 0\}$. A dimer touching sites $l$ and $m$ has the occupation variables $o_l$ and $o_m$ incremented. An occupation number in excess of one signals a violation of the hard-core condition.

Algorithm 6.5 (**naive-dimer(patch)**) implements base-4 or base-5 counting (base-10 is the decimal system), with $i_{\max} = 4$. With $i_{\min} = 1$, dimer configurations are generated, whereas $i_{\min} = 0$ allows us to enumerate monomer–dimer configurations. Remarkably, the algorithm leaves out no configuration, and generates no duplicates. We could spruce up Alg. 6.5 (**naive-dimer(patch)**) into a mixed-base enumeration routine with site-dependent values of $i_{\min}$ or $i_{\max}$. Then, the corner sites would go through only two values, other boundary sites would go through three values, and only inner sites would go from $i = 1$ to $i = 4$. Boundary conditions would thus be implemented in a natural way. However, the performance gain would be minor, and it is better to move straight ahead to more powerful tree-based enumerations.

```
procedure naive-dimer(patch)
{i₁,...,i₈} ← {i_min,...,i_min}
if ({i₁,...,i₈} legal) then
   { output {i₁,...,i₈}
   for n = 1,2... do
      ⎧ for k = 8,7,... do
      ⎪    ⎧ if (k = 0) stop (terminate program)
      ⎪    ⎪ if (i_k = i_max) then
      ⎪    ⎪    { i_k ← i_min
      ⎪    ⎨ else
      ⎪    ⎪    ⎧ i_k ← i_k + 1
      ⎪    ⎪    ⎨ if ({i₁,...,i₈} legal) then
      ⎨    ⎩    ⎩    { output {i₁,...,i₈}
      ⎪          goto 1
      ⎩ 1 continue
```

**Algorithm 6.5** naive-dimer(patch). Enumeration of numbers (digits from $i_{min}$ to $i_{max}$) and outputting of legal dimer configurations.

## 6.2.2   Breadth-first and depth-first enumeration

Algorithm 6.4 (naive-dimer), like computer enumeration in general, is more reliable than a pencil-and-paper approach, and of great use for testing Monte Carlo programs. Often, we even write discretized versions of programs with the sole aim of obtaining enumeration methods to help debug them. For testing purposes, it is a good strategy to stick to naive approaches. Nevertheless, the high rejection rate of Alg. 6.4 (naive-dimer) is due to a design error: the program goes through numbers, instead of running over dimer configurations. The present subsec-

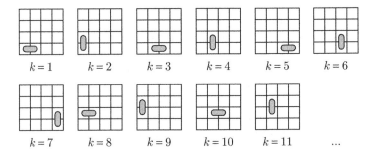

**Fig. 6.18** Single-dimer configurations on a $4 \times 4$ square lattice without periodic boundary conditions.

tion corrects this error and discusses direct enumeration methods for dimers (not numbers). These methods are as easy to implement as the naive approaches, and they are widely applicable. We start with single dimers, listed in Fig. 6.18 in a reasonable but arbitrary order.

We can say that the configuration $k = 3$ in Fig. 6.18 is smaller than the configuration $k = 7$, simply because it is generated first. The single-dimer ordering of Fig. 6.18 induces a lexicographic order of many-dimer configurations, analogous to the ordering of words that is induced by the alphabetical ordering of letters. There is one difference: within a dimer configuration, the ordering of the dimers is meaningless. We may arrange them in order of increasing $k$: the configuration on the left of Fig. 6.19 is thus called "1–5–9–10–14", rather than "5–1–9–10–14". (The problem arises from the indistinguishability of dimers, and has appeared already in the discussion of bosonic statistics in Section 4.1.2.) The configuration 1–5–9–10–14 is lexicographically larger than 1–5–9–11, on the right in Fig. 6.19.

Lexicographic order induces a tree structure among configurations (see Fig. 6.20). At the base of this tree, on its left, is the empty configuration (node 0, zero dimers), followed by the single-dimer configurations (nodes $a, b$, with $a < b$), two-dimer configurations (nodes $c < d < e < f$), etc.

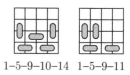

1–5–9–10–14  1–5–9–11

**Fig. 6.19** Ordering of dimer configurations in terms of the numbering scheme of Fig. 6.18.

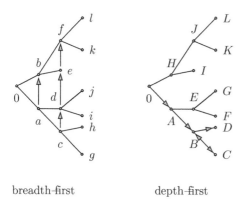

breadth-first          depth-first

**Fig. 6.20** A tree, and two different strategies for visiting all its nodes, starting from the root, 0.

A straightforward, "breadth-first", strategy for visiting (enumerating) all the nodes of a tree is implemented in Alg. 6.6 (**breadth-dimer**), with no more effort than for the naive routines at the beginning of this section.

Algorithm 6.6 (**breadth-dimer**) works on arbitrary lattices and is simplest if we can store the configurations on disk. An input file containing $n$-dimer configurations (such as $\{c, d, e, f\}$ in Fig. 6.20) yields an ordered output of $(n + 1)$-dimer configurations (such as $\{g, h, i, j, k, l\}$), which can be fed back into the program. Again, it is a great virtue of the method that configurations are never left out, nor are they produced twice.

For larger systems, we may still be able to go through all configurations but lack the time or space to write them out to a file. It is then preferable

**procedure** breadth-dimer
**input** $\{k_1, \ldots, k_n\}$
**for** $k = k_n + 1, \ldots, k_{\max}$ **do**
$\left\{ \begin{array}{l} \text{if } (\{k_1, \ldots, k_n, k\} \text{ legal}) \text{ then} \\ \quad \{ \text{ output } \{k_1, \ldots, k_n, k\} \end{array} \right.$

**Algorithm 6.6** breadth-dimer. Breadth-first enumeration. $k_{\max}$ is the number of single-dimer configurations.

**Table 6.3** Number of dimer configurations on a $6 \times 6$ square lattice without periodic boundary conditions, from Alg. 6.7 (depth-dimer)

| # of dimers | # of configurations |
|---|---|
| 0 | 1 |
| 1 | 60 |
| 2 | 1622 |
| 3 | 26 172 |
| 4 | 281 514 |
| 5 | 2 135 356 |
| 6 | 11 785 382 |
| 7 | 48 145 820 |
| 8 | 146 702 793 |
| 9 | 333 518 324 |
| 10 | 562 203 148 |
| 11 | 693 650 988 |
| 12 | 613 605 045 |
| 13 | 377 446 076 |
| 14 | 154 396 898 |
| 15 | 39 277 112 |
| 16 | 5 580 152 |
| 17 | 363 536 |
| 18 | 6728 |

to enumerate the nodes of a tree in depth-first fashion, as indicated on the right of Fig. 6.20. We go as deep into the tree as we can, visiting nodes in the order $A \to B \to C \ldots$, and backtracking in the order $C \to B \to D$ or $D \to \cdots \to A \to E$ if stuck. Algorithm 6.7 (depth-dimer) implements this strategy with the help of three subroutines, which attempt to put, move, or delete a dimer:

$$\text{put-dimer}: \{k_1, \ldots, k_n\} \to \{k_1, \ldots, k_n, k\} \text{ with next } k > k_n,$$
$$\text{move-dimer}: \{k_1, \ldots, k_n\} \to \{k_1, \ldots, k_{n-1}, k\} \text{ with next } k > k_n,$$
$$\text{delete-dimer}: \{k_1, \ldots, k_n\} \to \{k_1, \ldots, k_{n-1}\}.$$

**procedure** depth-dimer
$\mathcal{D} \leftarrow \{1\}$ (set of dimers)
next $\leftarrow$ "put"
**while** $(\mathcal{D} \neq \{\})$ **do**
$\left\{ \begin{array}{l} \text{if } (\text{next} = \text{"put"}) \text{ then} \\ \quad \left\{ \begin{array}{l} \text{call put-dimer}\,(\Upsilon) \\ \text{if } (\Upsilon = 1) \; \{\text{next} \leftarrow \text{"put"}\} \text{ else } \{\text{next} \leftarrow \text{"move"}\} \end{array} \right. \\ \text{else if } (\text{next} = \text{"move"}) \text{ then} \\ \quad \left\{ \begin{array}{l} \text{call move-dimer}\,(\Upsilon) \\ \text{if } (\Upsilon = 1) \; \{\text{next} \leftarrow \text{"put"}\} \text{ else } \{\text{next} \leftarrow \text{"delete"}\} \end{array} \right. \\ \text{else if } (\text{next} = \text{"delete"}) \text{ then} \\ \quad \left\{ \begin{array}{l} \text{call delete-dimer}\,(\Upsilon) \\ \text{next} \leftarrow \text{"move"} \end{array} \right. \end{array} \right.$

**Algorithm 6.7** depth-dimer. Depth-first enumeration. The flag $\Upsilon = 1$ signals a successful "put" or "move" (the set $\mathcal{D}$ contains the dimers).

The output of Alg. 6.7 (depth-dimer) for a $4 \times 4$ square lattice without periodic boundary conditions is shown in Fig. 6.21. The configurations $i = 1, \ldots, 8, 12, 13, 16$ result from a "put", those at iterations $i = 10, 17$ result from a pure "move", and $i = 9, 11, 14, 15$ are generated from a "delete" (indicated in dark) followed by a "move".

Algorithm 6.7 (depth-dimer) runs through the 23 079 663 560 configurations on a $6 \times 6$ square lattice with periodic boundary conditions with astounding speed (in a few minutes), but only cognoscenti will be able to make sense of the sequence generated. During the enumeration, no

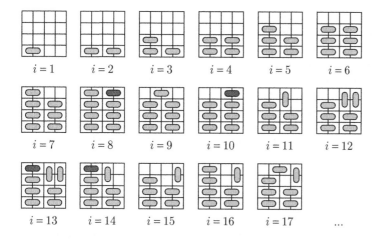

**Fig. 6.21** Dimer configurations $\mathcal{D}$ on a $4 \times 4$ square lattice without periodic boundary conditions, in the order generated by Alg. 6.7 (`depth-dimer`).

configuration is ever forgotten, none is generated twice, and we are free to pick out any information we want (see Table 6.3). We must choose the data formats carefully, as the numbers can easily exceed $2^{31}$, the largest integer that fits into a standard four-byte computer word.

We shall now analyze pair correlations of dimers, in order to study the existence of long-range order. For concreteness, we restrict our attention to the complete dimer model. In this case, there can be four different dimer orientations, and so the probability of any given orientation is $1/4$. If the dimer orientations on far distant sites are independent of each other, then each of the 16 possible pairs of orientations should appear with the same probability, $1/16$. Data for pair correlations of dimers on a $6 \times 6$ square lattice are shown in Table 6.3. There, we cannot go to distances larger than $\{\Delta_x, \Delta_y\} = \{3, 3\}$; we see, however, that the configuration $c$, with periodic boundary conditions, has a probability $5936/90\,176 = 0.0658$. This is very close to $1/16 = 0.0625$. On the basis of this very preliminary result, we can conjecture (correctly) that the dimer model on a square lattice is not in an ordered phase. Orientations of dimers on far distant sites are independent of each other.

**Table 6.4** Number of dimer configurations on a $6 \times 6$ square lattice, with dimers prescribed as in Fig. 6.22. The total number of complete coverings is $90\,176$ with periodic boundary conditions and $6728$ without.

| Case | # of configurations | |
|------|:--:|:--:|
| | Periodic boundary | |
| | With | Without |
| $a$ | 4888 | 242 |
| $b$ | 6184 | 1102 |
| $c$ | 5936 | 520 |
| $d$ | 6904 | 1034 |
| $e$ | 5800 | 1102 |
| $f$ | 5472 | 640 |

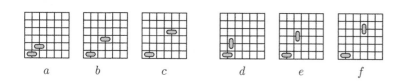

**Fig. 6.22** Pair correlation functions (the probability that two dimers are in the positions shown) for a $6 \times 6$ square lattice.

In the present subsection, we have approached the double limit of enumeration (in the sense of Subsection 5.1.4): not only is it impossible to generate a list of all dimer configurations on a large lattice (because of its sheer size), but we cannot even compute the size of the list without actually producing it. Enumerations of complete dimer configurations, the subject of Subsection 6.2.3, are the only exception to this rule: the configurations can be counted but not listed for very large system sizes, by methods related to the indirect enumeration methods for the Ising model. The genuine difficulty of monomer–dimer enumerations away from complete filling was first pointed out by Valiant (1979).

### 6.2.3   Pfaffian dimer enumerations

In this subsection, we enumerate complete dimer configurations on a square lattice by indirect methods, which are closely related to the determinantal enumerations used for the two-dimensional Ising model. To simplify matters, we leave aside the issue of periodic boundary conditions, even though they are easily included. For the same reason, we remain with small systems, notwithstanding the fact that the Pfaffian approach allows us to go to very large systems and count configurations, compute correlation functions, etc.

Obtaining the number of complete dimer configurations involves computing the partition function

$$Z = \sum_{\substack{\text{dimer} \\ \text{configurations}}} 1. \tag{6.10}$$

What constitutes a legal dimer configuration may be encoded in a matrix $A^+ = (a_{kl}^+)$ with indices $\{k, l\}$ running over the sites $\{1, \ldots, N\}$. On a $4 \times 4$ square lattice, $A^+$ is given by

$$A^+ = \begin{bmatrix} \cdot & + & \cdot & \cdot & + & \cdot & \cdot & \cdot & \cdot & \cdot & \cdot & \cdot & \cdot & \cdot & \cdot & \cdot \\ + & \cdot & + & \cdot & \cdot & + & \cdot & \cdot & \cdot & \cdot & \cdot & \cdot & \cdot & \cdot & \cdot & \cdot \\ \cdot & + & \cdot & + & \cdot & \cdot & + & \cdot & \cdot & \cdot & \cdot & \cdot & \cdot & \cdot & \cdot & \cdot \\ + & \cdot & + & \cdot & \cdot & \cdot & \cdot & + & \cdot & \cdot & \cdot & \cdot & \cdot & \cdot & \cdot & \cdot \\ \cdot & + & \cdot & \cdot & + & \cdot & + & \cdot & \cdot & + & \cdot & \cdot & \cdot & \cdot & \cdot & \cdot \\ \cdot & \cdot & + & \cdot & + & \cdot & + & \cdot & \cdot & \cdot & + & \cdot & \cdot & \cdot & \cdot & \cdot \\ \cdot & \cdot & \cdot & + & \cdot & + & \cdot & + & \cdot & \cdot & \cdot & + & \cdot & \cdot & \cdot & \cdot \\ \cdot & \cdot & \cdot & + & \cdot & + & \cdot & \cdot & \cdot & \cdot & + & \cdot & + & \cdot & \cdot & \cdot \\ \cdot & \cdot & \cdot & \cdot & + & \cdot & + & \cdot & \cdot & + & \cdot & \cdot & + & \cdot & \cdot & \cdot \\ \cdot & \cdot & \cdot & \cdot & \cdot & + & \cdot & + & \cdot & + & \cdot & \cdot & \cdot & + & \cdot & \cdot \\ \cdot & \cdot & \cdot & \cdot & \cdot & \cdot & + & \cdot & + & \cdot & + & \cdot & \cdot & \cdot & + & \cdot \\ \cdot & \cdot & \cdot & \cdot & \cdot & \cdot & \cdot & + & \cdot & + & \cdot & \cdot & \cdot & \cdot & \cdot & + \\ \cdot & \cdot & \cdot & \cdot & \cdot & \cdot & \cdot & \cdot & + & \cdot & \cdot & \cdot & \cdot & + & \cdot & \cdot \\ \cdot & \cdot & \cdot & \cdot & \cdot & \cdot & \cdot & \cdot & \cdot & + & \cdot & \cdot & + & \cdot & + & \cdot \\ \cdot & \cdot & \cdot & \cdot & \cdot & \cdot & \cdot & \cdot & \cdot & \cdot & + & \cdot & \cdot & + & \cdot & + \\ \cdot & \cdot & \cdot & \cdot & \cdot & \cdot & \cdot & \cdot & \cdot & \cdot & \cdot & + & \cdot & \cdot & + & \cdot \end{bmatrix}. \tag{6.11}$$

The elements of this matrix which are marked by a "$\cdot$" are zero and those marked by a "$+$" are equal to one. For example, the elements $a_{12}^+ = a_{21}^+ = 1$ signal the edge between sites 1 and 2, in our standard numbering scheme shown again in Fig. 6.23.

Complete (fully packed) dimer configurations correspond to permutations, as in

$$P = \begin{pmatrix} \overbrace{P_1 \quad P_2}^{\text{first dimer}} & \overbrace{P_3 \quad P_4}^{\text{second dimer}} & \cdots & \overbrace{P_{N-1} \quad P_N}^{\text{last dimer}} \\ 1 \quad 2 & 3 \quad 4 & \cdots & N-1 \quad N \end{pmatrix}.$$

| 13 | 14 | 15 | 16 |
| 9  | 10 | 11 | 12 |
| 5  | 6  | 7  | 8  |
| 1  | 2  | 3  | 4  |

**Fig. 6.23** Numbering scheme for the $4 \times 4$ square lattice.

This simply means that the first dimer lives on sites $\{P_1, P_2\}$, the second on $\{P_3, P_4\}$, etc. (permutations are written "bottom-up" $\left(\left(\begin{smallmatrix} P_1 & \cdots & P_N \\ 1 & \cdots & N \end{smallmatrix}\right)\right)$, rather than "top-down"; see Subsection 1.2.2). All dimer configurations are permutations, but not all permutations are dimer configurations: we need to indicate whether a permutation is compatible with the lattice structure. Using the matrix $A^+$ in eqn (6.11), we arrive at

$$ Z = \frac{1}{(N/2)!2^{N/2}} \sum_{\text{permutations } P} a^+_{P_1 P_2} a^+_{P_3 P_4} \cdots a^+_{P_{N-1} P_N}. \tag{6.12} $$

The combinatorial factor in this equation takes into account that we need to consider only a subset of permutations, the matchings $M$, where the lattice sites of each dimer are ordered (so that $P_1 < P_2$, $P_3 < P_4$, etc.), and where, in addition, dimers are ordered within each configuration, with $P_1 < P_3 < \cdots < P_{N-1}$. We can also write the partition function as

$$ Z = \sum_{\text{matchings } M} \underbrace{a^+_{P_1 P_2} a^+_{P_3 P_4} \cdots a^+_{P_{N-1} P_N}}_{\substack{\text{weight of matching:} \\ \text{product of elements of } A^+}}. \tag{6.13} $$

By construction, the weight of a matching is one if it is a dimer configuration, and otherwise zero.

On a $4 \times 4$ square lattice, there are $16! = 20\,922\,789\,888\,000$ permutations, and $16!/8!/2^8 = 2\,027\,025$ matchings, of which 36 contribute to $Z$ (without periodic boundary conditions), as we know from earlier enumerations. However, the straight sum in eqn (6.13) cannot be computed any more efficiently than through enumeration, so the representation in eqn (6.13) is not helpful.

To make progress, a sign must be introduced for the matching. This leads us to the Pfaffian,[1] defined for any antisymmetric matrix $A$ of even order $N$ by

$$ \text{Pf}\, A = \sum_{\text{matchings } M} \text{sign}(M) \underbrace{a_{P_1 P_2} a_{P_3 P_4} \cdots a_{P_{N-1} P_N}}_{N/2 \text{ terms}}. \tag{6.14} $$

The sign of the matching is given by the corresponding permutation $P = \left(\begin{smallmatrix} P_1 & \cdots & P_N \\ 1 & \cdots & N \end{smallmatrix}\right)$, and even though many permutations give the same matching, the product of the sign and the weight is always the same. Each permutation can be written as a cycle configuration. We remember that a permutation of $N$ elements with $n$ cycles has a sign $(-1)^{n+N}$, as was discussed in Subsection 1.2.2.

Pfaffians can be computed in $\text{O}(N^3)$ operations, much faster than straight sums over matchings, but the Pfaffian of $A^+$ is certainly different from the quantity defined in eqn (6.13). Therefore, the challenge in the present subsection is to find a matrix $A$ whose Pfaffian gives back the sum over matchings of $A^+$, and to understand how to evaluate $\text{Pf}\, A$. We

---

[1] J. F. Pfaff (1765–1825), was a professor in Halle, Saxony, and a teacher of C. F. Gauss.

must first understand the relationship between the Pfaffian in eqn (6.14) and the determinant

$$
\det A = \sum_{\substack{\text{permutations}}} \text{sign}(P) \underbrace{a_{1P_1}, a_{2P_2} \ldots a_{NP_N}}_{N \text{ terms}}
$$

$$
= \sum_{\substack{\text{cycle configs}}} (-1)^n \underbrace{(a_{P_1 P_2} a_{P_2 P_3} a_{P_3 P_4} a_{P_4 P_5} \cdots \cdot a_{P_k P_1})}_{\text{cycle 1}} \cdots \underbrace{\phantom{\cdots}}_{\text{cycle } n} . \quad (6.15)
$$

We see that the Pfaffian is a sum of products of $N/2$ terms, and the determinant a sum of products of $N$ terms. (Terms in the determinant that are missing in the Pfaffian are underlined in eqn (6.15).) The matchings of a second copy of $A$ or, more generally, the matchings of another matrix $B$ can be used to build an alternating cycle configuration. For concreteness, we illustrate in the case of a $4 \times 4$ matrix how two matchings (one of a matrix $A$ ($M_A$), the other of a matrix $B$ ($M_B$)) combine into an alternating cycle ($M_A \cup M_B$). (See Table 6.5; the products of the signs and weights of the matchings agree with the sign and weight of the alternating cycle.)

**Table 6.5** Two matchings giving an alternating cycle (see Fig. 6.24). The product of the signs of $M_A$ and $M_B$ equals the sign of $M_A \cup M_B$.

| Object | Sign | Weight | Dimers | Perm. | Cycle |
|---|---|---|---|---|---|
| $M_A$ | $-1$ | $a_{13} a_{24}$ | $\{1,3\}\{2,4\}$ | $\left(\begin{smallmatrix} 1 & 3 & 2 & 4 \\ 1 & 2 & 3 & 4 \end{smallmatrix}\right)$ | $(1)(2,3)(4)$ |
| $M_B$ | $1$ | $b_{12} b_{34}$ | $\{1,2\}\{3,4\}$ | $\left(\begin{smallmatrix} 1 & 2 & 3 & 4 \\ 1 & 2 & 3 & 4 \end{smallmatrix}\right)$ | $(1)(2)(3)(4)$ |
| $M_A \cup M_B$ | Product of the above | $a_{13} b_{34}$ $\times a_{42} b_{21}$ | Both of the above | $\left(\begin{smallmatrix} 3 & 1 & 4 & 2 \\ 1 & 2 & 3 & 4 \end{smallmatrix}\right)$ | $(1342)$ |

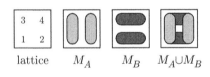

lattice    $M_A$    $M_B$    $M_A \cup M_B$

**Fig. 6.24** A $2 \times 2$ square lattice, a matching $M_A$ of a matrix $A$, a matching $M_B$ of a matrix $B$, and the alternating cycle generated from the combined matching $M_A \cup M_B$.

Generalizing from the above, we see that matchings of an antisymmetric matrix $A$ and those of another matrix $B$ combine to alternating cycle configurations:

$$
(\text{Pf } A)(\text{Pf } B) = \sum_{\substack{\text{alternating} \\ \text{cycle configs}}} (-1)^n
$$

$$
\times \underbrace{(a_{P_1 P_2} b_{P_2 P_3} a_{P_3 P_4} b_{P_4 P_5} \cdots \cdot b_{P_k P_1})}_{\text{alternating cycle 1}} \cdots \underbrace{\phantom{\cdots}}_{\text{alternating cycle } n} ,
$$

and give, for the special case $A = B$,

$$
\underline{(\text{Pf } A)^2} = \sum_{\substack{\text{cycle configs} \\ \text{of even length}}} (-1)^n \underbrace{(a_{P_1 P_2} \cdots \cdot a_{P_k P_1})}_{\text{cycle 1}} \cdots \underbrace{\phantom{\cdots}}_{\text{cycle } n} = \underline{\det A} \quad (6.16)
$$

In eqn (6.16), the restriction to cycles of even length is irrelevant because, for any antisymmetric matrix, the cycles of odd length add up to zero (for example, the two (distinct) cycles $(234)$ and $(243)$ correspond to $a_{23} a_{34} a_{42}$ and $a_{24} a_{43} a_{32}$—the sum of the two is zero). Equation (6.16) gives a first method for computing the Pfaffian of a matrix $A$, up to its sign, as the square root of its determinant.

We yet have to find the matrix $A$ enumerating the matchings of $A^+$. Instead of combining two matchings of the same matrix $A$, we consider as the partner of $A$ a matrix $B_0$, called the standard matrix,

$$
B_0 = \begin{bmatrix}
\cdot & + & \cdot & \cdot & \cdot & \cdot & \cdot & \cdot & \cdot & \cdot & \cdot & \cdot & \cdot & \cdot & \cdot & \cdot \\
- & \cdot & \cdot & \cdot & \cdot & \cdot & \cdot & \cdot & \cdot & \cdot & \cdot & \cdot & \cdot & \cdot & \cdot & \cdot \\
\cdot & \cdot & \cdot & + & \cdot & \cdot & \cdot & \cdot & \cdot & \cdot & \cdot & \cdot & \cdot & \cdot & \cdot & \cdot \\
\cdot & \cdot & - & \cdot & \cdot & \cdot & \cdot & \cdot & \cdot & \cdot & \cdot & \cdot & \cdot & \cdot & \cdot & \cdot \\
\cdot & \cdot & \cdot & \cdot & \cdot & + & \cdot & \cdot & \cdot & \cdot & \cdot & \cdot & \cdot & \cdot & \cdot & \cdot \\
\cdot & \cdot & \cdot & \cdot & - & \cdot & \cdot & \cdot & \cdot & \cdot & \cdot & \cdot & \cdot & \cdot & \cdot & \cdot \\
\cdot & \cdot & \cdot & \cdot & \cdot & \cdot & \cdot & + & \cdot & \cdot & \cdot & \cdot & \cdot & \cdot & \cdot & \cdot \\
\cdot & \cdot & \cdot & \cdot & \cdot & \cdot & - & \cdot & \cdot & \cdot & \cdot & \cdot & \cdot & \cdot & \cdot & \cdot \\
\cdot & \cdot & \cdot & \cdot & \cdot & \cdot & \cdot & \cdot & \cdot & + & \cdot & \cdot & \cdot & \cdot & \cdot & \cdot \\
\cdot & \cdot & \cdot & \cdot & \cdot & \cdot & \cdot & \cdot & - & \cdot & \cdot & \cdot & \cdot & \cdot & \cdot & \cdot \\
\cdot & \cdot & \cdot & \cdot & \cdot & \cdot & \cdot & \cdot & \cdot & \cdot & \cdot & + & \cdot & \cdot & \cdot & \cdot \\
\cdot & \cdot & \cdot & \cdot & \cdot & \cdot & \cdot & \cdot & \cdot & \cdot & - & \cdot & \cdot & \cdot & \cdot & \cdot \\
\cdot & \cdot & \cdot & \cdot & \cdot & \cdot & \cdot & \cdot & \cdot & \cdot & \cdot & \cdot & \cdot & + & \cdot & \cdot \\
\cdot & \cdot & \cdot & \cdot & \cdot & \cdot & \cdot & \cdot & \cdot & \cdot & \cdot & \cdot & - & \cdot & \cdot & \cdot \\
\cdot & \cdot & \cdot & \cdot & \cdot & \cdot & \cdot & \cdot & \cdot & \cdot & \cdot & \cdot & \cdot & \cdot & \cdot & + \\
\cdot & \cdot & \cdot & \cdot & \cdot & \cdot & \cdot & \cdot & \cdot & \cdot & \cdot & \cdot & \cdot & \cdot & - & \cdot
\end{bmatrix}. \tag{6.17}
$$

This matrix allows only a single matching with nonzero (unit) weight, namely $\{1,2\}, \{3,4\}, \ldots, \{N-1, N\}$, so that its Pfaffian is $\mathrm{Pf}\, B_0 = 1$. The standard matrix $B_0$ can be combined with a matrix $A$, due to Kasteleyn (1961), which differs from the matrix $A^+$ of eqn (6.11) only in its signs:

$$
A = \begin{bmatrix}
\cdot & + & \cdot & \cdot & - & + & \cdot & \cdot & \cdot & \cdot & \cdot & \cdot & \cdot & \cdot & \cdot & \cdot \\
- & \cdot & + & \cdot & + & - & \cdot & \cdot & \cdot & \cdot & \cdot & \cdot & \cdot & \cdot & \cdot & \cdot \\
\cdot & - & \cdot & + & \cdot & \cdot & + & \cdot & \cdot & \cdot & \cdot & \cdot & \cdot & \cdot & \cdot & \cdot \\
+ & \cdot & - & \cdot & \cdot & \cdot & - & + & \cdot & \cdot & \cdot & \cdot & \cdot & \cdot & \cdot & \cdot \\
\cdot & + & \cdot & \cdot & - & + & \cdot & \cdot & - & \cdot & \cdot & \cdot & \cdot & \cdot & \cdot & \cdot \\
\cdot & \cdot & \cdot & \cdot & + & \cdot & - & \cdot & \cdot & + & \cdot & - & \cdot & \cdot & \cdot & \cdot \\
\cdot & \cdot & \cdot & \cdot & \cdot & + & \cdot & - & \cdot & - & + & \cdot & + & \cdot & \cdot & \cdot \\
\cdot & \cdot & \cdot & \cdot & \cdot & \cdot & + & \cdot & - & \cdot & - & + & \cdot & + & \cdot & \cdot \\
\cdot & \cdot & \cdot & \cdot & \cdot & \cdot & \cdot & + & \cdot & - & \cdot & \cdot & - & + & \cdot & \cdot \\
\cdot & \cdot & \cdot & \cdot & \cdot & \cdot & \cdot & \cdot & + & \cdot & - & \cdot & \cdot & - & + & \cdot
\end{bmatrix} \tag{6.18}
$$

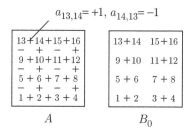

$a_{13,14} = +1, \ a_{14,13} = -1$

**Fig. 6.25** Construction rules for the matrices $A$ and $B_0$ in eqns (6.17) and (6.18).

(see Fig. 6.25). The matrix $A$ gives nonzero weights to matchings corresponding to dimer configurations. Some alternating cycle configurations of $A$ and $B_0$ are shown in Fig. 6.26. In view of eqn (6.15), it remains to be shown that the weight of each cycle is $-1$, to be sure that the Pfaffian of $A$ counts the number of dimer configurations, but this is easy to show for alternating cycles that are collapsed onto one edge (as in the configuration labeled $a$ in Fig. 6.26), or for alternating cycles of length 4 (as in the configuration $b$). The general case follows because general alternating cycles encircle an odd number of elementary squares of the lattice or, in other words, can be broken up into an odd number of cycles of length 4 (the alternating cycle in configuration $f$, in Fig. 6.26 again, encloses seven elementary squares). It follows that

$$
\underline{\mathrm{Pf}\, A} = \mathrm{Pf}\, A \ \mathrm{Pf}\, B_0 = \left\{ \begin{matrix} \text{sum of alternating} \\ \text{cycle configurations} \\ \text{(one per matching)} \end{matrix} \right\} = \underline{\left\{ \begin{matrix} \text{number of} \\ \text{matchings} \end{matrix} \right\}}.
$$

On the $4 \times 4$ lattice without periodic boundary conditions, the Pfaffian of the matrix $A$ in eqn (6.18), obtained via eqn (6.16) as the square root

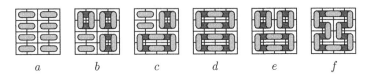

**Fig. 6.26** Alternating cycle configurations from dimers (light) and the unique matching of $B_0$ (dark). Some cycles collapse onto a single edge.

of its determinant, comes out to $\text{Pf}\, A = 36$. This agrees with the number of complete dimer configurations obtained in Table 6.2, by enumeration methods, but we can now compute the number of complete matchings on the $10 \times 10$, the $50 \times 50$, and the $100 \times 100$ lattice. However, we again have an enumeration of the second kind, as the Pfaffian approach allows us to count, but not to list, dimer configurations.

Pfaffians appear in many areas in physics and, for example, are a crucial ingredient in the fermionic path integral, a subject beyond the scope of this book. We shall only move ahead one step, and give an algorithm that computes a Pfaffian directly, without passing through the determinant and losing a sign. (For the dimer problem, we do not really need this algorithm, because we know that the Pfaffian is positive.) The algorithm illustrates that Pfaffians exist by themselves, independent of determinants. The relationship between Pfaffians and determinants mimics the relationship between fermion and bosons: two fermions can make a boson, just as two Pfaffians can combine into a determinant, but fermions also exist by themselves, and they also carry a sign.

As in the case of determinants, there are three linear operations on antisymmetric matrices of order $N = 2M$ which change the Pfaffian of a matrix $C$ in a predictable manner, while systematically transforming $C$ to the standard matrix $B_0$ of eqn (6.17). These transformation rules are the following.

First, the Pffaffian of a matrix is multiplied by $\mu$ if, for any constant $\mu$, both row $i$ and column $i$ are multiplied by $\mu$. This follows from the definition in eqn (6.13). This transformation multiplies the determinant by $\mu^2$.

Second, the Pfaffian of a matrix changes sign if, for $i \neq j$, both the rows $i$ and $j$ and the columns $i$ and $j$ are interchanged. (The terms in the sum in eqn (6.12) are the same, but they belong to a permutation which has one more transposition, that is, an opposite sign.) This transformation does not change the determinant.

Third, the Pfaffian of a matrix is unchanged if $\lambda$ times row $j$ is added to the elements of row $i$ and then $\lambda$ times column $j$ is added to column $i$. This third transformation also leaves the determinant unchanged. We verify this here for an antisymmetric $4 \times 4$ matrix $C = (c_{kl})$, and $i =$

$1, j = 3$:

$$
\mathrm{Pf}
\begin{pmatrix}
0 & c_{12} & c_{13} + \lambda c_{23} & c_{14} + \lambda c_{24} \\
-c_{12} & 0 & c_{23} & c_{24} \\
-c_{13} - \lambda c_{23} & -c_{23} & 0 & c_{34} \\
-c_{14} - \lambda c_{24} & -c_{24} & -c_{34} & 0
\end{pmatrix}
$$
$$
= c_{12}c_{34} - (c_{13} + \lambda c_{23})c_{24} + (c_{14} + \lambda c_{24})c_{23}
$$

is indeed independent of $\lambda$. More generally, the rule can be proven by constructing alternating paths with the standard matrix $B_0$ of eqn (6.17).

We now illustrate the use of the transformation rules for an antisymmetric $4 \times 4$ matrix $C$ (which is unrelated to the dimer problem),

$$
C = \left(
\begin{array}{cc|cc}
0 & 6 & 1 & 3 \\
-6 & 0 & 1 & -1 \\
\hline
-1 & -1 & 0 & -1 \\
-3 & 1 & 1 & 0
\end{array}
\right)
\qquad (\mathrm{Pf}\, C = -2).
$$

Because of its small size, we may compute its Pfaffian from the sum over all matchings:

$$
\mathrm{Pf}\, C = c_{12}c_{34} - c_{13}c_{24} + c_{14}c_{23} = 6 \times (-1) - (1) \times (-1) + 3 \times 1 = -2.
$$

Analogously to the standard Gaussian elimination method for determinants, the above transformation rules allow us to reduce the matrix $C$ to the standard matrix $B_0$. We first subtract ($\lambda = -1$) the third column $j = 3$ from the second column $i = 2$ and then the third row from the second:

$$
C' = \left(
\begin{array}{cc|cc}
0 & 5 & 1 & 3 \\
-5 & 0 & 1 & 0 \\
\hline
-1 & -1 & 0 & -1 \\
-3 & 0 & 1 & 0
\end{array}
\right)
\qquad (\mathrm{Pf}\, C' = \mathrm{Pf}\, C).
$$

We then add three times the third row (and column) to the first,

$$
C'' = \left(
\begin{array}{cc|cc}
0 & 2 & 1 & 0 \\
-2 & 0 & 1 & 0 \\
\hline
-1 & -1 & 0 & -1 \\
0 & 0 & 1 & 0
\end{array}
\right)
\qquad (\mathrm{Pf}\, C'' = \mathrm{Pf}\, C'),
$$

subtract the second row and column from the first,

$$
C''' = \left(
\begin{array}{cc|cc}
0 & 2 & 0 & 0 \\
-2 & 0 & 1 & 0 \\
\hline
0 & -1 & 0 & -1 \\
0 & 0 & 1 & 0
\end{array}
\right)
\qquad (\mathrm{Pf}\, C''' = \mathrm{Pf}\, C''),
$$

and finally subtract the fourth row from the second and the fourth column from the second,

$$C'''' = \begin{pmatrix} 0 & 2 & 0 & 0 \\ -2 & 0 & 0 & 0 \\ \hline 0 & 0 & 0 & -1 \\ 0 & 0 & 1 & 0 \end{pmatrix} \quad (\mathrm{Pf}\,C'''' = \mathrm{Pf}\,C''').$$

In the matrix $C''''$, it remains to multiply the fourth column and row by $-1$ and the second column and row by $\frac{1}{2}$, to arrive at the standard matrix $B_0$, so that $\mathrm{Pf}\,C'''' = \mathrm{Pf}\,C = -2 \cdot \mathrm{Pf}\,B_0 = -2$. A direct Gaussian elimination algorithm can compute the Pfaffian of any antisymmetric matrix in about $\propto N^3$ operations, as we have seen in the above example. This includes the calculation of the sign, which is not an issue for dimers, but which often appears in fermion problems.

The matrices $A$, $B$, and $C$ in this subsection are all integer-valued, and therefore their Pfaffians and determinants, being sums of products of matrix elements, are guaranteed to be integer. On the other hand, the Gaussian elimination algorithm for Pfaffians and the analogous method for determinants work with multiplications of rational numbers $\lambda$ and $\mu$, so that the rounding errors of real arithmetic cannot usually be avoided. Modern algorithms compute Pfaffians and determinants of integer matrices without divisions (see, for example, Galbiati and Maffioli (1994)). This is one of the many meeting points of research-level computational physics with modern discrete and applied mathematics, where they jointly attack problems as yet unsolved.

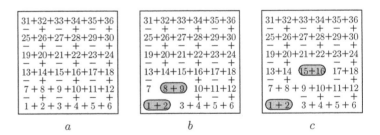

**Fig. 6.27** The original $A$-matrix ($a$), and modifications ($b, c$) with prescribed dimer orientations.

We now apply Pfaffians to the calculation of dimer–dimer correlation functions and then use them in a direct sampling algorithm. For concreteness, we consider first a $6 \times 6$ square lattice without periodic boundary conditions, with its 6728 complete dimer configurations, as we know from earlier enumerations (see Table 6.3). We can recover this number from the Pfaffian of the $36 \times 36$ matrix $A_a$ labeled $a$ in Fig. 6.27. We now use Pfaffians, rather than the direct enumeration methods used for Table 6.4, to count how many dimer configurations have one horizontal dimer on site 1 and site 2 and another dimer on sites 8 and 9 (see part $b$ in Fig. 6.27). The answer is given by the Pfaffian of the matrix $A_b$, obtained from $A_a$ by suppressing all connections of sites 1, 2, 8, and 9 to sites other than the ones forming the prescribed dimers (see Fig. 6.27). We readily obtain $\sqrt{\det A_b} = \operatorname{Pf} A_b = 242$, in agreement with case $a$ in Table 6.4. Other pair correlations can be obtained analogously, for example $\operatorname{Pf} A_c = 1102$ (corresponding to case $c$ in Fig. 6.27). Pfaffians of modified $A$-matrices can be computed numerically for very large $N$, and have even been determined analytically in the limit $N \to \infty$ (see Fisher and Stephenson (1963)).

As a second application of Pfaffian enumerations, and to provide a preview of Subsection 6.2.4 on Markov-chain Monte Carlo algorithms, we discuss a Pfaffian-based direct-sampling algorithm for complete dimer configurations on a square lattice. For concreteness, we consider the $4 \times 4$ square lattice without periodic boundary conditions. The algorithm is constructive. We suppose that four dimers have already been placed, in the configuration $a$ of Fig. 6.28. The matrix $A_a$ corresponding to this configuration is

$$
A_a =
\begin{bmatrix}
\cdot & + & \cdot & \cdot & - & \cdot & \cdot & \cdot & \cdot & \cdot & \cdot & \cdot & \cdot & \cdot & \cdot & \cdot \\
- & \cdot & \cdot & \cdot & \cdot & + & \cdot & \cdot & \cdot & \cdot & \cdot & \cdot & \cdot & \cdot & \cdot & \cdot \\
\cdot & \cdot & \cdot & + & \cdot & \cdot & \cdot & \cdot & \cdot & \cdot & \cdot & \cdot & \cdot & \cdot & \cdot & \cdot \\
+ & \cdot & - & \cdot & \cdot & + & \cdot & \cdot & - & \cdot & \cdot & \cdot & \cdot & \cdot & \cdot & \cdot \\
\cdot & - & \cdot & \cdot & - & \cdot & + & \cdot & \cdot & + & \cdot & \cdot & \cdot & \cdot & \cdot & \cdot \\
\cdot & \cdot & \cdot & - & \cdot & \cdot & \cdot & \cdot & \cdot & \cdot & \cdot & \cdot & \cdot & \cdot & \cdot & \cdot \\
\cdot & \cdot & \cdot & \cdot & + & \cdot & \cdot & \cdot & \cdot & \cdot & + & \cdot & \cdot & \cdot & \cdot & \cdot \\
\cdot & \cdot & \cdot & \cdot & \cdot & + & \cdot & \cdot & + & \cdot & \cdot & \cdot & \cdot & \cdot & \cdot & \cdot \\
\cdot & \cdot & \cdot & + & \cdot & \cdot & - & \cdot & \cdot & + & \cdot & \cdot & \cdot & \cdot & \cdot & \cdot \\
\cdot & \cdot & \cdot & \cdot & + & \cdot & \cdot & - & \cdot & \cdot & \cdot & \cdot & \cdot & \cdot & \cdot & \cdot \\
\cdot & \cdot & \cdot & \cdot & \cdot & \cdot & - & \cdot & \cdot & \cdot & \cdot & \cdot & \cdot & \cdot & \cdot & \cdot \\
\cdot & \cdot & \cdot & \cdot & \cdot & \cdot & \cdot & \cdot & \cdot & \cdot & \cdot & \cdot & + & \cdot & \cdot & \cdot \\
\cdot & \cdot & \cdot & \cdot & \cdot & \cdot & \cdot & \cdot & \cdot & \cdot & \cdot & - & \cdot & + & \cdot & \cdot \\
\cdot & \cdot & \cdot & \cdot & \cdot & \cdot & \cdot & \cdot & \cdot & \cdot & \cdot & \cdot & - & \cdot & + & \cdot \\
\cdot & \cdot & \cdot & \cdot & \cdot & \cdot & \cdot & \cdot & \cdot & \cdot & \cdot & \cdot & \cdot & - & \cdot & + \\
\cdot & \cdot & \cdot & \cdot & \cdot & \cdot & \cdot & \cdot & \cdot & \cdot & \cdot & \cdot & \cdot & \cdot & - & \cdot
\end{bmatrix}
\quad \text{(config. $a$ in Fig. 6.28).}
$$

It is generated from the matrix $A$ of eqn (6.18) by cutting all the links that would conflict with the dimers already placed. The Pfaffian of this matrix is $\operatorname{Pf} A_a = 4$. This means that four dimer configurations are compatible with the dimers that are already placed.

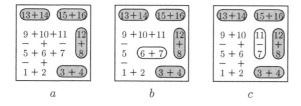

$\qquad a \qquad\qquad\qquad b \qquad\qquad\qquad c$

**Fig. 6.28** Pfaffian direct-sampling algorithm: the dimer orientation on site 7 is determined from $\pi(b)$ and $\pi(c)$.

To carry the construction one step further, we place a dimer on site 7, but have a choice between letting it point towards site 6 (configuration $b$ in Fig. 6.28) or site 11 (configuration $c$). Again, we cut a few links, compute Pfaffians, and find $\pi_b = \operatorname{Pf} A_b = 1$ and $\pi_c = \operatorname{Pf} A_c = 3$. We find that configuration $b$ has a probability $1/4$ and configuration $c$ a probability 3/4. These probabilities are used to sample the orientation of the dimer on site 7, and to conclude one more step in the construction of a direct sample. The Pfaffian direct-sampling algorithm illustrates the fact that an exact solution of a counting problem generally yields a direct-sampling algorithm. An analogous direct-sampling algorithm can be easily constructed for the two-dimensional Ising model. (To test the relative probabilities of having a new spin $k$ point parallel or antiparallel to a neighboring spin $l$ already placed, we compute the determinants of modified matrices with infinitely strong coupling $J_{kl} = +\infty$ or $J_{kl} = -\infty$.)

These correlation-free direct-sampling algorithms are primarily of theoretical interest, as they are slow and specific to certain two-dimensional lattices, which are already well understood. Nevertheless, they are intriguing: they allow one to estimate, by sampling, quantities from the analytic solutions, which cannot be obtained by proper analytic methods. For the Ising model, we can thus extract, with statistical uncertainties, the histogram of magnetization and energy $\mathcal{N}(M, E)$ from the analytic solution (see Fig. 5.7), the key to the behavior of the Ising model in a magnetic field, which cannot be obtained analytically. In the dimer model, we are able to compute complicated observables for which no analytic solution exists.

### 6.2.4    Monte Carlo algorithms for the monomer–dimer problem

As discussed in the previous section, the enumeration problem for monomers and dimers is generally difficult. It is therefore useful to consider Monte Carlo sampling algorithms. We first discuss a local Markov-chain Monte Carlo algorithms using flips of a pair of neighboring dimers (see Fig. 6.30). These moves are analogous to the spin flips in the Ising model (see Subsection 5.2.1). The local Monte Carlo algorithm satisfies detailed balance because we move with the same probability from configuration $a$ to configuration $b$ as from $b$ back to $a$. However, it is not completely evident that this algorithm is also ergodic, that is, that we can connect any two configurations by a sequence of moves.

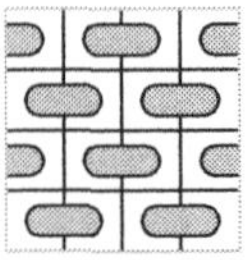

**Fig. 6.29** Dimer configuration without flippable dimer pairs proving nonergodicity of the local algorithm in the presence of periodic boundary conditions.

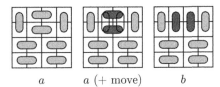

$a$        $a$ (+ move)        $b$

**Fig. 6.30** A local Monte Carlo move flipping a pair of neighboring dimers in the square lattice.

For concreteness, we discuss the ergodicity for a $6 \times 6$ lattice without periodic boundary conditions. (The local algorithm is trivially nonergodic in the presence of periodic boundary conditions because of the existence of winding-number sectors similar to those for path integrals, in Chapter 3, see Fig. 6.29.) We first show that any configuration must contain a flippable pair of neighboring dimers. Let us try to construct a configuration without such a pair. Without restricting our argument, we can suppose that the dimer in the lower right corner points upwards (see the gray dimer in Fig. 6.31). In order not to give a flippable pair, the dimer on site $a$ must then point in the direction shown, and this in turn imposes the orientation of the dimer on site $b$, on site $c$, etc., until we have no choice but to orient the dimers on site $g$ and $h$ as shown. These two dimers are flippable.

**Fig. 6.31** Proof that any complete dimer configuration contains a flippable pair of dimers.

The argument proving the existence of flippable dimer pairs can be extended to show that any configuration of dimers can be connected to the standard configuration of eqn (6.17), where all dimers are in horizontal orientation. (If any two configurations can be connected to the standard configuration, then there exists a sequence of moves going from the one configuration to the other, and the algorithm is ergodic.) We first suppose that the configurations contains no rows of sites where all the dimers are already oriented horizontally. The lower part of the configuration then again resembles the configuration shown in Fig. 6.31. After flipping the dimers $g$ and $h$, we can flip dimer $c$, etc., and arrange the lowest row of sites in Fig. 6.31 to be all horizontal. We can then continue with the next higher row, until the whole configuration contains only horizontal dimers. We conclude that the local dimer algorithm is ergodic. However, it is painfully slow.

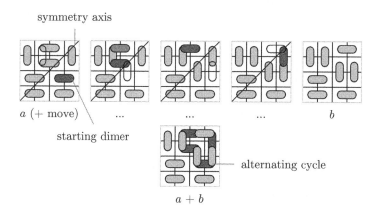

**Fig. 6.32** Pivot cluster algorithm for dimers on a square lattice with periodic boundary conditions.

Much more powerful algorithms result from the fact that the combination of any two dimer configurations $a$ and $b$ (two matchings) gives rise to an alternating cycle configuration (see the discussion in Subsection 6.2.3). In any nontrivial alternating cycle, we can simply replace dimers of configuration $a$ with those of configuration $b$. If we choose as

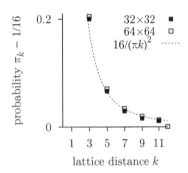

**Fig. 6.33** "Ladder" correlations for the complete dimer model in square lattices.

configuration $b$ the configuration $a$ itself, transformed with respect to a symmetry operation of the lattice, we arrive at the pivot cluster algorithm, that was already used for binary mixtures in Subsection 6.1.3 (Krauth and Moessner 2003). This algorithm can be directly implemented from Alg. 6.3 (`pocket-binary`) (see Fig. 6.32). It gives one of the fastest simulation methods for dimer models and monomer–dimer models in two or more dimensions. It can be adapted to a variety of lattices. Snapshots of configurations generated with this algorithm are shown in Fig. 6.34. On these large systems, we can compute dimer–dimer correlation functions, and see explicitly that they do not give long-range order. In Fig. 6.33, we show the probability of having a horizontal dimer at one site and another horizontal dimer $k$ sites above. This "ladder" correlation tends towards the value $1/16$ indicating that the dimers become uncorrelated in the limit $k \to \infty$. However, the correlation goes to zero as a power law in $k$, not exponentially, showing that the dimer model, while lacking long-range order, is critical. (The dimer correlations in Fig. 6.33 have been computed analytically by Fisher and Stephenson (1963).) In contrast, the monomer–dimer model has exponentially decaying correlations, which cannot be obtained analytically.

**Fig. 6.34** Dimers (*left*) and monomer–dimers (*right*) in a lattice with periodic boundary conditions (from adapted Alg. 6.3 (`pocket-binary`)).

At the end of Subsection 6.2.3, we discussed that exact solutions for enumeration problems yield direct-sampling methods. We can now discuss the relationship between enumeration and sampling under another angle. It is known that the monomer–dimer problem is computationally hard: while Pfaffian methods allow us to count the number of complete dimer configurations, say on the left of Fig. 6.34, is it impossible to determine the exact number of monomer–dimer configurations for large systems, say, on the right of Fig. 6.34. On the other hand, sampling configurations of the monomer–dimer model is easy, and we can use sampling methods for precision estimates for the number of monomer–dimer configurations on large lattices. Clearly and remarkably, sampling does not meet the same limitations as enumeration.

## 6.2.5 Monomer–dimer partition function

Monte Carlo simulations and Pfaffian computations allow us to understand, and even to prove rigorously, that the monomer–dimer model on the square lattice has no phase transition, and that dimer–dimer correlations become critical at complete packing. Remarkably, in this model, no order–disorder transition takes place. In a typical configuration on a large lattice, the proportions of horizontal and vertical dimers are thus equal. This result is a consequence of the mapping of the dimer model onto the two-dimensional Ising model at the critical point. Rather than to follow this direction, we highlight in this subsection a powerful theorem due to Heilmann and Lieb (1970) which shows that on any lattice (in two, three, or higher dimensions or on any irregular graph), the monomer–dimer model never has a phase transition. The constructive proof of this theorem has close connections to enumeration methods.

We first generalize our earlier partition function, which counted the number of complete packings, to variable densities of dimers and suppose that each dimer has zero energy, whereas a monomer, an empty lattice site, costs $E > 0$. Different contributions are weighted with a Boltzmann factor

$$Z(\beta) = \sum_{M=0}^{N/2} \underbrace{\mathcal{N}(M)}_{\substack{\text{number of} \\ \text{dimer confs}}} \underbrace{\left(e^{-\beta E}\right)^{N-2M}}_{\text{monomer weight}},$$

where $M$ represents the number of dimers in a given configuration. At zero temperature, the penalty for placing monomers becomes prohibitive, and on lattices which allow complete packings, the partition function $Z(\beta = \infty)$ thus gives back the sum over the partition function (6.10), the number of complete packings.

For concreteness, we consider the partition function $Z_{L\times L}(\beta)$ for an $L \times L$ square lattice with periodic boundary conditions. $Z_{L\times L}(\beta)$ is a polynomial in $x = e^{-\beta E}$ with positive coefficients which, for small lattices, are given by

$$Z_{2\times 2}(x) = x^4 + 8x^2 + 8,$$
$$Z_{4\times 4}(x) = x^{16} + 32x^{14} + 400x^{12} + \cdots + 3712x^2 + 272, \qquad (6.19)$$
$$Z_{6\times 6}(x) = x^{36} + 72x^{34} + \cdots + 5\,409\,088\,488x^{12} + \cdots + 90\,176.$$

The coefficients of $Z_{2\times 2}$ correspond to the fact that in the $2 \times 2$ lattice without periodic boundary conditions, we can build one configuration with zero dimers and eight configurations each with one and with two dimers. The partition function $Z_{4\times 4}$ follows from our first enumeration in Table 6.2, and the coefficients of the $6 \times 6$ lattice are generated by depth-first enumeration (see Table 6.6).

In a finite lattice, $Z(\beta)$ is always positive, and for no value of the temperature (or of $x$ on the positive real axis) can the free energy $\log Z$, or any of its derivatives, generate a divergence. The way a transition can nevertheless take place in the limit of an infinite system limit, was clarified by Lee and Yang (1952). It involves considering the partition function as a function of $x$, taken as a complex variable.

**Table 6.6** Number of dimer configurations in a $6 \times 6$ square lattice with periodic boundary conditions, from Alg. 6.7 (`depth-dimer`)

| $M$ (# dimers) | $\mathcal{N}(M)$ (# configs) |
|---|---|
| 0 | 1 |
| 1 | 72 |
| 2 | 2340 |
| 3 | 45 456 |
| 4 | 589 158 |
| 5 | 5 386 752 |
| 6 | 35 826 516 |
| 7 | 176 198 256 |
| 8 | 645 204 321 |
| 9 | 1 758 028 568 |
| 10 | 3 538 275 120 |
| 11 | 5 185 123 200 |
| 12 | 5 409 088 488 |
| 13 | 3 885 146 784 |
| 14 | 1 829 582 496 |
| 15 | 524 514 432 |
| 16 | 81 145 872 |
| 17 | 5 415 552 |
| 18 | 90 176 |

Standard algorithms allow us to compute the (complex-valued) zeros of these polynomials, that is, the values of $x$ for which $Z_{L \times L}(x) = 0$, etc. For all three lattices, the zeros remain on the imaginary axis, and the partition function can be written as

$$Z_{2 \times 2}(x) = \underbrace{(x^2 + 1.1716)(x^2 + 6.828)}_{2 \text{ terms}},$$

$$Z_{4 \times 4}(x) = \underbrace{(x^2 + 0.102)(x^2 + 0.506) \dots (x^2 + 10.343)}_{8 \text{ terms}}, \qquad (6.20)$$

$$Z_{6 \times 6}(x) = \underbrace{(x^2 + 0.024)(x^2 + 0.121) \dots (x^2 + 10.901)}_{18 \text{ terms}}.$$

We note that the two eqns (6.19) and (6.20) feature the same polynomials, and that the factorized representation allows us to read off the zeros. For example, the polynomial $Z_{2 \times 2}(x) = 0$ for $x^2 = -1.1716$ and for $x^2 = -6.828$, that is, for $x = \pm i\sqrt{1.1716}$ and $x = \pm i\sqrt{6.828}$. The zeros of the above partition functions are all purely imaginary. The generalization of this finding constitutes the Heilmann–Lieb theorem.

Using enumeration, we cannot go much beyond a $6 \times 6$ lattice. To nevertheless refine our heuristics, we can compute the partition functions for all the lattices shown in Fig. 6.35. A single run of Alg. 6.7

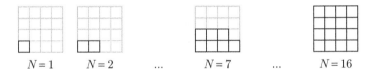

$N = 1$ $\qquad$ $N = 2$ $\qquad$ ... $\qquad$ $N = 7$ $\qquad$ ... $\qquad$ $N = 16$

**Fig. 6.35** Building up a $4 \times 4$ square lattice without periodic boundary conditions from sublattices containing $N = 1, \dots, 16$ sites.

**Table 6.7** Number of configurations $\mathcal{N}(M, N)$ with $M$ dimers on $N$-site lattices of Fig. 6.35

| $M$ | $N = 6$ | 7 | 8 | 9 | 10 | 11 | 12 | 13 | 14 | 15 | 16 |
|---|---|---|---|---|---|---|---|---|---|---|---|
| 0 | 1 | 1 | 1 | 1 | 1 | 1 | 1 | 1 | 1 | 1 | 1 |
| 1 | 6 | 8 | 10 | 11 | 13 | 15 | 17 | 18 | 20 | 22 | 24 |
| 2 | 8 | 16 | 29 | 37 | 55 | 76 | 102 | 117 | 149 | 184 | 224 |
| 3 | 2 | 7 | 26 | 42 | 90 | 158 | 267 | 343 | 524 | 746 | 1044 |
| 4 | . | . | 5 | 12 | 52 | 128 | 302 | 460 | 908 | 1545 | 2593 |
| 5 | . | . | . | . | 7 | 29 | 123 | 251 | 734 | 1572 | 3388 |
| 6 | . | . | . | . | . | . | 11 | 40 | 232 | 682 | 2150 |
| 7 | . | . | . | . | . | . | . | . | 18 | 88 | 552 |
| 8 | . | . | . | . | . | . | . | . | . | . | 36 |

(`depth-dimer`) on the $4 \times 4$ lattice suffices to compute the numbers of $M$-dimer configurations for all these lattices (see Table 6.7). (We simply retain for each configuration on the $4 \times 4$ lattice the largest occupied site, $N'$. This configuration contributes to all lattices $N$ in Table 6.7 with $N \geq N'$.) We then write down partition functions and compute roots of polynomials, as before, but for all the 16 lattices of Fig. 6.35:

$$Z_8(x) = (x^2 + 0.262)(x^2 + 1.151)(x^2 + 2.926)(x^2 + 5.661),$$
$$Z_9(x) = x(x^2 + 0.426)(x^2 + 1.477)(x^2 + 3.271)(x^2 + 5.826),$$
$$Z_{10}(x) = (x^2 + 0.190)(x^2 + 0.815)(x^2 + 1.935)(x^2 + 3.623)(x^2 + 6.436).$$

These partition functions are again as in eqn (6.20), slightly generalized to allow lattices with odd number of sites:

$$Z_N(x) = x \prod (x^2 + b_i) \text{ for } N \text{ odd},$$
$$Z_N(x) = \prod (x^2 + b_i) \text{ for } N \text{ even}.$$

Furthermore, the zeros of the polynomial $Z_8$ are sandwiched in between the zeros of $Z_9$, which are themselves sandwiched in between the zeros of $Z_{10}$. (For the zeros of $Z_8$ and $Z_9$, for example, we have that $0 < 0.262 < 0.426 < 1.151$, etc.) This observation can be generalized. Let us consider more lattices differing in one or two sites, for example lattices containing sites $\{1, \ldots, 7\}$, $\{1, \ldots, 6\}$, and also $\{1, \ldots, 3, 5, \ldots, 7\}$. The partition function on these lattices are related to each other (see Fig. 6.36):

$$Z_{1,\ldots,8}(x) = x Z_{1,\ldots,7}(x) + Z_{1,\ldots,6}(x) + Z_{1,\ldots,3,5,\ldots,7}(x), \tag{6.21}$$

simply because site 8 on the lattice $\{1, \ldots, 8\}$ hosts either a monomer or a dimer, which must point in one of a few directions.

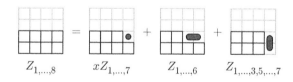

$$Z_{1,\ldots,8} \qquad x Z_{1,\ldots,7} \qquad Z_{1,\ldots,6} \qquad Z_{1,\ldots,3,5,\ldots,7}$$

**Fig. 6.36** Monomer–dimer partition function on a lattice of 8 sites expressed through partition functions on sublattices.

If the zeros both of $Z_{1,\ldots,6}$ and of $Z_{1,\ldots,3,5,\ldots,7}$ are sandwiched in by the zeros of $Z_{1,\ldots,7}$, then it follows, by simply considering eqn (6.21), that the zeros of $Z_{1,\ldots,7}$ are sandwiches in by the zeros of $Z_{1,\ldots,8}(x)$. We see that a relation for lattices of six and seven sites yields the same relation for lattices between seven and eight sites. More generally, the zeros of the partition function for any lattice stay on the imaginary axis, and are sandwiched in by the zeros of partition functions for lattices with one more site. It also follows from the absence of zeros of the partition function in the right half plane of the complex variable $x$ that the logarithm of the partition function, the free energy, is an analytic function in this same region. It nowhere entered the argument that the lattice was planar, or the sublattices of the special form used in our example. Whatever the lattice (2-dimensional lattice, three-dimensional, an irregular graph, etc.), there can thus never be a phase transition in the monomer–dimer model.

# Exercises

## (Section 6.1)

(6.1) To sample configuration of the one-dimensional clothes-pin model, implement Alg. 6.1 (`naive-pin`) and the rejection-free direct-sampling algorithm (Alg. 6.2 (`direct-pin`)). Check that the histograms for the single-particle density $\pi(x)$ agree with each other and with the analytic expression in eqn (6.7). Generalize the two programs to the case of periodic boundary conditions (pins on a ring). Compute the periodically corrected pair correlation function on the ring (the histogram of the distance between two pins) and check the equivalence in eqn (6.9) between the pair correlations on a ring and the boundary correlations on a line.
NB: Implement periodic distances as in Alg. 2.6 (`diff-vec`).

(6.2) Show by direct analysis of eqn (6.7) that the single-particle probability $\pi(x)$ of the clothes-pin model has a well-defined limit $L \to \infty$ at constant covering density $\eta = 2N\sigma/L$. Again, from an evaluation of eqn (6.7) for finite $N$, show that, for $\eta < \frac{1}{2}$, the function $\pi(x)$ is exactly constant in an inner region of the line, more than $(2N-1)\sigma$ away from the boundaries. Prove this analytically for three or four pins, and also for general values of $N$.
NB: The general proof is very difficult—see Leff and Coopersmith (1966).

(6.3) Generalize Alg. 6.2 (`direct-pin`) to the case of $N_l$ large and $N_s$ small particles on a segment of length $L$. Compute the single-particle probability distribution $\pi(x)$ for the large particles. Generalize the program to the case of particles on a ring with periodic boundary conditions (a "binary necklace"). Determine the pair-correlation function, as in Exerc. 6.1.
NB: After sampling the variables $\{y_1, \ldots, y_N\}$, with $N = N_l + N_s$, use Alg. 1.12 (`ran-combination`) to decide which ones of them are the large particles.

(6.4) (Relates to Exerc. 6.3). Implement a local Monte Carlo algorithm for the binary necklace of $N_l$ large and $N_s$ small beads (particles on a line with periodic boundary conditions). To sample all possible arrangements of small and large beads, set up a local Monte Carlo algorithm with the two types of move shown in Fig. 6.37. Compare results for the

pair-correlation function with Exerc. 6.3. Comment on the convergence rate of this algorithm.

**Fig. 6.37** Two types of move on a binary necklace.

(6.5) Implement Alg. 6.3 (`pocket-binary`), the cluster algorithm for hard squares and disks in a rectangular region with periodic boundary conditions. Test your program in the case of equal-size particles in one dimension, or in the binary necklace problem of Exerc. 6.3. Generate phase-separated configurations of binary mixtures of hard squares, as in Fig. 6.14. NB: Use Algs 2.5 and 2.6 to handle periodic boundary conditions. Run the program for several minutes in order to generate configurations as in Fig. 6.10 without using grid/cell schemes. If possible, handle initial conditions as in Exerc. 1.3 (see also Exerc. 2.3). The legal initial configuration at the very first start of the program may contain all of the small particles in one half of the box, and all of the large particles in the other.

## (Section 6.2)

(6.6) Implement Alg. 6.4 (`naive-dimer`) on a $4 \times 4$ square lattice. Use an occupation-number vector for deciding whether a configuration is legal. Check your program against the data in Table 6.2. Modify it to allow you to choose the boundary conditions, and also to choose between the enumeration of complete dimer configurations or of monomers and dimers. Implement Alg. 6.5 (`naive-dimer(patch)`). Can you treat lattices larger than $4 \times 4$?

(6.7) Consider the numbers of configurations with $M > 0$ dimers on the $4 \times 4$ lattice (see Table 6.2). Explain why these numbers are all even, except for four dimers on the $4 \times 4$ lattice without periodic boundary conditions, where the number of configurations is odd.

(6.8) Implement Alg. 6.7 (`depth-dimer`) for dimers on the square lattice. Test it with the histogram of

Table 6.2. Use it to compute the analogous histograms on the $6 \times 6$ lattice with or without periodic boundary conditions, but make sure that your representation allows you to treat sufficiently large numbers without overflow. Store all the complete dimer configurations in a file and compute the pair correlation functions of Table 6.6.

(6.9) Generalize Alg. 6.6 (breadth-dimer) for tetris molecules (see Fig. 6.38) on an $L \times L$ square lattice without periodic boundary conditions. Start from a list of single-tetris configurations analogous to Fig. 6.18. How many configurations of eight tetris molecules are there on a $7 \times 7$ lattice?

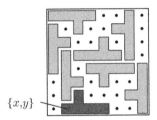

**Fig. 6.38** Tetris molecules on an $8 \times 8$ lattice.

NB: The enumeration program relies on a list of single-molecule configurations, and on a list of sites touched by each of them (there are 24 such configurations on the $4 \times 4$ lattice, and 120 configurations on the $6 \times 6$ lattice). Write down explicitly that a molecule based at $\{x, y\}$ (where $1 \leq x, y \leq L$), in the same orientation as the dark molecule in Fig. 6.38, exists if $x + 3 \leq L$ and if $y + 1 \leq L$ and then touches also the positions $\{x+1, y\}$ $\{x+2, y\}$ $\{x+3, y\}$, and $\{x+1, y+1\}$. All coordinates $\{x, y\}$ should then be translated into site numbers.

(6.10) Implement the matrix $A$ of eqn (6.18), generalized for an $L \times L$ square lattice without periodic boundary conditions (see Fig. 6.25). Use a standard linear algebra routine to compute det $A$ and $\mathrm{Pf}\, A = \sqrt{\det A}$. Test your implementation against the enumeration results of Table 6.2 and Table 6.3. Extend the program to compute Pfaffians of modified matrices, as $A'$ and $A''$ in Fig. 6.27. Then implement a Pfaffian-based rejection-free direct sampling algorithm for complete dimer configurations. At each step during the construction of the configuration, the algorithm picks a site that is not yet covered by a dimer, and must compute probabilities for the different orientations of the dimer on that site, as discussed in Fig. 6.28. Check your algorithm in the $4 \times 4$ square lattice. It should generate the 36 different complete dimer configurations with

approximately equal frequency (see Table 6.2).
NB: An analogous algorithm exists for the two-dimensional Ising model, and the Ising spin glass. In that latter case, unlike for dimers, there are no good Markov-chain sampling methods.

(6.11) The matrices $\{A_1, \ldots, A_4\}$ in Fig. 6.39 allow one to count complete dimer configurations with periodic boundary conditions: consider an arbitrary alternating cycle, either planar or winding around the lattice (in $x$ or $y$ direction, or both). Compute its weight for all four matrices (see Kasteleyn (1961)). Show that the number of complete dimer configurations on the square lattice with periodic boundary conditions is

$$Z = \frac{1}{2}(-\mathrm{Pf}\, A_1 + \mathrm{Pf}\, A_2 + \mathrm{Pf}\, A_3 + \mathrm{Pf}\, A_4).$$

Implement matrices analogous to $\{A_1, \ldots, A_4\}$ for $L \times L$ square lattices with periodic boundary conditions, and recover the results for complete dimer configurations in Tables 6.2 and 6.3.

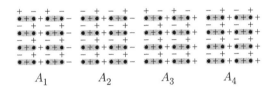

**Fig. 6.39** Standard matrices for the $4 \times 4$ lattice (compare with Fig. 6.25).

(6.12) Implement the pivot cluster algorithm for dimers on the $L \times L$ square lattice with periodic boundary conditions. Test it by writing out complete dimer configurations on file: on the $4 \times 4$ lattice, the 272 complete dimer configurations should be generated equally often. Sample configurations with $M < L^2/2$ dimers and store them on file. Now show how a single iteration of Alg. 6.6 (breadth-dimer) allows one to estimate $\mathcal{N}(M+1)/\mathcal{N}(M)$. Estimate $\mathcal{N}(M)$ from several independent Monte Carlo runs. Test your procedure in the $4 \times 4$ lattice and the $6 \times 6$ lattices against enumeration data, try it out on much larger lattices.
NB: Thus it is easy to estimate the number of dimer configurations for any $M$, because Markov-chain Monte Carlo algorithms converge very well. We note that Valiant (1979) has rigorously established that counting the number of monomer–dimer configurations without statistical errors is difficult: estimation is easy, precise counting difficult.

# References

Asakura S., Oosawa F. (1954) On interaction between 2 bodies immersed in a solution of macromolecules, *Journal of Chemical Physics* **22**, 1255–1256

Buhot A., Krauth W. (1999) Phase separation in two-dimensional additive mixtures, *Physical Review E* **59**, 2939–2941

Dinsmore A. D., Yodh A. G., Pine D. J. (1995) Phase-diagrams of nearly hard-sphere binary colloids, *Physical Review E* **52**, 4045–4057

Fisher M. E., Stephenson J. (1963) Statistical mechanics of dimers on a plane lattice. II. Dimer correlations and monomers, *Physical Review* **132**, 1411–1431

Galbiati G., Maffioli F. (1994) On the computation of Pfaffians, *Discrete Applied Mathematics* **51**, 269–275

Heilmann O. J., Lieb E. H. (1970) Monomers and dimers, *Physical Review Letters* **24**, 1412–1414

Kasteleyn P. W. (1961) The statistics of dimers on a lattice I. The number of dimer arrangements on a quadratic lattice, *Physica* **27**, 1209–1225

Krauth W., Moessner R. (2003) Pocket Monte Carlo algorithm for classical doped dimer models, *Physical Review B* **67**, 064503

Lee T. D., Yang C. N. (1952) Statistical theory of equations of state and phase transitions. 2. Lattice gas and Ising model, *Physical Review* **87**, 410–419

Leff H. S., Coopersmith M. H. (1966) Translational invariance properties of a finite one-dimensional hard-core fluid, *Journal of Mathematical Physics* **8**, 306–314

Valiant L. G. (1979) Complexity of enumeration and reliability problems, *SIAM Journal on Computing* **8**, 410–421

# Dynamic Monte Carlo methods

In the first six chapters of this book, we have concentrated on equilibrium statistical mechanics and related computational-physics approaches, notably the equilibrium Monte Carlo method. These and other approaches allowed us to determine partition functions, energies, superfluid densities, etc. Physical time played a minor role, as the observables were generally time-independent. Likewise, Monte Carlo "time" was treated as of secondary interest, if not a nuisance: we strove only to make things happen as quickly as possible, that is, to have algorithms converge rapidly.

The moment has come to reach beyond equilibrium statistical mechanics, and to explore time-dependent phenomena such as the crystallization of hard spheres after a sudden increase in pressure or the magnetic response of Ising spins to an external field switched on at some initial time. The local Monte Carlo algorithm often provides an excellent framework for studying dynamical phenomena.

The conceptual difference between equilibrium and dynamic Monte Carlo methods cannot be overemphasized. In the first case, we have an essentially unrestricted choice of a priori probabilities, since we only want to generate independent configurations $x$ distributed with a probability $\pi(x)$, in whatever way we choose, but as fast as possible. In dynamic calculations, the time dependence becomes the main object of our study. We first look at this difference between equilibrium and dynamics in the case of the random-sequential-deposition problem of Chapter 2, where a powerful dynamic algorithm perfectly implements the faster-than-the-clock paradigm. We then discuss dynamic Monte Carlo methods for the Ising model and encounter the main limitation of the faster-than-the-clock approach, the futility problem.

In the final section of this chapter, we apply a Monte Carlo method called simulated annealing, an important tool for solving difficult optimization problems, mostly without any relation to physics. In this approach, a discrete or continuous optimization problem is mapped onto an artificial physical system whose ground state (at zero temperature or infinite pressure) contains the solution to the original task. This ground state is slowly approached through simulation. Simulated annealing will be discussed for monodisperse and polydisperse hard disks on the surface of a sphere, under increasing pressure. It works prodigiously in one case, where the disks end up crystallizing, but fails in another case, where they settle into a glassy state.

Disks are dropped randomly into a box (Fig. 7.1), but they stay put only if they fall into a free spot. Most of the time, this is not the case, and the last disk must be removed again. It thus takes a long time to fill the box. In this chapter, we study algorithms that do this much faster: they go from time $t = 4262$ to $t = 20332$ in one step, and also find out that the box is then full and that no more disks can be added. All problems considered in this chapter treat dynamic problems, where time dependence plays an essential role.

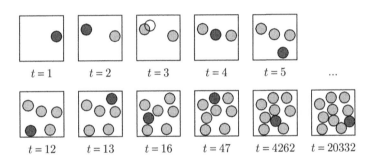

**Fig. 7.1** Random sequential deposition of disks in a box. Any disk generating overlaps (as at time $t = 3$) is removed.

# 7.1   Random sequential deposition

A number of dynamic models in statistical physics do not call on equilibrium concepts such as the Boltzmann distribution and equiprobability. From Chapter 2, we are already familiar with one of these models,
that of random sequential deposition. This model describes hard disks
which are deposited, one after another, at random positions in an initially empty square. Disks stick to the region in which they have been
deposited if they do not overlap with any other disk placed earlier; otherwise, they are removed, leaving the state of the system unchanged. It
is instructive to implement random sequential deposition (see Alg. 7.1
(`naive-deposition`)) and to compare this dynamic Monte Carlo process
with Alg. 2.7 (`direct-disks`), the equilibrium direct-sampling method
for hard disks.

**procedure** `naive-deposition`
$k \leftarrow 1$
**for** $t = 1, 2, \ldots$ **do**
$\Big\{$
$\quad x_k \leftarrow \mathtt{ran}\,(x_{\min}, x_{\max})$
$\quad y_k \leftarrow \mathtt{ran}\,(y_{\min}, y_{\max})$
$\quad$ **if** $\big( \min_{l<k}[\mathtt{dist}(\mathbf{x}_k, \mathbf{x}_l)] > 2r \big)$ **then**
$\quad\quad\Big\{$
$\quad\quad\quad$ **output** $\{\mathbf{x}_k, t\}$
$\quad\quad\quad k \leftarrow k + 1$

——

**Algorithm 7.1** `naive-deposition`. Depositing hard disks of radius $r$ on
a deposition region delimited by $\{x_{\min}, y_{\min}\}$ and $\{x_{\max}, y_{\max}\}$.

As seen in earlier chapters, equilibrium Monte Carlo problems possess
a stationary probability distribution $\pi(x)$. In contrast, dynamic processes such as random sequential deposition are defined through a rule.
In fact, the rule is all there is to the model.

Random sequential deposition raises several questions. For example,
we would like to compute the stopping time $t_{\mathrm{s}}$ for each sample, the
time after which it becomes impossible to place an additional disk in
the system. We notice that Alg. 7.1 (`naive-deposition`) is unable to
decide whether the current simulation time is smaller or larger than the
stopping time.

We would like to investigate the structure of the final state, after $t_{\mathrm{s}}$.
More generally, we would like to compute the ensemble-averaged density
of the system as a function of time, up to the stopping time, for different
sizes of the deposition region. To compare the time behavior of different-
sized systems, we must rescale the time as $\tau = t/(\text{deposition area})$. This
lets us compare systems which have seen the same number of deposition
attempts per unit area. In the limit of a large area $L \times L$, the rescaled
stopping time $\tau_s$ diverges, and it becomes difficult to study the late
stages of the deposition process.

### 7.1.1  Faster-than-the-clock algorithms

In dynamic Monte Carlo methods, as in equilibrium methods, algorithm design does not stop with naive approaches. In the late stages of Alg. 7.1 (`naive-deposition`), most deposition attempts are rejected and do not change the configuration. This indicates that better methods can be found. Let us first rerun the simulation of Fig. 7.1, but mark in dark the accessible region (more than two radii away from any disk center and more than one radius from the boundary) where new disks can still be placed successfully (see Fig. 7.2). The accessible region has already appeared in our discussion of entropic interactions in Chapter 6.

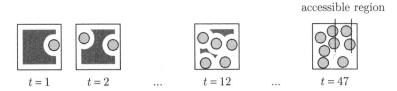

accessible region

$t=1$   $t=2$   ...   $t=12$   ...   $t=47$

**Fig. 7.2** Some of the configurations of Fig. 7.1, together with their accessible regions, drawn in dark.

At time $t = 47$, the remaining accessible region—composed of two tiny spots—is hardly perceptible: it is no wonder that the waiting time for the next successful deposition is very large. In Fig. 7.2, this event occurs at time $t = 4263$ ($\Delta_t = 4215$; see Fig. 7.1), but a different set of random numbers will give different results. Clearly, the waiting time $\Delta_t$ until the next successful deposition is a random variable whose probability distribution depends on the area of the accessible region. Using the definition
$$\lambda = 1 - \frac{\text{area of accessible region}}{\text{area of deposition region}},$$
we can see that a single deposition fails (is rejected) with a probability $\lambda$. The rejection rate increases with each successful deposition and reaches 1 at the stopping time $\tau_s$.

The probability of failing once is $\lambda$, and of failing twice in a row is $\lambda^2$. $\lambda^k$ is thus the probability of failing at least $k$ times in a row, in other words, the probability for the waiting time to be larger than $k$.

The probability of waiting exactly $\Delta_t$ steps is given by the probability of having $k$ rejections in a row, multiplied by the acceptance probability

$$\pi(\Delta_t) = \underbrace{\lambda^{\Delta_t-1}}_{\substack{\Delta_t-1 \\ \text{rejections}}} \overbrace{(1-\lambda)}^{\text{acceptance}} = \lambda^{\Delta_t-1} - \lambda^{\Delta_t}.$$

The distribution function $\pi(\Delta_t)$—a discretized exponential function—can be represented using the familiar tower scheme shown in Fig. 7.3 (see Subsection 1.2.3 for a discussion of tower sampling).

We note that the naive algorithm samples the distribution $\pi(\Delta_t)$ and at the same time places a disk center inside the accessible region, that is, it mixes two random processes concerning the when and the where of the next successful deposition. Faster-than-the-clock methods, the subject of the present subsection, nicely disentangle this double sampling problem. Instead of finding the waiting time the hard way (by trying and trying), starting at time $t$, these methods sample $\Delta_t$ directly (see Fig. 7.3). After deciding when to place the disk, we have to find out where to put it; this means that a disk is placed (at the predetermined time $t+\Delta_t$) anywhere in the accessible region.

To actually determine the waiting time, we do not need to implement tower sampling with Alg. 1.14 (`tower-sample`), but merely solve the inequality (see Fig. 7.3)

$$\lambda^{\Delta_t} < \mathtt{ran}\,(0,1) < \lambda^{\Delta_t-1},$$

which yields

$$(\Delta_t - 1)\,\log\lambda < \log\mathtt{ran}\,(0,1) < \Delta_t\,\log\lambda,$$

$$\Delta_t = 1 + \mathtt{int}\left[\frac{\log\mathtt{ran}\,(0,1)}{\log\lambda}\right]. \tag{7.1}$$

Of course, we can sample $\Delta_t$ (with the help of eqn (7.1)) only after computing $\lambda$ from the area of the accessible region. This region (see Fig. 7.4) is very complicated in shape: it need not be simply connected, and connected pieces may have holes.

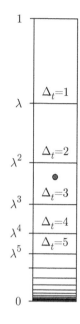

**Fig. 7.3** Tower sampling in random sequential deposition: a pebble $\mathtt{ran}\,(0,1)$ samples the waiting time.

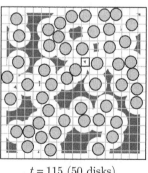 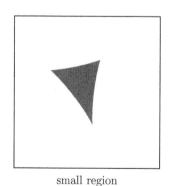

$t = 115$ (50 disks)  small region

**Fig. 7.4** Accessible region, in dark, and the method of cutting it up into small regions (*left*). A small region, spanned by a convex polygon (*right*).

We can considerably simplify the task of computing the area of the accessible region by cutting up the accessible region into many small regions $\{\mathcal{R}_1,\ldots,\mathcal{R}_K\}$, each one confined to one element of a suitable grid. We let $A(\mathcal{R}_k)$ be the area of the small region $\mathcal{R}_k$, so that the total accessible area is $A_{\mathrm{acc}} = \sum_k A(\mathcal{R}_k)$.

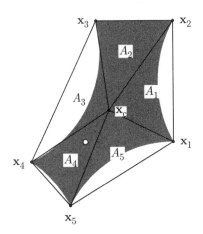

**Fig. 7.5** Convex polygon with five vertices $\{\mathbf{x}_1, \ldots, \mathbf{x}_5\}$ and a central point $\mathbf{x}_c$. A random sample inside the fourth triangle is indicated.

For a sufficiently fine grid, the small regions $\mathcal{R}_k$ have no holes, but there may be more than one small region within a single grid element. The shapes of the small regions are typically as shown in Fig. 7.5.

The small region $\mathcal{R}_k$ is spanned by vertices $\{\mathbf{x}_1, \ldots, \mathbf{x}_n\}$, where $\mathbf{x}_k = \{x_k, y_k\}$. The polygon formed by these vectors has an area

$$A_{\text{polygon}} = \frac{1}{2}\left(x_1 y_2 + \cdots + x_n y_1\right) - \frac{1}{2}\left(x_2 y_1 + \cdots + x_1 y_n\right), \qquad (7.2)$$

as we might remember from elementary analytic geometry. We must subtract segments of circles (the nonoverlapping white parts of the polygon in Fig. 7.5) from $A_{\text{polygon}}$ to obtain $A(\mathcal{R}_k)$. The sum of all the resulting areas, $A_{\text{acc}}$, allows us to compute $\lambda$ and to sample $\Delta_t$.

After sampling the waiting time, we must place a disk inside the accessible region. It falls into the small region $\mathcal{R}_k$ with probability $A(\mathcal{R}_k)$, and the index $k$ is obtained by tower sampling in $\{A(\mathcal{R}_1), \ldots, A(\mathcal{R}_K)\}$. Knowing that the disk center will fall into small area $k$, we must sample a random position $\mathbf{x}$ inside $\mathcal{R}_k$. This process involves tower sampling again, in $n$ triangles formed by neighboring pairs of vertices and a center point $\mathbf{x}_c$ (which, for a convex polygon, is inside its boundary) (see Alg. 7.2 (`direct-polygon`)). In this program, Alg. 7.3 (`direct-triangle`) samples a random point inside an arbitrary triangle; any sampled point in the white segments in Fig. 7.5 instead of in the small area is rejected, and the sampling repeated. Eventually, a point $\mathbf{x}$ inside the gray region in Fig. 7.5 will be found.

The algorithm for random deposition must keep track of the small regions $\mathcal{R}_k$ as they are modified, cut up, and finally eliminated during the deposition process. It is best to handle this task using oriented surfaces, by putting arrows on the boundaries of $\mathcal{R}_k$ and of the exclusion disk (see Fig. 7.6). We let the accessible region be on the left-hand side of

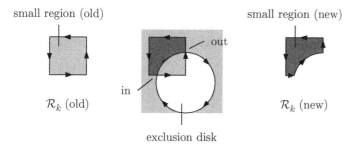

small region (old)    small region (new)

$\mathcal{R}_k$ (old)    in    out    $\mathcal{R}_k$ (new)

exclusion disk

**Fig. 7.6** Intersection of a small region $\mathcal{R}_k$ with an exclusion disk.

```
procedure direct-polygon
input {x₁,...,xₙ}
x_c ← ∑ x_k/n
x_{n+1} ← x₁
for k = 1,...,n do
    { A_k ← (x_c y_k + x_k y_{k+1} + x_{k+1} y_c
      -x_k y_c - x_{k+1} y_k - x_c y_{k+1})/2  (see eqn (7.2))
    k ← tower-sample(A₁,...,Aₙ)
    x ← direct-triangle(x_c, x_k, x_{k+1})
output x
```

**Algorithm 7.2** `direct-polygon`. Uniformly sampling a random position inside a convex polygon with $n > 3$ vertices.

the edges delimiting it. Arrows then go around $\mathcal{R}_k$ in an anticlockwise sense, and they circle the exclusion disk in a clockwise sense (other disk centers can be placed on its outside, again on the left-hand side of the boundary). The explicit trigonometric computations must produce an ordered list of "in" and "out" intersections, which will be new vertices. All the pieces from "out" to "in" intersections are part of the new boundary, in addition to the arcs of the exclusion disks from "in" to the next "out". An example of this calculation is shown in Fig. 7.6. The rules also apply when the exclusion disk cuts the small area $\mathcal{R}_k$ into several pieces. Algorithm 7.4 (`fast-deposition`) contains little more than the routines Alg. 7.2 (`direct-polygon`), Alg. 7.3 (`direct-triangle`), in addition to the tower-sampling algorithm. They should all be incorporated as subroutines and written and tested independently.

## 7.2 Dynamic spin algorithms

The faster-than-the-clock approach is ideally suited to random sequential deposition because disks are placed once and never move again. Keeping track of the accessible area generates overhead computations only in the

**procedure** direct-triangle
**input** $\{\mathbf{x}_1, \mathbf{x}_2, \mathbf{x}_3\}$
$\{\Upsilon_1, \Upsilon_2\} \leftarrow \{\mathtt{ran}\,(0,1), \mathtt{ran}\,(0,1)\}$
**if** $(\Upsilon_1 + \Upsilon_2 > 1)$ **then**
$\quad \{\ \{\Upsilon_1, \Upsilon_2\} \leftarrow \{1 - \Upsilon_1, 1 - \Upsilon_2\}$
$\mathbf{x} \leftarrow \mathbf{x}_1 + \Upsilon_1 \cdot (\mathbf{x}_2 - \mathbf{x}_1) + \Upsilon_2 \cdot (\mathbf{x}_3 - \mathbf{x}_1)$
**output** $\mathbf{x}$
——

**Algorithm 7.3** direct-triangle. Sampling a random pebble $\mathbf{x}$ inside a triangle with vertices $\{\mathbf{x}_1, \mathbf{x}_2, \mathbf{x}_3\}$.

**procedure** fast-deposition
**input** $\{\mathcal{R}_1, \ldots, \mathcal{R}_K\}, t$
$\lambda \leftarrow 1 - [\sum_l A(\mathcal{R}_l)] / A_{\text{tot}}$ (probability of doing nothing)
$\Delta_t \leftarrow 1 + \mathtt{int}\,[(\log \mathtt{ran}\,(0,1)) / \log \lambda]$ (see eqn (7.1))
$k \leftarrow \mathtt{tower\text{-}sample}\,(A(\mathcal{R}_1), \ldots, A(\mathcal{R}_K))$
1 $\quad \mathbf{x} \leftarrow \mathtt{direct\text{-}polygon}\,(\mathcal{R}_k)$
**if** ($\mathbf{x}$ not legal point in $\mathcal{R}_k$) **then goto** 1
**output** $t + \Delta_t, \{\mathcal{R}_1, \ldots, \mathcal{R}_K\}$
——

**Algorithm 7.4** fast-deposition. Faster-than-the-clock sequential deposition of disks. Small regions $\mathcal{R}_k$ are described by vertices and edges.

neighborhood of a successful deposition. Small regions $\mathcal{R}_k$ are cut up, amputated, or deleted but otherwise never change shape, nor do they pop up in unexpected places. The management of the "in" and "out" intersections in Fig. 7.6 contains the bulk of the programming effort.

Algorithm 7.4 (fast-deposition) is thus exceptionally simple, because it lacks the dynamics of the usual models, in short because disks do not move. Rather than pursue further the subject of moving disks, in this section we consider dynamic spin models which have the same problem. First, a single-spin Ising model in an external field lets us revisit the basic setup of the faster-than-the-clock approach. We then apply the faster-than-the-clock approach to the Ising model on a lattice, with its characteristic difficulty in computing the probability of rejecting all moves. Finally, we discuss the futility problem haunting many dynamic Monte Carlo schemes, even the most insightful ones.

## 7.2.1 Spin-flips and dice throws

We consider a single Ising spin $\sigma = \pm 1$ in a magnetic field $h$, at a finite temperature $T = 1/\beta$ (see Fig. 7.7). The energy of the up-spin configuration ($\sigma = 1$) is $-h$, and the energy of the down-spin configuration ($\sigma = -1$) is $+h$, in short,

$$E_\sigma = -h\sigma. \tag{7.3}$$

We model the time evolution of this system with the Metropolis algo-

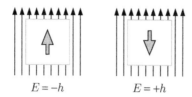

**Fig. 7.7** A single Ising spin in a magnetic field $h$.

rithm. If we write the energy change of a spin-flip as $\Delta_E = E_{-\sigma} - E_\sigma = 2h\sigma$, the probability of flipping a spin is

$$p(\sigma \rightarrow -\sigma) = \begin{cases} 1 & \text{if } \sigma = -1 \\ e^{-2\beta h} & \text{if } \sigma = +1 \end{cases}. \tag{7.4}$$

This transition probability satisfies detailed balance and ensures that at large times, the two spin configurations appear with their Boltzmann weights. (As usual, rejected moves leave the state of the system unchanged.) We keep in mind that our present goal is to simulate time evolution, and not only to generate configurations with the correct equilibrium weights.

At low temperature, and much as in Alg. 7.1 (**naive-deposition**), attempted flips of an up spin are rejected most of the time; $\sigma_{t+1}$ is then the same as $\sigma_t$. Again, we can set up the tower-of-probabilities scheme to find out how many subsequent rejections it will take until the up spin finally gets flipped. It follows from eqn (7.4) that any down spin at time $t$ is immediately flipped back up at the subsequent time step (see Fig. 7.8).

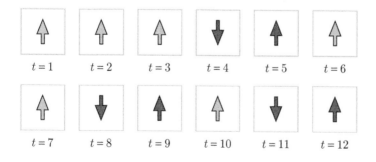

**Fig. 7.8** Time evolution of an Ising spin in a field. Flips from "+" to "−" at time $t$ are followed by back-flips to "+" at time $t + 1$.

For concreteness, we discuss an Ising spin at the parameter value

$h\beta = \frac{1}{2}\log 6$, where

$$\pi(+1) = e^{+\beta h} = e^{+\log\sqrt{6}} = \sqrt{6},$$
$$\pi(-1) = e^{-\beta h} = e^{-\log\sqrt{6}} = 1/\sqrt{6},$$

and where the magnetization comes out as

$$m = \langle\sigma\rangle = \frac{1\cdot\pi(1) + (-1)\cdot\pi(-1)}{\pi(1) + \pi(-1)} = \frac{\sqrt{6} - 1/\sqrt{6}}{\sqrt{6} + 1/\sqrt{6}} = \frac{5}{7}.$$

Because of eqn (7.4) ($\exp(-\log 6) = 1/6$), any up configuration has a probability $1/6$ to flip. This is the same as throwing a "flip-die" with five blank faces and one face that has "flip" written on it (see Fig. 7.9). This child's game is implemented by Alg. 7.5 (`naive-throw`): one random number is drawn per time step, but $5/6$ of the time, the flip is rejected. Rejected spin-flips do not change the state of the system. They only increment a counter and generate a little heat in the computer.

**Fig. 7.9** A child playing with a flip-die (see Alg. 7.5 (`naive-throw`)).

**procedure** `naive-throw`
**for** $t = 1, 2, \ldots$ **do**
$\left\{ \begin{array}{l} \Upsilon \leftarrow \mathbf{nran}(1, 6) \\ \mathbf{if}\ (\Upsilon = 1)\ \mathbf{output}\ t \end{array} \right.$

**Algorithm 7.5** `naive-throw`. This program outputs times at which the die in Fig. 7.9 shows a "flip".

Flip-die throwing can be implemented without rejections (see Alg. 7.6 (`fast-throw`)). The probability of drawing a blank face is $5/6$, the probability of drawing two blank faces in a row is $(5/6)^2$, etc. As in Subsection 7.1.1, the probability of drawing $k$ blank faces in a row, followed by drawing the "flip" face, is given by the probability of waiting at least $k$ times minus the probability of waiting at least $k+1$ throws:

$$\pi(k) = (5/6)^k - (5/6)^{k+1}.$$

$\pi(k)$ can be sampled, with rejections, by Alg. 7.5 (naive-throw), and without them by Alg. 7.6 (fast-throw).

**procedure fast-throw**
$\lambda \leftarrow 5/6$ (probability of doing nothing)
$t \leftarrow 0$
**for** $i = 1, 2, \ldots$ **do**
$\left\{\begin{array}{l} \Delta_t \leftarrow 1 + \text{Int}\left\{\log\left[\text{ran}\,(0,1)\right] / \log \lambda\right\} \\ t \leftarrow t+ \\ \textbf{output } t + \Delta_t \end{array}\right.$

**Algorithm 7.6** fast-throw. Faster-than-the-clock implementation of flip-die throwing.

The rejection-free Alg. 7.6 (fast-throw) generates one flip per random number and runs faster than Alg. 7.5 (naive-throw), with statistically identical output. The flip-die program is easily made into a simulation program for the Ising spin, spending exactly one step at a time in the down configuration and on average six steps at a time in the up configuration, so that the magnetization comes out equal to 5/7.

## 7.2.2  Accelerated algorithms for discrete systems

From the case of a single spin in a magnetic field, we now pass to the full-fledged simulation of the Ising model on $N$ sites, with time-dependent configurations $\sigma = \{\sigma_1, \ldots, \sigma_N\}$. We denote by $\sigma^{[k]}$ the configuration obtained from $\sigma$ by flipping the spin $k$. The energy change from a spin-flip, $\Delta_E = E_{\sigma^{[k]}} - E_\sigma$, enters into the probability of flipping a spin in the Metropolis algorithm:

$$p(\sigma \to \sigma^{[k]}) = \frac{1}{N} \min\left(1, e^{-\beta \Delta_E}\right).$$

This equation in fact corresponds to the local Metropolis algorithm (it is implemented in Alg. 5.7 (markov-ising)). The first term, $1/N$, gives the probability of selecting spin $k$, followed by the Metropolis probability of accepting a flip of that spin. In the Ising model and its variants at low temperature, most spin-flips are rejected. It can then be interesting to implement a faster-than-the-clock algorithm which first samples the time of the next spin-flip, and then the spin to be flipped, just as in the earlier deposition problem. Rejections are avoided altogether, although the method is not unproblematic (see Subsection 7.2.3). The probability $\lambda = 5/6$ of drawing a blank face in Alg. 7.6 (fast-throw) must now be generalized into the probability of doing nothing during one iteration of

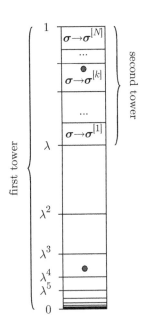

**Fig. 7.10** Two-pebble tower sampling in the Ising model. The first pebble, $\text{ran}\,(0,1)$, determines the waiting time $\Delta_t$, as in Fig. 7.3. Then, $\text{ran}\,(\lambda,1)$ samples the spin $k$ to be flipped.

the Metropolis algorithm,

$$\lambda = 1 - \sum_{k=1}^{N} p(\boldsymbol{\sigma} \to \boldsymbol{\sigma}^{[k]}). \tag{7.5}$$

This equation expresses that to determine the probability of doing nothing, we must know all the probabilities for flipping spins. Naively, we can recalculate $\lambda$ from eqn (7.5) after each step, and sample the waiting time as in Alg. 7.6 (`fast-throw`). After finding out when to flip the next spin, we must decide on which of the $N$ spins to flip. This problem is solved through a second application of tower sampling (see Alg. 7.7 (`dynamic-ising`), and Fig. 7.10). Output generated by this program is statistically indistinguishable from that of Alg. 5.7 (`markov-ising`). However, each spin-flip requires of the order of $N$ operations.

**procedure** `dynamic-ising`
**input** $t, \{\sigma_1, \ldots, \sigma_N\}$
**for** $k = 1, \ldots, N$ **do**
$\quad \left\{ \; p_k \leftarrow p(\boldsymbol{\sigma} \to \boldsymbol{\sigma}^{[k]}) \right.$
$\lambda \leftarrow 1 - \sum_k p_k$
$\Delta_t \leftarrow 1 + \text{int}[\log\,[\text{ran}\,(0,1)] \,/ \log \lambda]$
$l \leftarrow \texttt{tower-sample}(p_1, \ldots, p_N)$
$\sigma_l \leftarrow -\sigma_l$
$t \leftarrow t + \Delta_t$
**output** $t, \{\sigma_1, \ldots, \sigma_N\}$

——

**Algorithm 7.7** `dynamic-ising`. Simulation of the Ising model using the faster-than-the-clock approach (see Alg. 7.8 (`dynamic-ising(patch)`)).

The recalculation from scratch of $\lambda$, the probability of doing nothing, is easily avoided, because flipping a spin only changes the local environment of nearby spins, and does not touch most other spins. These other spins have as many up and down neighbors as before. The possible environments (numbers of up and down neighbors) fall into a finite number $n$ of classes. We must simply perform bookkeeping on the number of members of each class (see Alg. 7.8 (`dynamic-ising(patch)`)). The first paper on accelerated dynamic Monte Carlo algorithms, by Bortz, Kalos, and Lebowitz (1975), coined the name "$n$-fold way" for this strategy.

For concreteness, we consider a two-dimensional Ising model with isotropic interactions and periodic boundary conditions. Spin environments may be grouped into ten classes, from class 1 (for an up spin surrounded by four up spins) to class 10 (for a down spin surrounded by four down spins) (see Fig. 7.11). The tower of probabilities of all spin-flips can be reduced to a tower of 10 classes of spins, if we only know the number $\mathcal{N}_k$ of spins in each class $k$.

Flipping a spin thus involves concerted actions on the classes, as shown in the example in Fig. 7.12: the central up spin, being surrounded by only one up spin, is in class $k = 4$. Flipping it brings it into class $f(4) = 9$

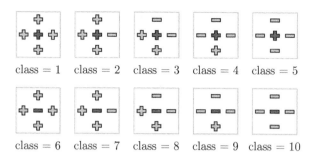

class = 1   class = 2   class = 3   class = 4   class = 5

class = 6   class = 7   class = 8   class = 9   class = 10

**Fig. 7.11** Classes of the two-dimensional Ising model with periodic boundary conditions (permuting neighbors does not change the class).

**Table 7.1** The 10 classes of the $n$-fold way: a flip of a spin moves it from class $k$ to $f(k)$. The flip of a neighboring spin, itself in class $j$, moves it from class $k$ to $g_j(k)$, see Fig. 7.12.

| | Site | Neighbor flip | |
|---|---|---|---|
| $k$ | $f(k)$ | $g_{1-5}(k)$ | $g_{6-10}(k)$ |
| 1 | 6 | 2 | – |
| 2 | 7 | 3 | 1 |
| 3 | 8 | 4 | 2 |
| 4 | 9 | 5 | 3 |
| 5 | 10 | 6 | 4 |
| 6 | 1 | 7 | 5 |
| 7 | 2 | 8 | 6 |
| 8 | 3 | 9 | 7 |
| 9 | 4 | 10 | 8 |
| 10 | 5 | – | 9 |

(see Table 7.1). Likewise, flipping the central spin (in class $l = 4$) transfers its right-hand neighbor from class $k = 1$ to class $g_4(1) = 2$. All these operations on the classes are encoded into functions $f(k)$ and $g_j(k)$ in Table 7.1. To be complete, the bookkeeping has to act on the sets $\{\mathcal{S}_1, \ldots, \mathcal{S}_{10}\}$ (set $\mathcal{S}_k$ contains all the spins (sites) in class $k$): changing a spin on site $k$ from class $l$ to class $j$ implies that we have to move $k$ from set $l$ ($\mathcal{S}_l \to \mathcal{S}_l \setminus \{k\}$) to set $j$ ($\mathcal{S}_j \to \mathcal{S}_j \cup \{k\}$). These computationally cheap operations should be outsourced into subroutines. As a consequence, the single-spin-flip algorithm runs on the order of const$\cdot N$ times faster than Alg. 7.7 (`dynamic-ising`), although the constant is quite small. However, at all but the smallest temperatures, the bookkeeping involved makes it go slower than the basic Alg. 5.7 (`markov-ising`).

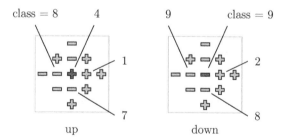

class = 8    4          9    class = 9

1          2

7          8

up          down

**Fig. 7.12** Consequences of a spin-flip of the central spin for the classes of the spin itself and its neighbors (see Table 7.1).

### 7.2.3   Futility

In previous chapters, we rated algorithms with a high acceptance probability as much better than methods which rejected most moves. Rejections appeared wasteful, and indicated that the a priori probability was inappropriate for the problem we were trying to solve. On the other hand,

**procedure** dynamic-ising(patch)
**input** $t, \{c_1, \ldots, c_N\}, \{\mathcal{S}_1, \ldots, \mathcal{S}_{10}\}$
**for** $k = 1, \ldots, 10$ **do**
$\{\; p_k \leftarrow \mathcal{N}(\mathcal{S}_k) p(\mathcal{S}_k)$
$\lambda \leftarrow 1 - \sum_{k=1}^{10} p_k$
$\Delta_t \leftarrow 1 + \text{int}[\log[\text{ran}(0,1)] / \log \lambda]$
$k \leftarrow \text{tower-sample}(p_1, \ldots, p_{10})$
$l \leftarrow$ random element of $\mathcal{S}_k$
**for** all neighbors $m$ of $l$ **do**
$\left\{ \begin{array}{l} \mathcal{S}_{c_m} \leftarrow \mathcal{S}_{c_m} \setminus \{m\} \\ c_m \leftarrow f_{\text{neigh}}(c_m, c_l) \\ \mathcal{S}_{c_m} \leftarrow \mathcal{S}_{c_m} \uplus \{m\} \end{array} \right.$
$c_l \leftarrow f_{\text{site}}(c_m, c_l)$
**output** $t + \Delta_t, \{c_1, \ldots, c_N\}, \{\mathcal{S}_1, \ldots, \mathcal{S}_{10}\}$

——

**Algorithm 7.8** dynamic-ising(patch). Improved version of Alg. 7.7 (dynamic-ising) using classes. Each spin-flip involves bookkeeping.

algorithms with a large acceptance probability allowed us to increase the range of moves (the throwing range), thereby moving faster through phase space. From this angle, faster-than-the-clock methods may appear to be of more worth than they really are: with their vanishing rejection rate, they seem to qualify for an optimal rating!

In a dynamic simulation with Alg. 7.7 (dynamic-ising), and in many other dynamic models which might be attacked with a faster-than-the-clock approach, we may soon be disappointed by the futility of the system's dynamics, even though it has no rejections. Intricate behind-the-scenes bookkeeping makes improbable moves or flips happen. The system climbs up in energy, but then takes the first opportunity to slide back down in energy to where it came from. We have no choice but to diligently undo all bookkeeping, before the same unproductive back-and-forth motion starts again elsewhere in the system.

Low-temperature dynamics, starting at configuration $a$ in Fig. 7.13, will drive the system to either $b$ or $c$, from where it will almost certainly fall back at the next time step. At low temperature, the system will take a very long time (and, more importantly, a very large number of operations) before hopping over one of the potential barriers and getting to either $d$ or $e$. In these cases, the dynamics is extremely repetitive, and futile. It may be wasteful to recompute the tower of probabilities, with or without bookkeeping tricks (the classes in Fig. 7.12). We may be better off saving much of the information about the probabilities for convenient reuse and lookup. After a move $\sigma \to \sigma^{[k]}$, we may look in an archive to quickly decide whether we have seen the current configuration before, and whether we can recycle an old tower of probabilities. There are many variations of this scheme.

To overcome problems such as the above futility, we may be tempted to implement special algorithms, but this cannot be achieved without effort.

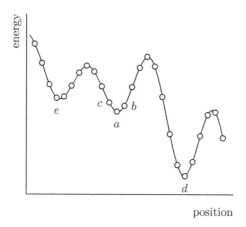

**Fig. 7.13** Futile dynamics: motion between points $a$, $b$, and $c$ is rejection-free, but it still takes a long time to get to $d$ or to $e$.

Before embarking on a special programming project, we should lucidly evaluate the benefits, hidden problems, programming skills required, etc. The computer time saved by insightful algorithms is easily overspent on human time in writing and debugging them. Naive implementations, such as those used throughout this book, keep us from getting trapped by the development of overcomplicated procedures which, after all, may be less powerful than we suspected.

## 7.3 Disks on the unit sphere

The Monte Carlo method that has accompanied us throughout this book has outgrown its origins in statistical physics. Nowadays, Monte Carlo methods appear wherever there is computing and modeling. Some specific tools of the trade are geared towards improving sampling efficiency, for example for integration in intermediate dimensions, when Riemann discretization and its variants no longer work.

Monte Carlo methods also help in solving general optimization problems in science and engineering. These problems can generally be formulated as artificial statistical-physics models whose ground state at zero temperature or at infinite pressure contains the sought-after solution. The simulation of the artificial model starts from high temperature or low pressure, with temperature gradually going down or pressure slowly increasing until the ground state is reached. This procedure, called simulated annealing, was introduced by Kirkpatrick, Gelatt, and Vecchi (1983), and constitutes an outstanding naive approach to hard optimization problems for which no other good solution is available, following the motto: "If all else fails, try statistical mechanics!"

In the present section, we study simulated annealing in a familiar context, again involving hard spheres, namely the close packing of disks on the surface of a sphere. This problem and its generalizations have been studied for a long time. Hundreds of papers dedicated to the problem are spread out over the mathematics and natural-sciences literature (see Conway and Sloane (1993)). The packing of disks has important applications, and it is closely connected with J. J. Thomson's problem of finding the energetically best position of equal charges on a sphere.

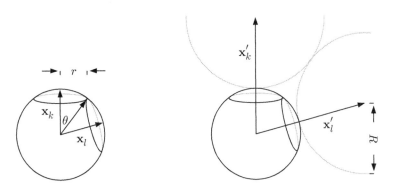

**Fig. 7.14** Two disks of radius $r$ (opening angle $\theta$) on the unit sphere (*left*). Spheres of radius $R$ touching the unit sphere (*right*).

We consider disks $k$ with an opening angle $\theta$ around a central vector $\mathbf{x}_k$, where $|\mathbf{x}_k| = 1$. The opening angle is related to the radius of the disks by $\sin \theta = r$ (see Fig. 7.14). Two disks with central vectors $\mathbf{x}_k$ and $\mathbf{x}_l$ overlap if

$$(\mathbf{x}_k \cdot \mathbf{x}_l) > \cos(2\theta)$$

or, equivalently, if

$$|\mathbf{x}_k - \mathbf{x}_l| < 2r.$$

The covering density of $N$ nonoverlapping disks, that is, the surface area of the spherical caps of opening angle $\theta$, is

$$\eta = \frac{2\pi N \int_0^\theta d\theta' \sin \theta'}{4\pi} = \frac{N}{2}(1 - \cos \theta),$$

where $\cos \theta = \sqrt{1 - r^2}$.

Nonoverlapping disks with central vectors $\mathbf{x}_k$, $|\mathbf{x}_k| = 1$, must satisfy $\min_{k \neq l} |\mathbf{x}_k - \mathbf{x}_l| > 2r$. The close-packing configuration $\{\mathbf{x}_1, \ldots, \mathbf{x}_N\}$ maximizes this minimum distance between the $N$ vectors, that is, it realizes the maximum

$$\max_{\substack{\{\mathbf{x}_1,\ldots,\mathbf{x}_N\} \\ |\mathbf{x}_k|=1}} \left( \min_{k \neq l} |\mathbf{x}_k - \mathbf{x}_l| \right). \tag{7.6}$$

The disk-packing problem is equivalent to the problem of the closest packing of spheres of radius

$$R = \frac{1}{1/r - 1} \tag{7.7}$$

touching the surface of the unit sphere. This follows from Fig. 7.14 because the touching spheres have $|\mathbf{x}'_k - \mathbf{x}'_l| = 2R$, where $|\mathbf{x}'_l| = 1 + R$. It was in this formulation, and for the special case $R = 1$ and $N = 13$, that the packing problem of 13 spheres was first debated, more than 300 years ago, by Newton and Gregory: Newton suspected, and Schütte and van der Waerden (1951) proved a long time later, that $R$ must be slightly below one. This means that 13 spheres can be packed around a central unit sphere only if their radius is smaller than 1. The best solution known has $r = 0.4782$, which corresponds to $R = 0.9165$ (see eqn (7.7)).

Disks with central vectors $\{\mathbf{x}_1, \ldots, \mathbf{x}_N\}$ are most simply placed on a sphere with $N$ calls to Alg. 1.22 (direct-surface). The minimum distance between vectors gives the maximum disk radius

$$r_{\max} = \frac{1}{2} \min_{k<l} |\mathbf{x}_k - \mathbf{x}_l|,$$

and must be compared with the disk radius. We reject the sample and try again if $r_{\max} < r$. In the limit of infinite time, this approach will come up with the optimal solution, but it is by no means practical, as was discussed in Chapter 2. In Fig. 7.15, a configuration of 16 disks and density $\eta = 0.3$ obtained with the direct-sampling algorithm is compared with a configuration of disks in a rectangular box with periodic boundary conditions and with same area. We note that both the surface of a sphere and a torus are homogeneous substrates (all bulk, no boundary, all points equivalent) and that the surface of a sphere is isotropic. On a sphere, all directions are equivalent, but not on a torus.

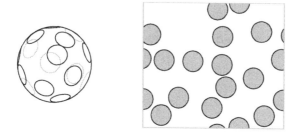

**Fig. 7.15** Equal disks (density $\eta = 0.3$, $N = 16$) on a sphere (*left*) and in a rectangular box with periodic boundary conditions (*right*).

For a long time, the problem of the closest packing on the unit sphere was formulated as a continuous minimization problem of the potential energy of particles with a two-body interaction

$$E_\alpha(\mathbf{x}_1, \ldots, \mathbf{x}_N) = \sum_{k<l} \frac{1}{|\mathbf{x}_k - \mathbf{x}_l|^\alpha},$$

in the limiting case $\alpha \to \infty$. For $\alpha = 1$, the minimum of the potential energy finds the equilibrium positions of equal static charges with a "1/distance" Coulomb interaction. This problem first appeared in the context of J. J. Thomson's "plum pudding" model of the atom (the precursor of Rutherford's atomic model) with electrons spread out on the surface of a sphere of uniform positive charge. For large $\alpha$, short distances dominate the energy more and more. (Note that, for example, $1/0.44^{1000} \ggg 1/0.45^{1000}$.) In the limit $\alpha \to \infty$, only the shortest interparticle distances contribute to the energy, and the minimum-energy configuration in the limit $\alpha \to \infty$ solves the disk-packing problem.

Continuous minimization routines (such as the Newton–Raphson algorithm) have been applied to find local minima of $E_\alpha(\mathbf{x}_1, \ldots, \mathbf{x}_N)$ for given $\alpha$, and to follow these minima with increasing $\alpha$. This strategy of continuous minimization was carried furthest by Kottwitz (1991), with $\alpha$ covering the astonishing range from 80 to more than a million. Running such a sophisticated Newton–Raphson program is a delicate task. Nevertheless, the approach proved to be successful, and most of the configurations found have not been improved upon. The empirical close-packed configurations for small $N$ are almost as firmly established as the mathematically proven optimal configurations for $N \le 12$ and $N = 24$, which have been known for many decades.

In this section, we use the much simpler approach of simulated annealing to find data as good as were ever discovered before. After discussing the method, we analyze the results in the asymptotic limit $N \to \infty$ and monitor the performance of variants of the model. Finally, in Subsection 7.3.4, we analyze the close-packing problem from the point of view of graph theory.

## 7.3.1  Simulated annealing

The configurations in Fig. 7.15 were obtained by direct sampling, but a Markov-chain algorithm allows us to go further. It is a simple matter to spruce up the Markov-chain sampling of pebbles on the surface of the unit sphere, Alg. 1.24 (`markov-surface`), into a hard-disk algorithm in which moves $\mathbf{x} \to \mathbf{x}'$ are proposed with the same probability as $\mathbf{x}' \to \mathbf{x}$, so that detailed balance is satisfied (see Alg. 7.9 (`markov-sphere-disks`) and Fig. 7.16).

**Fig. 7.16** Monte Carlo move (from $\mathbf{x}$ to $\mathbf{x}'$) on the sphere (see Alg. 7.9 (`markov-sphere-disks`)).

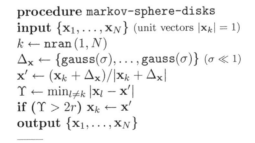

**procedure** `markov-sphere-disks`
**input** $\{\mathbf{x}_1, \ldots, \mathbf{x}_N\}$ (unit vectors $|\mathbf{x}_k| = 1$)
$k \leftarrow \mathrm{nran}\,(1, N)$
$\Delta_{\mathbf{x}} \leftarrow \{\mathrm{gauss}(\sigma), \ldots, \mathrm{gauss}(\sigma)\}$ ($\sigma \ll 1$)
$\mathbf{x}' \leftarrow (\mathbf{x}_k + \Delta_{\mathbf{x}})/|\mathbf{x}_k + \Delta_{\mathbf{x}}|$
$\Upsilon \leftarrow \min_{l \ne k} |\mathbf{x}_l - \mathbf{x}'|$
**if** $(\Upsilon > 2r)$ $\mathbf{x}_k \leftarrow \mathbf{x}'$
**output** $\{\mathbf{x}_1, \ldots, \mathbf{x}_N\}$
———

**Algorithm 7.9** `markov-sphere-disks`. Markov-chain algorithm for disks of radius $r$ on the unit sphere (with $\sigma \ll 1$).

During a Markov-chain simulation, disks almost never touch. This suggests that we should slightly swell the disks at certain times during the simulation, by a small fraction $\gamma$ of some maximum possible increase that would still keep the configuration legal (see Alg. 7.10 (resize-disks)). This program should be sandwiched in between long runs of the Markov-chain simulation. Combining Markov-chain simulation with careful resizing is reminiscent of the annealing procedure in metallurgy in which a metal is slowly cooled from high temperature in order to drive out grain boundaries and other imperfections to make it less brittle. On the computer, the approach is called simulated annealing. In our example, we approach infinite pressure rather than zero temperature.

**procedure** resize-disks
**input** $\{x_1, \ldots, x_N\}, r$
$\Upsilon \leftarrow \min_{k \neq l} |x_k - x_l|/2$
$r \leftarrow r + \gamma \cdot (\Upsilon - r)$
**output** $\{x_1, \ldots, x_N\}, r$

————

**Algorithm 7.10** resize-disks. Resizing disks by a factor $0 < \gamma \ll 1$ (the minimum is over $1 \leq k, l \leq N$, with $k \neq l$.)

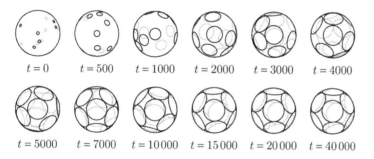

$t = 0$ $\quad$ $t = 500$ $\quad$ $t = 1000$ $\quad$ $t = 2000$ $\quad$ $t = 3000$ $\quad$ $t = 4000$

$t = 5000$ $\quad$ $t = 7000$ $\quad$ $t = 10\,000$ $\quad$ $t = 15\,000$ $\quad$ $t = 20\,000$ $\quad$ $t = 40\,000$

**Fig. 7.17** Simulated-annealing run for 13 disks on the unit sphere. The final density is $\eta = 0.791393$.

Simulated annealing is easily tried for 13 equal disks on a sphere (see Fig. 7.17, where one time step ($\Delta_t = 1$) consists of a single resizing of the disks with $\gamma = 0.01$, and 10 000 iterations of Alg. 7.9 (markov-sphere-disks), so that the whole simulation in Fig. 7.17 contains $4 \times 10^8$ moves). The step width, set by the standard deviation $\sigma$ of the Gaussians, is automatically adjusted after each resizing in order to keep the acceptance probability on the order of $\frac{1}{2}$. The packing density $\eta$ slowly increases during the run until the disks settle into a jammed configuration, and the step width goes to zero.

Naturally, there are many inequivalent jammed configurations of disks on the unit sphere, and most of them are not global but local minima (see Fig. 7.18). Even local minima can trap Alg. 7.9 (markov-sphere-disks) forever (for a small step size). However, running the simulated-annealing

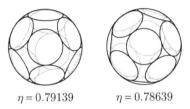

$\eta = 0.79139$         $\eta = 0.78639$

**Fig. 7.18** Optimal (*left*) and nonoptimal (*right*) jammed configurations for $N = 13$ (both obtained by simulated annealing).

**Table 7.2** Densities obtained in the runs shown in Fig. 7.19 ($\gamma = 0.01$). The configurations agree with proven or empirical optima.

| $N$ | Density $\eta$ |
|-----|----------------|
| 5 | 0.73223 |
| 6 | 0.87868 |
| 8 | 0.82358 |
| 12 | 0.89609 |
| 13 | 0.79139 |
| 15 | 0.80716 |
| 19 | 0.81096 |
| 24 | 0.86170 |
| 48 | 0.85963 |

algorithm many times (with small $\gamma$ and different random numbers) we shall notice that the solution with the highest density, up to global rotations, is obtained in the vast majority of runs. This indicates that the configuration is probably the optimal packing—it certainly has a large basin of attraction.

Results of simulated-annealing runs for different values of $N$ are shown in Fig. 7.19 (see also Table 7.2). The density depends very much on the number of disks; it increases slowly with $N$ but clearly stays below the close-packing density in two dimensions, $\eta_{max} = \pi/(2\sqrt{3}) = 0.907$.

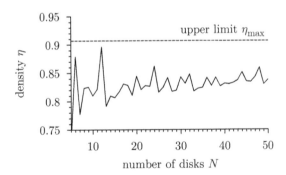

**Fig. 7.19** Density obtained by simulated annealing of disks on the surface of a sphere, for $5 \leq N \leq 50$. Most of the configurations are optimal.

Roughly speaking, simulated annealing works because an equilibrium physical system approaches the ground state in the limit of zero temperature. Likewise, the system of hard disks reaches the state of maximum compression (minimum volume) in the limit of infinite pressure. For infinitely slow annealing, the system remains in equilibrium, and almost always ends up in the ground state. In order to rigorously conform to this description, we would need to perform simulated annealing at constant pressure, similar to what we considered in Section 2.3.4. It would be best to take disks of fixed radius $r = \frac{1}{2}$ on a central sphere of variable radius $R$. The surface of the sphere generates a Boltzmann factor $\exp\left(-\beta PR^2\right)$. During the simulation, the pressure $P$, and thus the density, would increase slowly, and the outcome would be similar to that of Alg. 7.10 (`resize-disks`). This constant-pressure routine can escape from any local minimum. It is ergodic, as opposed to our simplified routine in the constant-volume ensemble, which is ergodic for some densities, and for others can be trapped by local minima, in the limit of small step size (see Fig. 7.20). A finite ergodic system must reach the most compressed state in the limit of infinitely slow pressure increase. This is the theoretical basis of simulated annealing.

However, the above argument is purely formal—practical annealing cannot proceed infinitely slowly, and usually not even slowly enough for applications. In the language of Subsection 1.4.2, the constant-pressure algorithm is practically nonergodic: it would spend weeks of computer times before escaping from local minima. In fact, both the constant-pressure algorithm and our simplified implementation rarely fall into local minima, because the basins of attractions of those minima are very small; in other words, simulating annealing rarely ever gets dangerously close to local minima. This nice behavior is specific to disks, spheres, and central potentials, and disappears for polydisperse (unequal) disks where the simulated annealing, with the local algorithm, is less successful. Good solutions are still found, but they are no longer the optimal ones (see Subsection 7.3.3).

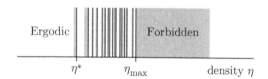

**Fig. 7.20** Jammed disk configurations, and their basins of attraction, for densities between $\eta^*$ and $\eta_{\mathrm{max}}$ (schematic).

## 7.3.2  Asymptotic densities and paper-cutting

As discussed, we have reasons to believe that the packings in Fig. 7.19 are the best possible. Irritatingly, however, the packing densities (except for $N = 12$) are substantially smaller than the close-packing density in the plane $\eta_{\mathrm{max}} = \pi/(2\sqrt{3}) = 0.907$, and we see no clear tendency driving $\eta(N)$ all the way to this upper limit. We must find out whether the close-packing solution for $N$ disks on a sphere has the same packing density as the close-packing solution for $N$ disks on a torus for $N \to \infty$.

We could employ general arguments to settle this point, but, rather, we shall construct hard-disk configurations that actually achieve $\eta \to \eta_{\mathrm{max}}$ in the large-$N$ limit (see Habicht and van der Waerden (1951)). This proves that in Fig. 7.19, we have simply not gone to large enough systems to see that the upper limit is eventually reached. The idea behind the

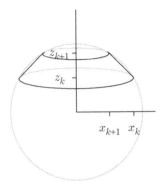

**Fig. 7.21** Cone section on the inside of the unit sphere, formed by one of the strips in Fig. 7.22.

construction is to build a cone section model that we can imagine cut out in paper and glued together inside the unit sphere (see Figs 7.21 and 7.22). Hexagonally close-packed configurations are drawn on the paper strips, and centers of disks that do not cut a boundary (the gray disks in Fig. 7.22) are projected onto the sphere. As the disks get smaller, fewer and fewer of them are on the boundaries (the white disks in the figure), and the packing fraction on the strip approaches the hexagonal close-packing density. Moreover, as the strips get smaller, their total area approaches the area of the sphere. It suffices to let the disks decrease in size more quickly than the strips to reach $\eta_{max}$. (See Exerc. 7.13 for a paper-cutting competition.)

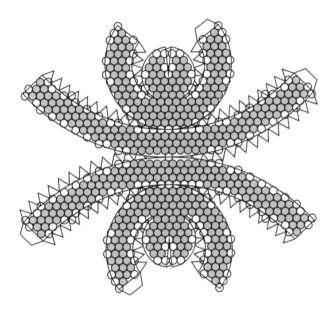

**Fig. 7.22** Cone section model consisting of close-packed strips of equal width, to be glued together and assembled inside the unit sphere.

The area of one strip in Fig. 7.22 is

$$\left\{ \begin{array}{c} \text{area} \\ \text{(one strip)} \end{array} \right\} = 2\pi \underbrace{\frac{x_{k+1} + x_k}{2}}_{\text{average strip length}} \underbrace{\sqrt{(z_{k+1} - z_k)^2 + (x_{k+1} - x_k)^2}}_{\text{strip width}}.$$

It is better to use angles $\phi_k$, where $x_k = \cos\phi_k$ and $z_k = \sin\phi_k$ ($\phi_{k+1} - \phi_k = \pi/(2n)$). The strip width is $2\sin[\pi/(4n)]$, and we find the following for the total area of a cone section model with $2n$ strips:

$$\frac{S_n}{4\pi} = \sum_{k=0}^{n-1} \frac{1}{2} \left( \cos\phi_{k+1} + \cos\phi_k \right) \left( 2\sin\frac{\pi}{4n} \right)$$

(where $\phi_k = k\pi/(2n)$), which may be rewritten and expanded in terms of $1/n$:

$$\frac{S_n}{4\pi} = \underbrace{\sum_{k=0}^{n-1} \cos\frac{(2k+1)\pi}{4n} \sin\frac{\pi}{2n}}_{\frac{1}{2}\sin^{-1}[\pi/(4n)]} \simeq 1 - \frac{\pi^2}{32n^2} + \cdots.$$

As the disks get smaller (in the large-$N$ limit), the fraction of cut-up disks that cannot be transferred from the strips to the sphere goes to zero. We may suppose the disks to have identical size on the strip and on the surface of the sphere, and then the packing density on a cone section model with only 12 strips ($n = 6$) already approaches 99% of the close-packing density in the plane (see Table 7.3).

We now increase the number of strips, in order to reach 100% of the close-packing density. This obliges us to think about the strip boundaries, and to exclude a zone of width $\simeq 2r$ from the strip. The total length of the boundaries is proportional to $n$, and we find

$$\left\{\begin{matrix} \text{free area} \\ \text{with } 2n \text{ strips} \end{matrix}\right\} = 4\pi\left(1 - \frac{c}{n^2} - c'\cdot nr\right).$$

This free area is maximized for $n = c''/r^{1/3}$; it varies according to

$$\left\{\begin{matrix} \text{free area} \\ \text{with } \propto r^{-1/3} \text{ strips} \end{matrix}\right\} = 4\pi\left(1 - \text{const}\cdot r^{2/3}\right). \qquad (7.8)$$

Equation (7.8) proves that the disposable area, in the limit $r \to 0$, goes to $4\pi$, although this limit is reached extremely slowly, and at the price of introducing a large number of grain boundaries separating regions with mutually incompatible hexagonal close-packed crystals. Nevertheless, in the large-$N$ limit, these crystallites contain infinitely many disks.

The conclusion of our paper-cutting exercise is that the best packing of $N$ disks on the unit sphere, in the limit $N \to \infty$, reaches the hexagonal close-packing density of disks in the plane. Furthermore, this optimal packing density grows very slowly with $N$. From eqn (7.8), it follows that the density of the homogeneous planar system is approached as

$$\simeq 1 - \frac{\text{const}}{N^{1/3}}.$$

The best packing thus has long grain boundaries, regions where the hexagonal ordering is perturbed. The number of grains increases with $N$, but less than proportionally, so that, in the limit $N \to \infty$, the grains contain more and more particles. Their extension, measured in multiples of the disk radius, diverges. Evidently, however, we do not expect the grain boundaries of the true optimal packing of $N$ disks to form concentric circles around the $z$-axis.

In the thermodynamic limit, packing densities on a sphere and in a plane thus become equivalent. In retrospect, however, we were well-advised in Chapter 2 to study the liquid–solid phase transition with periodic boundary conditions on an abstract torus, rather than on a

**Table 7.3** Surface area $S_n$ of a cone section model with $2n$ strips compared with the surface area of the unit sphere

| $n$ | $S_n/(4\pi)$ | $1 - \pi^2/(32n^2)$ |
|---|---|---|
| 1 | 0.70711 | 0.69157 |
| 2 | 0.92388 | 0.92289 |
| 3 | 0.96593 | 0.96573 |
| 6 | 0.99144 | 0.99143 |

sphere: in that case, the system sizes available for simulation are too small to capture the difference between a polycrystalline material and an amorphous block of matter.

### 7.3.3   Polydisperse disks and the glass transition

Unerringly, the simulated-annealing algorithm in Section 7.3.1 reaches the globally optimal solution, sidestepping the many local minima on its way. Notwithstanding this success, we must understand that simulated annealing is more a great "first try" than a prodigious scout of ground states and a true solver of general optimization problems. To see this in an example, we simply consider unequal (polydisperse) disks on the surface of the unit sphere instead of the equal disks studied so far (see Fig. 7.23). For concreteness, let us assume that disk $k$ has an opening angle

$$\theta_k = \theta \cdot (1 + \delta_k), \tag{7.9}$$

so that the density is equal to

$$\eta = \sum_{k=1}^{N} \left( \frac{1}{2} - \cos \theta_k \right).$$

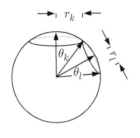

**Fig. 7.23** Polydisperse disks (with central vectors $\mathbf{x}_k$ and $\mathbf{x}_l$, and opening angles $\theta_k$ and $\theta_l$) on the surface of the unit sphere. Disks overlap if their scalar product $(\mathbf{x}_k \cdot \mathbf{x}_l)$ is greater than $\cos(\theta_k + \theta_l)$.

The optimization problem, a generalization of eqn (7.6), now consists in maximizing $\theta$ for a fixed ratio of the opening angles. Disks $k$ and $l$ overlap if $\arccos(\mathbf{x}_k \cdot \mathbf{x}_l) > \theta \cdot (2 + \delta_k + \delta_l)$, and we must solve the following optimization problem:

$$\max_{\substack{\{\mathbf{x}_1,\ldots,\mathbf{x}_N\} \\ |\mathbf{x}_1|=\cdots=|\mathbf{x}_N|=1}} \left[ \min_{k<l} \frac{\arccos(\mathbf{x}_k \cdot \mathbf{x}_l)}{2 + \delta_k + \delta_l} \right].$$

It is a simple matter to modify Alg. 7.9 (`markov-sphere-disks`) for polydisperse disks—it is best to work directly with opening angles—and attempt increases of $\theta$ in eqn (7.9). This modified code gets trapped in a different final configuration virtually each time it is run, even if the annealing is very slow (see Fig. 7.24, where $\delta_k = 0.2/N \cdot [k - \frac{1}{2}(N+1)]$ has been used).

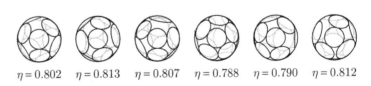

$\eta = 0.802$   $\eta = 0.813$   $\eta = 0.807$   $\eta = 0.788$   $\eta = 0.790$   $\eta = 0.812$

**Fig. 7.24** Inequivalent jammed configurations of 13 unequal disks on the unit sphere found by slow simulated annealing.

In Subsection 7.3.1, we described the motivation for the slow pressure increase by analogy with the physical process of annealing, where imperfections are driven out of a material, and the system is brought to the ground state. Annealing (in real life) must be done with care, especially in the presence of imperfections and disorder, because otherwise the system falls out of equilibrium and gets stuck in a metastable state. If the imperfections are too pronounced, the final result of annealing is very often a glass, rather than crystalline matter. This also happens in the simulated annealing process of polydisperse disks on a sphere. Disorder prevents the system from finding its ground state in a reasonable time interval, and keeps it blocked in a random jammed configuration, which now has a much larger basin of attraction than in the monodisperse case. The relations between this phenomenon (in Monte Carlo algorithms) and the glass transition (in nature) have been widely discussed in the literature (see, for example, Santen and Krauth (2000)).

We stress again that both the ease with which the monodisperse system falls into the optimal solution and the difficulty that the polydisperse system has in doing the same are remarkable features of the hard-disk system, and can also be found in three-dimensional hard-sphere systems and more generally in systems interacting with central potentials. These phenomena are intimately linked to the dynamics near phase transitions and the physics of glasses, and many fundamental problems are still unresolved.

## 7.3.4   Jamming and planar graphs

Up to now, we have studied the packing of equal disks on the unit sphere from an empirical point of view, mainly as a case study of simulated annealing. We can reach a deeper theoretical understanding of the problem by introducing the concept of a contact graph. In the case of a jammed configuration on the sphere, we draw an edge between the centers of disks $k$ and $l$ if they are in contact, that is, if $|\mathbf{x}_k - \mathbf{x}_l| = 2r$. The vertices (vectors $\mathbf{x}_k$) and edges form a graph drawn on the surface of the unit sphere. The edges do not intersect each other if we construct them as the shortest paths on the sphere between the vertices they connect, from $\mathbf{x}_k$ to $\mathbf{x}_l$.

The contact graph can also be drawn on the plane—not respecting distances, but respecting the general topology—without edge intersections: the contact graph is planar. To concretely draw the graph, we single out one special face of the graph and pull it to the outside of a convex polygon made up of the edges of that face. In simple terms, we imagine the unit sphere as a balloon, with the opening hole inside the special face (indicated by a cross in Fig. 7.25). We flatten the balloon and the graph drawn on its surface by dilating the hole to infinity. In the two examples in Fig. 7.25, the special face is made of vectors $\{\mathbf{x}_1, \ldots, \mathbf{x}_4\}$. On the sphere, it corresponds to the inside of vertices 1–4. In the plane, it represents the outside of the same vertices, including infinity. The rest of the graph is drawn inside the special face. Every contact graph is three-connected, which means that it does not fall apart into two disconnected pieces if two arbitrary vertices are suppressed together with

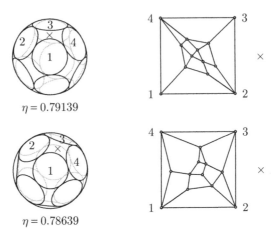

$\eta = 0.79139$

$\eta = 0.78639$

**Fig. 7.25** Optimal (*top*) and nonoptimal (*bottom*) jammed configurations for 13 equal disks. The contact graphs are shown on the *right*.

all the incident edges. This property implies that the drawing of the graph (the neighbor relations of faces) is essentially unique.

In a classic paper, Schütte and van der Waerden (1951) used the graph representation of jamming to compute the optimal configurations for small numbers of vertices. Going through detailed distinctions of cases, they were able to determine the optimal configurations for $N \leq 12$. (The optimal configuration for $N = 24$ was determined by Robinson (1964)). All these configurations are instantly reproduced by our simulated-annealing algorithm.

More generally, we see that any given jammed configurations of disks corresponds to a planar three-connected graph. The reverse problem consists in deciding whether a given planar three-connected graph can correspond to a jammed configuration of disks. This problem is most interesting, as its solution allows us to reduce the optimal-packing problem to an enumeration problem of planar graphs. The problem has been partially solved (see Krauth and Loebl (2004)), and work on this fascinating subject is continuing at the very moment that we are finishing writing this book....

# Exercises

## (Section 7.1)

(7.1) Implement Alg. 7.1 (`naive-deposition`). Include the use of periodic boundary conditions (use subroutine Alg. 2.6 (`diff-vec`)), and implement a patch to identify the stopping time $t_s$. (You may also speed up the program by using a grid/cell scheme, which is particularly simple for this problem (see Subsection 2.4.1).) Use this program to determine the probability distribution of the covering density at the stopping time $t_s$ for systems of different size. If possible, compare the disk–disk pair-correlation function of the final state of random sequential deposition (at $t_s$) with the one of typical hard disk configurations (from Alg. 2.9 (`markov-disks`)) at comparable densities.
NB: For the stopping test, note that the accessible region is bounded by exclusion disks. For any pair of disks $\{k, l\}$, determine a forbidden interval of angles for which the boundary of the exclusion disk of $k$ is covered by the exclusion disk of $l$. The stopping time is reached if the boundaries of all exclusion disks are completely covered by forbidden intervals of angles. Use a height representation of all the forbidden intervals to determine whether an exclusion disk is completely covered (see Fig. 7.26).

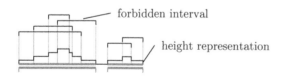

**Fig. 7.26** Height representation of forbidden intervals.

(7.2) Prove that Alg. 7.3 (`direct-triangle`) indeed samples uniformly distributed pebbles inside a triangle. Implement the routine. Test that the center of mass of pebbles converges to the geometrically computed center of mass of the triangle. Use this routine to implement the triangle algorithm of Subsection 1.1.6, a basic application of a priori probabilities. Finally, use Alg. 7.2 (`direct-polygon`) to sample a pebble uniformly distributed in a convex polygon with $n$ vertices. Check that for a regular (symmetric) hexagon, the center of mass of sampled pebbles converges to the center of the hexagon.

(7.3) Implement Alg. 7.4 (`fast-deposition`), not for disks, but for the simpler case of equal oriented squares (see Section 6.1.3). Work with rectangular small regions $\mathcal{R}_k$. Run your system up to the stopping time $t_s$. Compute the average covering density at time $t_s$ for different ratios of the area of the squares to the deposition area.

## (Section 7.2)

(7.4) Implement the Algs 7.5 (`naive-throw`) and 7.6 (`fast-throw`) and convince yourself that they produce equivalent output. Modify the two programs so that their output (spin state $\sigma(t)$) describes the physical time in a simulation of the single-spin model of eqn (7.4).

(7.5) Implement Alg. 7.7 (`dynamic-ising`) for the two-dimensional Ising model on a small lattice with periodic boundary conditions (use Alg. 1.14 (`tower-sample`)). Check your implementation by computing the mean energy (compare with Table 5.3). Modify the program so that it updates spin configurations after a time interval $\Delta_t$ different from 1. Can you realize the limiting case $\Delta_t \to 0$? Implement Alg. 7.8 (`dynamic-ising(patch)`) (the $n$-fold-way program) for the two-dimensional Ising model with periodic boundary conditions on a small lattice. Check your program by comparing again the mean energy with the exact results.

## (Section 7.3)

(7.6) Consider the close packing of 12 equal disks on the unit sphere, with the centers $\mathbf{x}_i$ of the disks forming a regular dodecahedron. Using polar coordinates, compute the central vectors $\{\mathbf{x}_1, \ldots, \mathbf{x}_{12}\}$, and the largest possible disk radius (use $\cos(2\pi/5) = \frac{1}{4}(-1+\sqrt{5})$). Determine the covering density $\eta$ and compare with Table 7.2. What is the maximum radius $R$ of spheres arranged as a regular dodecahedron on the surface of the unit sphere?

(7.7) Modify Alg. 1.22 (`direct-surface`) into a direct-sampling algorithm of equal disks on the unit sphere (this program was used to generate Fig. 7.15). Compute the densest configurations obtained during very long runs for a small number of disks ($N = 5, 8, 12, 13, 16, \dots$) and compare with the data in Table 7.2. Does this program allow you to estimate the close packing densities?

(7.8)  Implement Alg. 7.9 (`markov-sphere-disks`) and run it for small values of $N$. If possible, handle initial conditions as in Exerc. 1.3; use Gaussian unit random vectors for constructing an initial configuration (see Exerc. 7.7). During a long run at small density of disks, compute polar angles $\theta_k$ and $\phi_k$ of central vectors $\mathbf{x}_k$ and check that two-dimensional histograms of $\cos\theta_k$ and of $\phi_k$ are flat. This shows that the simulation is isotropic. Compute the acceptance probability of moves as a function of the standard deviation $\sigma$ in Alg. 7.9 (`markov-sphere-disks`), and find a rule to automatically set $\sigma$ such that it is of the order of $\frac{1}{2}$ (that is, smaller than 0.9, and larger than 0.1).

NB: When sampling integrals by the Monte Carlo method, adaptive choice of the step-size is strictly forbidden, because it interferes with the detailed-balance condition.

(7.9)  (Uses Exerc. 7.8.) Use Alg. 7.10 (`resize-disks`) together with Alg. 7.9 (`markov-sphere-disks`) of Exerc. 7.8 to implement simulated annealing for hard disks on the unit sphere. Adapt the step-size $\sigma$ in order to keep an acceptance rate of order $\frac{1}{2}$. Check that for $N = 12$ disks, your program converges towards the perfect dodecahedron arrangement of disks. Then, implement the 13-sphere problem: recover the configurations shown in Fig. 7.18. Study the influence of the annealing rate $\gamma$ on the quality of the final configuration (on the probability that it is not optimal). List all nonequivalent jammed configurations generated.

(7.10)  Perform further experiments with the simulated annealing algorithm of Exerc. 7.9. First, run the program for $N = 15$. Convince yourself that the best configuration has density $\eta = 0.80731$. Show that there are in fact two nonequivalent optimal solutions, with different contact graphs. Run the simulated annealing algorithm for $N = 19$. Show that in the configuration with highest density ($\eta = 0.81096$), one of the disks is free to move. It "rattles" in a free spot formed by its neighbors. Finally, modify the annealing program for polydisperse disks. Show that many nonequivalent solutions are obtained even for very small annealing rates.

NB: The two nonequivalent minima for $N = 15$ and the rattling solutions for $N = 19$ were found by Kottwitz (1991).

(7.11)  Write a Markov-chain simulation program for $N$ disks on a sphere at constant pressure (see Subsection 2.3.4). For a sphere radius $R$, the Boltzmann factor is $\exp\left(-\beta P \cdot 4\pi R^2\right)$. Determine the equation of state (covering density $\eta$ vs. pressure $P$) of the system for finite $N$. Interpret your findings in the light of the discussion of Subsection 7.3.2. Do you expect a liquid–solid phase transition to take place in this system for large $N$?

(7.12)  Generate $N$ randomly distributed "cities" in the unit square and use the simulated annealing algorithm to find a good solution to the traveling salesman problem: a round-trip tour of shortest length visiting all the cities. Implement local Monte Carlo moves as shown in Fig. 7.27 by rearranging the tour at two cities $k$ and $l$ (instead of connecting city $k$ with city $k'$ and city $l$ with city $l'$, connect $k$ with $l$ and $k'$ with $l'$). Use as energy the total length of the tour. Start the simulated annealing algorithm at high temperature and gradually lower the temperature. Compare the final solution found for different runs.

NB: This is a historic application of simulated annealing. The solutions found by this method are usually nonoptimal, and are considerably less accurate than those found by other heuristic methods (for $N \lesssim 100$ cities, even visual inspection usually gives better tours). However, no other method can be implemented as quickly.

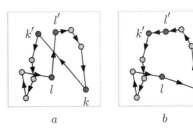

**Fig. 7.27** Local Monte Carlo move in the traveling salesman problem.

(7.13)  (Paper-cutting competition.) Write a portable computer program implementing the cone section model of Subsection 7.3.2 (see Fig. 7.22) or any other nonrandom algorithm for placing $N$ equal disks on the surface of the unit sphere (use values of $N$ between 12 and at least $1\,000\,000$). Note the highest covering densities $\eta(N)$ obtained. Communicate your program, and a sketch of the algorithm used, to the author. The best solutions found will be included in subsequent editions of this book.

# References

Bortz A. B., Kalos M. H., Lebowitz J. L. (1975) A new algorithm for Monte Carlo simulation of Ising spin systems, *Journal of Chemical Physics* **17**, 10–18

Conway J. H., Sloane N. J. A. (1993) *Sphere Packings, Lattices, and Groups, 2nd edn*, Springer, New York

Habicht W., van der Waerden B. L. (1951) Lagerung von Punkten auf der Kugel [in German], *Mathematische Annalen* **96**, 223–234

Kirkpatrick S., Gelatt C. D., Vecchi M. P. (1983) Optimization by simulated annealing, *Science* **220**, 671–680

Kottwitz D. A. (1991) The densest packing of equal circles on a sphere, *Acta Crystallographica* **A47**, 158–165

Krauth W. (2002) Disks on a sphere and two-dimensional glasses, *Markov Processes and Related Fields* **8**, 215–219

Krauth W., Loebl M. (2004) Jamming and geometric representations of graphs, preprint, math.CO/0406166

Robinson R. M. (1961) Arrangement of 24 points on a sphere, *Mathematische Annalen* **144**, 17–48

Santen L., Krauth W. (2000) Absence of thermodynamic phase transition in a model glass former, *Nature* **405**, 550–551

Schütte K., van der Waerden B. L. (1951) Auf welcher Kugel haben 5, 6, 7, 8 oder 9 Punkte mit Mindestabstand Eins Platz? [in German], *Mathematische Annalen* **123**, 96

# Acknowledgements

The cover illustration, taken from a work by Robert Filliou (1923–1987), is used with kind permission of Marianne Filliou. I thank Françoise Ninghetto for making the cover project possible.

The illustration of Count Buffon (Fig. 1.6) first appeared in an article *Les mathématiciens jouent à la roulette pour comprendre le hasard*, in *Le Monde* (Paris), edition of 13 december 1996. Used with kind permission of *Le Monde*.

Figures 1.1, 1.2, 1.3, and 7.9 were first published as Figures 1, 2, 7, and 16; in Krauth, W. "Introduction to Monte Carlo Algorithms" in Advances in Computer Simulation, Lectures Held at the Eötvös Summer School in Budapest, Hungary, 16–20 July 1996, *Springer Lecture Notes in Physics*, Vol. 501, Kertesz J., Kondor I. (Eds.), Copyright Springer 1998. With kind permission of Springer Science and Business Media.

Material in this book was shaped through essential discussions and collaborations with C. Bouchiat, A. Buhot, M. Caffarel, D. M. Ceperley, C. Dress, O. Duemmer, S. Grossmann, M. Holzmann, J. L. Lebowitz, P. Le Doussal, M. Loebl, M. Mézard, R. Moessner, A. Rosso, L. Santen, and M. Staudacher.

I thank P. Zoller for the initial suggestion that it was time to start writing, M. Staudacher for encouragement throughout the project, and P. M. Goldbart for help.

I am indebted to A. Barrat, O. Duemmer, J.-G. Malherbe, R. Moessner, and A. Rosso for reading through many versions of the text, generously offering advice and suggesting improvements in the content and the presentation of the material. I thank C. T. Pham for expert help with many technical questions.

It was a pleasure working with S. Adlung, E. Robottom, and the editorial team at OUP.

# Index

Printed and bound by CPI Group (UK) Ltd, Croydon, CR0 4YY